INTERNATIONAL CENTRE FOR MECHANICAL SCIENCES

COURSES AND LECTURES - No. 261

DYNAMICAL SYSTEMS AND MICROPHYSICS

EDITED BY

A. BLAQUIERE, F. FER, A. MARZOLLO

FOREWORD BY

Louis de BROGLIE

SPRINGER-VERLAG WIEN GMBH

Originally published by Springer-Verlag Wien New York in 1980

ISBN 978-3-211-81533-5 ISBN 978-3-7091-4330-8 (eBook)
DOI 10.1007/978-3-7091-4330-8

LIST OF CONTRIBUTORS

L. Accardi, Istituto di Matematica, Università di Milano, Via Saldini 50, Milano, Italy.

A. Avez, Département de Mécanique, Université Paris VI, 2 Place Jussieu, 75005, Paris, France.

A. Blaquiére, Laboratoire d'Automatique Théorique, Université de Paris VII, 2 Place Jussieu, 75009 Paris, France.

P. Claverie, Laboratoire de Chimie Quantique, Université de Paris VI, Institut de Biologie Physico-Chimique, 13, rue P. et M. Curie, 75005, Paris, France.

O. Costa de Beauregard, Institute Henri Poincaré, 11 rue P. et M. Curie, 75231 Paris Cedex 05, France.

M. Courbage, Bld. du Triomphe, C.P. 231, Bruxelles, Belgium.

G. Della Riccia, Dept. of Mathematics, The Ben Gurion University of the Negev, Beer Sheva, Israel.

S. Diner, Institut de Biologie Physico-Chimique, Laboratoire de Chimie Quantique, 13 rue P. et M. Curie, 75005 Paris, France.

E. Etim, Laboratori Nazionali INFN, Frascati (Roma), Italy.

D. Fargue, Ecole des Mines, 60 Bld. Saint-Michel, 75006 Paris, France.

F. Fer, Ecole des Mines, 60 Bld. Saint-Michel, 75006 Paris, France.

C. Galles, Mathematiques, Université Paris VI, 4, Place Jussieu, 75005 Paris, France.

G. Ghirardi, International Centre for Theoretical Physics, Strada Costiera 11, 34014 Trieste, Italy.

P. Glansdorff, Université libre, 206, Avenue Monjoie, 1180 Bruxelles, Belgium.

M. Grmela, Centre de recherche de mathematiques appliquées, Université de Montréal, Montréal H3C 3J7, Canada.

F. Langouche, Instituut voor Theoretische Fysica, Universiteit Leuven, Celestijnenlaan 200D, 3030 Leuven, Belgium.

G. Leitmann, Mechanical Engineering Dept., University of California, Berkeley, California, 94720, USA.

G. Lochak, Fondation Louis de Broglie, 1, rue Montgolfier, F-75003 Paris, France.

T.W. Marshall, Department of Mathematics, Manchester University, Manchester M13 9PL, Great Britain.

T.A. Minelli, Istituto di Fisica dell'Università, Via Marzolo 8, Padova, Italy.

L. Morato, LADSEB, C.N.R., C.P. 1075, Padova, Italy.

A. Pascolini, Istituto di Fisica dell'Università, Via Marzolo 8, Padova, Italy.

L. Pesquera, Laboratoire de Chimie Quantique, Institut de Biologie Physico-Chimique, 13, rue P. et M. Curie, 75005 Paris, France.

D. Roekaerts, Instituut voor Theoretische Fysica, Universiteit Leuven, Celestijnenlaan 200D, B-3030 Leuven, Belgium.

C. Ruiz Garrido, Mathematiques, Université Paris VII, 4, Place Jussieu, 75005 Paris, France.

J. Salmon, Conservatoire National des Arts et Métiers, 292, rue Saint-Martin, 75141 Paris CEDEX 03, France.

F. Selleri, Istituto di Fisica, Università di Bari, Via Amendola 173, 70126 Bari, Italy.

R. Thom, Institut des Hautes Etudes Scientifiques, 91440 Bures sur Yvette, France.

E. Tirapegui, Institut de Physique Theorique, Université Catholique de Louvain, Chemin du Cyclotron 2, 1348 Louvain-la-Neuve, Belgium.

J. Vassalo Pereira, Faculty of Sciences, University of Lisbon, Rua da Escola Politecnica, Lisboa-2, Portugal.

J.C. Willems, Mathematics Institute, P.O. Box 800, 9700 AV Groningen, The Netherlands.

CONTENTS

Je me réjouis de voir paraître ce livre, car il est le premier à résulter de la réunion de tant de spécialistes des diverses branches de la Physique et des Mathématiques unissant leurs efforts pour tenter de résoudre les problèmes que posent l'interprétation et les difficultés actuelles de la Microphysique.

Je suis particulièrement heureux de retrouver parmi les noms de ceux qui ont participé à la rédaction de ce volume les noms de plusieurs de mes anciens élèves. Mais c'est aussi pour moi un grand plaisir de voir les noms de mathématiciens et de physiciens venus de nombreux pays étrangers et représentants des tendances d'esprit très diverses, voire même opposées, qui ont librement ~~exprimé~~ exposé leurs idées durant ces deux semaines de travail.

Evidemment les opinions développées dans ce livre sont souvent très différentes, mais c'est souvent du choc des idées opposées que sont sorties les idées nouvelles et fructueuses.

On doit exprimer de bien vifs remerciements aux organisateurs de ces réunions et surtout au C.I.S.M. et à la ville d'Udine qui a accueilli le Séminaire avec tant de cordialité

Louis de Broglie

I am particularly glad that this volume appears. It is actually the first one to be issued as a result of the meeting of so many specialists from various areas of physics and mathematics striving together towards solving the problems and difficulties of present microphysics.

I am personally pleased to see the names of several pupils of mine among the contributors to this volume. Also, I am delighted that mathematicians and physicists from many countries and representing different, sometimes even opposite trends of thought, freely discussed their ideas during this two working weeks. It often happened that new and fruitful theories came out from the contrast of antagonist ideas.

Thanks should be expressed to the organizers of this meeting and above all to C.I.S.M. and to the town of Udine which has so warmly welcomed the seminar.

Louis de Broglie

PREFACE

In spite of a half century of experimental verification of quantum predictions in an incredibly large variety of physical phenomena, with a very high accuracy, a number of features of quantum mechanics are still quite generally considered as unsatisfactory. According to what is reported among physicists, we find for example in the list: can Quantum Mechanics really account for anything else that stationary states and reversible processes? Did it succeed in the explanation of intricate properties of high energy particles? What about the highly controversed question of quantum non-separability?

As the concern of physicists for the above mentioned aspects of quantum mechanics increased, a new mathematical discipline has emerged from the bustling scientific activity of the last two decades, namely the mathematical theory of dynamical systems, or more concisely system theory. Its applications increased in range and importance with each passing year and rapidly reached various areas such as mathematical physics, economics, biology, operations research, and related fields.

Among other topics, system theory is concerned with: optimal control and differential games; that is, more generally, optimization theory, stability theory, linear and nonlinear oscillators, stochastic and hereditary dynamical systems.

This bunch of new techniques and concepts has thrown a new light on dynamic analysis of systems, in the areas on which we shall focus our attention in this book. It has thus disclosed new paths for approaching modern physics, and made questionable some of the arguments which led to a discontinuity in the course of physics, about fifty years ago.

With the objectives of exploring the current and potential areas of interactions of system theory with modern physics and of fostering active exchange of ideas among people with different backgrounds, a two-week seminar was organized in September 1979 with the theme "Mathematical Theory of Dynamical Systems and Microphysics", at the International Centre for Mechanical Sciences (CISM) in Udine, Italy. We record here some of the results of fruitful discussions among the participants of the seminar.

The material published in this volume can be ideally divided into three parts, though the reader will find several overlaps between them.

In Part I we present papers dealing with optimization theory and stochastic approaches to quantum mechanics. This set of papers is, itself, subdivided into three groups. In the first group, optimization theory is presented in introductory lectures by Leitmann. They prepare the ground for the lectures by Blaquière which make the junction between optimization theory (e.g. an extension of the least action principle and of the theory of Hamilton-Jacobi) and a relativistic stochastic approach to quantum mechanics. The second group to which the papers by Accardi, Etim-Etim, Mitter and Morato belong, is more specifically concerned with the existing parallelism between the theory of stochastic processes (in the mathematical sense) and quantum theory. The third group, represented by the papers of Claverie, Diner, Marshall and Pesquera deals with Stochastic Electrodynamics (S.E.D.). In contrast to the theories above, S.E.D. is basically classical electrodynamics with radiation damping and a non-white stochastic electromagnetic field.

For instance stochasticity enters thermodynamics, and causality is obviously concerned with time symmetry or asymmetry as in some discussions of the EPR paradox.

Part II is devoted to "Evolution and irreversibility", which is maybe a too broad subject to keep a thorough homogeneity; but no classification is perfect. This part may be divided in four groups. The first one bears on general aspects of the question: a lecture of Thom on reversibility versus irreversibility; one of Lochak on stability and irreversibility; and a paper of Vassalo-Pereira about relations between stationarity and adiabatic invariance.

The second group is a more thermodynamical one: a review by Glansdorff of the modern views of non-equilibrium stable processes; a paper of Courbage on irreversibility and stochasticity in dynamical systems; papers of Fer on the comparison of irreversibility and hamiltonian character, Della Riccia on the Hamilton-Jacobi treatment of one sort of dissipative systems, of Salmon on the irreversibility axioms in Thermodynamics, and of Grmela on an inversion structure of non-hamiltonian dynamics.

The third group is centered about the notion of state: papers of Willems on System theoretic foundations for modelling physical systems, of Avez on a new method of quantization starting from symplectic dynamics, of Galles et al. on euclidean Markov fields of integer spin, of Tirapegui et al. on a W.K.B. expansion.

The last group deals with models which go out of the frame of classical quantum mechanics: a paper of Fargue on the hereditary evolution of continuous electric media, and of Minelli and Pascolini on a solition-like solution of a non-linear Schrödinger equation.

Part III contains a contribution to the question of E.P.R. paradox and related topics. The reader will find a paper of Ghirardi on quantum non-separability, another of Selleri on the einsteinian view of locality, a paper of Costa de Beauregard on the same debate, and on experiment proposals by Gutkowski.

We take this opportunity to express our gratitude to CISM which originated the idea of this seminar and shared the major part in its scientific and practical organization. We wish to thank Dr. Vinicio Turello, president of CISM for his generous hospitality, and all CISM staff, and in particular Mr. Carlo Tasso, for their able and competent assistance.

We are also indebted to UNESCO and IFAC (International Federation for Automatic Control) for their sponsorships. In particular, the former provided financial assistance to several researchers from developing countries and made possible their participation. Also thanks are due to the Fondation Louis de Broglie for its participation to the seminar.

A. Blaquière, F. Fer, A. Marzollo

January 1980

PART I

OPTIMIZATION THEORY AND STOCHASTIC APPROACHES TO QUANTUM MECHANICS

Optimal Feedback Control for Dynamical Systems with One or Two Controllers†

G. Leitmann
University of California
Berkeley, California

PREFACE

This chapter contains a discussion of conditions that must be satisfied in order that a dynamical system be "controlled" in an "optimal" fashion. In particular, the treatment is restricted to systems governed by ordinary differential equations involving functions that are to be chosen so as to control the system's response in a desired fashion. In the case of a single controller, the desired goal is attainment of the minimum of a prescribed functional; in the case of two controllers, the desired goal is attainment of a saddlepoint of a prescribed functional.

†Based in part on research supported by the National Science Foundation under Grant ENG 78-13931.

In either case, the choice of the optimal control value at a given instant is based on the system's current state; that is, we deal with necessary conditions for optimal state-feedback control.

1. Dynamical System

We are concerned with a subset of the universe that is defined by n real numbers, loosely referred to as the state of the system, $x = (x_1, x_2, \ldots, x_n)^T \in R^n$, which changes with the passing of time $t \in (-\infty, \infty)$; the evolution of the state is governed by a controller. Such a system is called a dynamical system. We deal with a dynamical system whose state evolution is determined by the solution of an ordinary differential equation, the state equation.

Given an initial state x^o, we are interested in the motion of the state due to the controller's decisions. It is convenient to let time be one of the state components, say $x_n \equiv t$, and to introduce relative time $\tau = t - t_o$ where $t_o = x_n^o$, the initial time. Now consider a bounded Lebesgue measurable function

$$u(\cdot) : [0, \tau_f] \to R^m$$

and a prescribed C^1-function

$$f(\cdot) : R^n \times R^m \to R^n .$$

The state equation is

$$\dot{x}(\tau) = f(x(\tau), u(\tau)) \qquad (1.1)$$

where the state's evolution is given by an absolutely continuous function

$$x(\cdot) : [0, \tau_f] \to R^n , \; x(0) = x^o \qquad (1.2)$$

satisfying (1.1) for almost all $\tau \in [0, \tau_f]$.

2. Feedback Control

The controller governs the state's evolution through his choice of the control value $u(\tau)$. This control decision is based on information. Classically, the only information used is time, or relative time; we speak of open loop control. Here we consider a control choice based on knowledge of the system's state at relative time τ, that is, on $x(\tau)$; we speak of closed loop or feedback control. Thus, a feedback control is a function

$$p(\cdot) : R^n \to R^m .$$

With such a choice of the controller, state equation (1.1) is replaced by

$$\dot{x}(\tau) = f(x(\tau),p(x(\tau))) \tag{2.1}$$

where, provided it exists, $x(\cdot) : [0,\tau_f] \to R^n$, with $x(0) = x^o$, is a solution of (2.1). Indeed, without further restrictions on the class of admissible feedback controls no solution might exist or, if solutions exist, they need not be unique. In fact, one may wish to consider solutions of (2.1) in some generalized sense; for instance, see References 1-5. We shall not be concerned with questions of existence and hence shall not dwell any more on the subject.

Given a feedback control $p(\cdot)$ and initial state x^o, suppose there exists a solution $x(\cdot) : [0,\tau_f] \to R^n$, with $x(0) = x^o$. Then at $\tau \in [0,\tau_f]$ there is a control value

$$u(\tau) = p(x(\tau)) . \tag{2.2}$$

Let $K_x(x^o)$ denote the set of all solutions generated by $p(\cdot)$ from initial state x^o, and let $K_u(x^o)$ denote the corresponding set of control functions $u(\cdot) : [0,\tau_f] \to R^m$ generated via relation (2.2);

that is,

$$K_u(x^o) \triangleq \{u(\cdot) \mid x(\cdot) \in K_x(x^o)\} .$$

Now we restrict feedback controls to a set of admissible ones.

Definition 2.1.[†] Set P is the set of admissible feedback controls if and only if for all $x^o \in R^n$ and $p(\cdot) \in P$,

i) $u(\cdot) \in K_u(x^o)$ is bounded and Lebesgue measurable,

and

ii) given $p'(\cdot)$, $p''(\cdot) \in P$,

$t' = x_n' \in (-\infty,\infty)$, and

$p(\cdot)$ such that

$p(x) = p'(x)$ if $x_n \leqslant x_n'$

$p(x) = p''(x)$ if $x_n > x_n'$

implies that

$p(\cdot) \in P$.

3. Playability[††]

Among other motivations, we suppose that the controller wishes to steer the system's state from a given initial state x^o to one belonging to a prescribed target set $\theta \subset R^n$.

Definition 3.1. A feedback control $p(\cdot)$ is playable at x^o if and only if

i) $p(\cdot) \in P$

and

[†] To satisfy the conditions of this definition it suffices to admit bounded Borel measurable functions; see Stalford and Leitmann.[6]

[††] This nomenclature is a holdover from game theory.

ii)[†] there exists at least one solution $x(\cdot) : [0,\tau_f] \to R^n$ of (2.1) such that $x(0) = x^o, x(\tau) \notin \theta$ for $\tau < \tau_f$, and $x(\tau_f) \in \theta$. Such a solution is termed a terminating one and τ_f is then the corresponding terminal (relative) time. A corresponding triplet $\{x^o, p(\cdot), x(\cdot)\}$ is also called terminating; such a triplet is sometimes called a play.

4. Performance Index

In order to permit consideration of another one of the controller's motivations, namely optimization, we introduce a measure of the "goodness" of the chosen feedback control, a performance index.

Let

$f_o(\cdot) : R^n \times R^m \to R$

be a prescribed C^1-function. Then for given $x^o \in R^n$, $p(\cdot) \in P$ and a corresponding solution $x(\cdot) : [0,\tau_f] \to R^n$, $x(0) = x^o$, the performance index is defined by

$$V(x^o, p(\cdot), x(\cdot)) = \int_0^{\tau_f} f_o(x(\tau), p(x(\tau)))d\tau \ . \tag{4.1}$$

If there is also a terminal cost, namely,

$$V(x^o, p(\cdot), x(\cdot)) = h(x(\tau_f)) + \int_0^{\tau_f} f_o(x(\tau), p(x(\tau)))d\tau \ , \tag{4.2}$$

[†] In the case of open loop control, the requirement that $x(\tau) \notin \theta$ for $\tau < \tau_f$ is sometimes not imposed; e.g., see Leitmann and Stalford.[7]

where $h(\cdot) : R^n \to R$ is a given C^1-function, one may consider an integral performance index

$$\tilde{V}(x^o,p(\cdot),x(\cdot)) = \int_0^{\tau_f} [\text{grad } h(x(\tau))\ f(x(\tau),p(x(\tau))) + f_o(x(\tau),p(x(\tau)))]d\tau \ . \tag{4.3}$$

For given initial state $x(0) = x^o$, (4.2) and (4.3) differ only by the constant term $h(x^o)$.

5. Optimality

In addition to requiring termination on the prescribed target, the controller desires to minimize the value of the performance index (4.1).

Definition 5.1. A feedback control $p^*(\cdot)$ is optimal on $X \subseteq R^n$ if and only if

i) it is playable at all $x^o \in X$,

and for all $x^o \in X$

ii) $V(x^o,p^*(\cdot),x^*(\cdot)) \leqslant V(x^o,p(\cdot),x(\cdot))$

for all terminating triplets $\{x^o,p^*(\cdot),x^*(\cdot)\}$ and $\{x^o,p(\cdot),x(\cdot)\}$.

We note that condition ii) of Definition 3.1, together with relation 4.1, implies that

$$V(x^o,p(\cdot),x(\cdot)) = 0 \quad \forall\, x^o \in \theta$$

and, in particular,

$$V(x^o,p^*(\cdot),x^*(\cdot)) = 0 \quad \forall\, x^o \in \theta \ . \tag{5.1}$$

Consequently, the conditions of Definition 5.1 are met trivially if $x^o \in \theta$, and so $p^*(\cdot)$ is optimal on $X^* = X \cup \theta$. Thus, in view of the uniqueness of the minimum (but not necessarily of the optimal control or the terminating solutions generated by it), for given $p^*(\cdot)$ there exists a function

$$V^*(\cdot) : X^* \to R$$

such that for all $x^o \in X^*$

$$V^*(x^o) = V(x^o, p^*(\cdot), x^*(\cdot)) \tag{5.2}$$

for all terminating triplets $\{x^o, p^*(\cdot), x^*(\cdot)\}$. Note that $V^*(\cdot)$ is defined for a given $p^*(\cdot)$. This dependence can be removed in case of non-unique optimal feedback controls. Suppose $\{p^{i*}(\cdot),\ i \in I\}$ is the set of all optimal feedback controls, where I is an index set and $p^{i*}(\cdot)$ is optimal on X_i . Then

$$V^*(\cdot) : \bigcap_I X_i \to R$$

such that, for all $x^o \in \bigcap_I X_i$,

$$V^*(x^o) = V(x^o, p^{i*}(\cdot), x^{i*}(\cdot))$$

for all terminating $\{x^o, p^{i*}(\cdot), x^{i*}(\cdot)\}$, is in fact independent of $p^{i*}(\cdot)$ and $x^{i*}(\cdot)$.

The relation between optimal feedback control and optimal open loop control is discussed in Leitmann.[8] In the following sections we derive <u>necessary</u> conditions for optimality of a feedback control. This derivation is based on a geometric approach that was utilized earlier for optimal open loop control; for example, see References 9-12.

6. Trajectories, Limiting and Isocost Surfaces

Now we introduce some additional concepts.

Let

$$y = (x_o, x_1, \ldots, x_n)^T \in R^{n+1}$$

denote the augmented state. Then, given a triplet $\{x^o, p(\cdot), x(\cdot)\}$ where $x(\cdot) : [0,\tau_f] \to R^n$, $x(0) = x^o$, is a solution, not necessarily terminating, of (2.1), consider the absolutely continuous function

$$x_o(\cdot) : [0,\tau_f] \to R \quad , \quad x_o(0) = x_o^o$$

satisfying

$$\dot{x}_o(\tau) = f_o(x(\tau), p(x(\tau))) \tag{6.1}$$

for almost all $\tau \in [0,\tau_f]$. Thus

$$x_o(\tau) = \int_0^\tau f_o(x(s), p(x(s)))ds + x_o^o \tag{6.2}$$

and, in particular,

$$x_o(\tau_f) - x_o^o = V(x^o, p(\cdot), x(\cdot)) \ . \tag{6.3}$$

Now consider C^1-function

$$g(\cdot) : R^n \times R^m \to R^{n+1}$$

whose components are

$$g_i(\cdot) = f_i(\cdot) \quad , \quad i = 0,1,\ldots,n \ .$$

Then one can combine state equation (2.1) with (6.1) to obtain the augmented state equation

$$\dot{y}(\tau) = g(x(\tau), p(x(\tau))) \quad , \quad y(0) = y^o \ . \tag{6.4}$$

The target set in R^{n+1} is simply

$$\Theta \triangleq R \times \theta \ .$$

Next we introduce trajectories in R^n and R^{n+1}, respectively.

Definition 6.1. Let $\{x^o, p(\cdot), x(\cdot)\}$ be a triplet with solution $x(\cdot) : [0,\tau_f] \to R^n$, $x(0) = x^o$, not necessarily terminating. A trajectory in R^n generated by this triplet is the ordered set of points

$$\pi \triangleq \{x \in R^n \mid x = x(\tau) \ , \ \tau \in [0,\tau_f]\}$$

In Definition 6.1, the ordering is the "natural" one induced by time $x_n \equiv t$; other orderings are sometimes admitted as, for instance, in Blaquière.[13]

Here we note that condition ii) of Definition 2.1 assures that trajectories can be "joined". Consider two triplets $\{x^{o'}, p'(\cdot), x'(\cdot)\}$ and $\{x^{o''}, p''(\cdot), x''(\cdot)\}$ with solutions $x'(\cdot) : [0,\tau_f'] \to R^n$, $x''(\cdot) : [0,\tau_f''] \to R^n$ and trajectories π', π'', respectively, such that $x'(\tau_f') = x''(0) = x^{o''}$. Then there exists a triplet $\{x^o, p(\cdot), x(\cdot)\}$, $x^o = x^{o'}$, that generates trajectory $\pi = \pi' \cup \pi''$.

Definition 6.2. Let $\{x^o, p(\cdot), x(\cdot)\}$ be a triplet with solution $y(\cdot) = (x_o(\cdot), x_1(\cdot), \ldots, x_n(\cdot))^T : [0,\tau_f] \to R^{n+1}$, $y(0) = y^o$, not necessarily terminating, where

$$x_o(\tau) + \int_\tau^{\tau_f} f_o(x(s), p(x(s)))ds = x_o(\tau_f)$$

and $x_o(\tau_f) = C$. A trajectory in R^{n+1} generated by this triplet is an ordered set of points

$$\pi(C) \triangleq \{y \in R^{n+1} \mid y = y(\tau) \quad , \quad \tau \in [0,\tau_f]\} \ .$$

Thus, given a triplet $\{x^o, p(\cdot), x(\cdot)\}$, there is one trajectory π in R^n and a one-parameter family of trajectories $\{\pi(C)\}$ in R^{n+1}; see Figure 6.1.

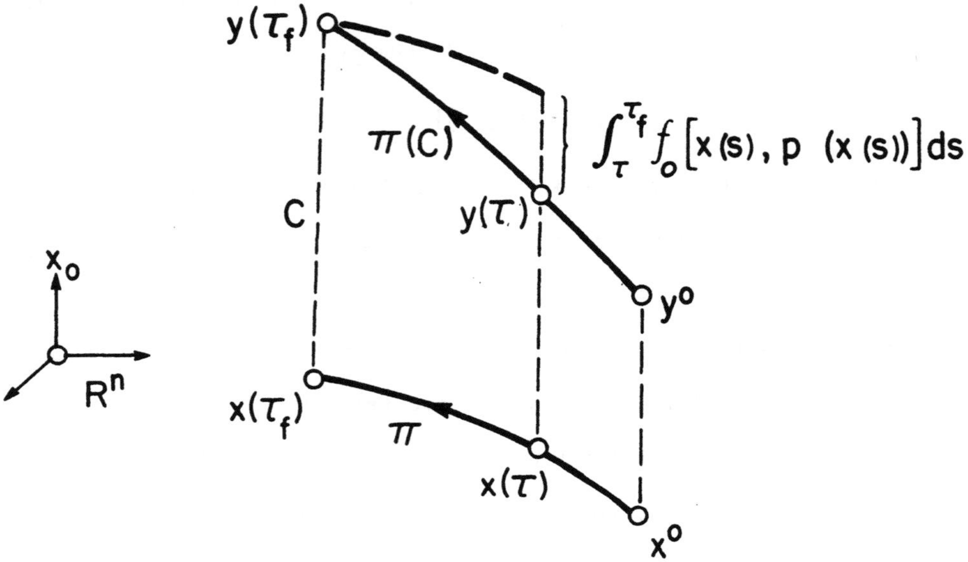

Figure 6.1, Trajectories

If $p(\cdot) = p^*(\cdot)$ and a corresponding solution of (6.4) is $y^*(\cdot) : [0,\tau_f] \to R^{n+1}$, then we denote a corresponding trajectory by $\pi^*(C)$. If such a solution terminates, namely,

$$y^*(\cdot) : [0,\tau_f^*] \to R^{n+1} \quad , \quad y^*(\tau_f^*) \in \Theta ,$$

then we term it and a corresponding trajectory <u>optimal</u>.

Suppose that optimal feedback control $p^*(\cdot)$ generates a terminating solution $x^*(\cdot) : [0,\tau_f^*] \to R^n$. Then all members of $\{\pi^*(C)\}$.

possess the same projection onto R^n, namely, the optimal trajectory in R^n

$$\pi^* \triangleq \{x \in R^n \mid x = x^*(\tau) , \tau \in [0,\tau_f^*]\} .$$

Next we employ function $V^*(\cdot)$ defined on X^* via relation (5.2) to define so-called <u>limiting surfaces</u> and <u>isocost surfaces</u>.

<u>Definition 6.3.</u> A limiting surface corresponding to constant C is the set

$$\sum(C) \triangleq \{y \in R^{n+1} \mid x_o + V^*(x) = C , \quad x \in X^*\} .$$

An isocost surface corresponding to constant C is given by $\sum(C) \cap R^n$; hence, it is the set

$$S(C) \triangleq \{x \in X^* \mid V^*(x) = C\} .$$

The reason for the terminology "limiting surface" will emerge subsequently. An isocost surface is so named because it is the locus of all initial states for which the <u>minimum</u> value of the performance index (cost) is the same.

$\sum(C)$ is a set of points in R^{n+1} which are in one-to-one correspondence with the points of X^*. In terms of C, there is a one-parameter family $\{\sum(C)\}$ of limiting surfaces and a corresponding family of isocost surfaces $\{S(C)\}$. The members of $\{\sum(C)\}$ are deduced from one another by translation along the x_o-direction and are ordered by parameter C. Thus, one and only one member of $\{\sum(C)\}$ passes through a given point of $R \times X^*$.

A given limiting surface $\sum(C)$ separates $R \times X^*$ into two disjoint sets

$$A / \sum(C) \triangleq \{y \in R^{n+1} \mid x_o > C - V^*(x) \ , \ x \in X^*\}$$
$$B / \sum(C) \triangleq \{y \in R^{n+1} \mid x_o < C - V^*(x) \ , \ x \in X^*\} \tag{6.5}$$

denoting points "above" and "below" $\sum(C)$, respectively. Thus,

$$R \times X^* = \sum(C) \cup A / \sum(C) \cup B / \sum(C) \ .$$

A limiting surface $\sum(C)$ that contains $y^o \in R^{n+1}$ is defined by

$$\Phi(y) = C \quad , \quad C = x_o^o + V^*(x^o)$$

where function

$$\Phi(\cdot) : R \times X^* \to R$$

is given by

$$\Phi(y) = x_o + V^*(x) \ . \tag{6.6}$$

7. A Basic Theorem

Now we are ready to state a fundamental property of limiting surfaces.

Theorem 7.1 Let $p^*(\cdot)$ be a feedback control that is optimal on X , and let $p(\cdot)$ be an admissible feedback control. Let $\sum(\tilde{C})$ be a limiting surface corresponding to $p^*(\cdot)$. Let $\pi^*(C^*)$ be a trajectory in R^{n+1} generated by $p^*(\cdot)$ and emanating from $y^*(0) \in \sum(\tilde{C})$. Let $\pi(C)$ be a trajectory in R^{n+1} generated by $p(\cdot)$ and emanating from $y(0) \in \sum(\tilde{C})$. Then

(i) $\pi(C) \cap B / \sum(\tilde{C}) = \phi$, and

(ii) $[\pi^*(C^*) \cap R \times X^*] \subset \sum(\tilde{C})$ where $C^* = \tilde{C}$.

In other words, no trajectory emanating from a given limiting surface dips "below" that surface; the points (in $R \times X^*$) of a trajectory generated by an optimal control lie on the limiting surface that contains the initial point of the trajectory. These properties account for the terminology "limiting surface".

Proof. Let us first prove (i) of the theorem. Let π denote the projection of $\pi(C)$ on R^n. Let y' be a point of $\pi(C)$, and let π' denote the portion of π that ends at x' and corresponds to solution $x'(\cdot) : [0,\tau_f'] \to R^n$, $x'(0) = x(0) = x^o$.

Note that $y' \notin B / \sum(\tilde{C})$ if $x' \notin X^*$ since $\sum(\tilde{C}) \subset R \times X^*$. So suppose that $x' \in X^*$. Since $p^*(\cdot)$ is optimal on X^*, it generates a terminating solution $x''(\cdot) : [0,\tau_f''] \to R^n$ with $x''(0) = x'$; let π'' denote the corresponding trajectory.

In view of condition ii) of Definition 2.1, there exists an admissible feedback control $\hat{p}(\cdot)$ given by

$$\hat{p}(x) = p(x) \quad \text{for} \quad x_n \leqslant x_n'$$

$$\hat{p}(x) = p^*(x) \quad \text{for} \quad x_n > x_n'$$

that generates a solution $\hat{x}(\cdot) : [0,\hat{\tau}_f] \to R^n$, $\hat{x}(0) = x'(0) = x(0) = x^o$, with trajectory $\hat{\pi} = \pi' \cup \pi''$; see Figure 7.1. The value of the performance index for transfer along $\hat{\pi}$ generated by $\hat{p}(\cdot)$ is $V(x^o,\hat{p}(\cdot),\hat{x}(\cdot))$. Now let $x^*(\cdot) : [0,\tau_f^*] \to R^n$, $x^*(0) = x^o$, be a terminating solution generated by $p^*(\cdot)$. Then by ii) of Definition 5.1, we have

$$V^*(x^o) = V(x^o,p^*(\cdot),x^*(\cdot)) \leqslant V(x^o,\hat{p}(\cdot),\hat{x}(\cdot)) . \qquad (7.1)$$

But†

†Note that the "additivity" property of the performance index along a trajectory is required here; of course, integrals possess this property.

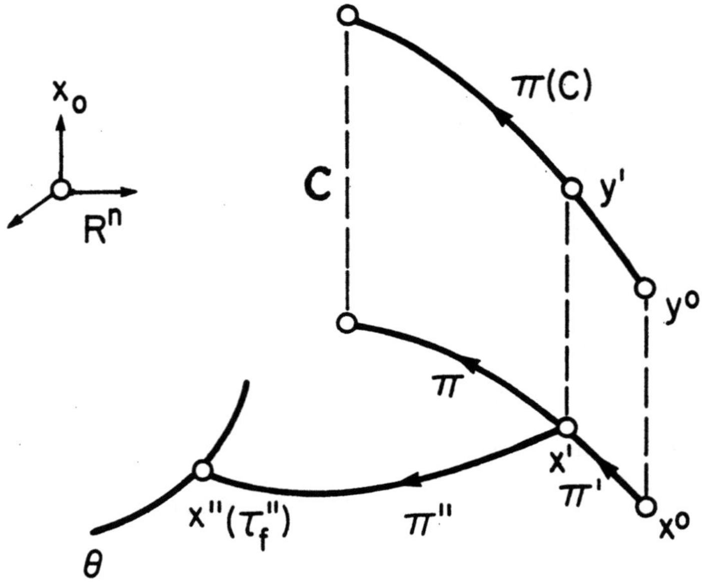

Figure 7.1, Trajectories in Theorem 7.1

$$V(x^o,\hat{p}(\cdot),\hat{x}(\cdot)) = V(x^o,p(\cdot),x'(\cdot)) + V(x',p^*(\cdot),x''(\cdot)) \qquad (7.2)$$

where

$$V(x',p^*(\cdot),x''(\cdot)) = V^*(x') \ . \qquad (7.3)$$

Then, by (7.1) and (7.2),

$$V^*(x^o) - V^*(x') \leq V(x^o,p(\cdot),x'(\cdot)) \qquad (7.4)$$

where, according to (6.3),

$$V(x^o,p(\cdot),x'(\cdot)) = x_o' - x_o^o \ . \qquad (7.5)$$

However, since $y(0) \in \sum(\tilde{C})$,

$$x_o^o + V^*(x^o) = \tilde{C} \ . \qquad (7.6)$$

Then, on setting $y = y'$ in (6.5), relations (7.4)-(7.6) lead to

$$y' \in A \,/\, \sum(\tilde{C}) \cup \sum(\tilde{C}) \ .$$

Since y' is any point of $\pi(C)$, this concludes the proof of (i).

Proposition (ii) of the theorem follows at once from the definitions of $\pi^*(C^*)$ and of $\sum(\tilde{C})$. Suppose that $p^*(\cdot)$ generates a solution $y^*(\cdot) : [0,\tau_f] \to R^{n+1}$, $y^*(0) \in \sum(\tilde{C})$ and $x^*(0) = x^o$, with trajectory $\pi^*(C)$. It follows from the definition of $\pi^*(C^*)$ that

$$x_o^*(\tau) + V^*(x^*(\tau)) = C^* . \tag{7.7}$$

But $y^*(0) \in \sum(\tilde{C})$ implies that

$$x_o^*(0) + V^*(x^o) = \tilde{C} . \tag{7.8}$$

Thus, in view of (7.7) and (7.8), we have $C^* = \tilde{C}$. Then, on letting $x = x^*(\tau)$ in Definition 6.3, we obtain $x_o = x_o^*(\tau)$. This concludes the proof of (ii).

Finally it should be pointed out that Theorem 7.1 is predicated on but two assumptions: the group property of admissible feedback controls allowing them to be "joined", and the additivity property of the performance index.

8. Regular Optimal Trajectories

Before proceeding to a discussion of conditions necessary for optimality, we restrict the class of optimal trajectories.

Definition 8.1. Let $p^*(\cdot)$ be an optimal feedback control on X, which generates a terminating solution $y^*(\cdot) : [0,\tau_f^*] \to R^{n+1}$ with corresponding optimal trajectory $\pi^*(C)$ whose projection on R^n is π^*. Trajectory $\pi^*(C)$ is regular if and only if

(i) int $X \neq \phi$;

(ii) all points of π^*, with the possible exception

of terminal point $x^*(\tau_f^*)$ belong to the interior of X ;

(iii) every point of π^*, with the possible exception of terminal point $x^*(\tau_f^*)$, possesses a neighborhood in R^n on which $V^*(\cdot)$ is of class C^2 ;

(iv) every point of π^* possesses a neighborhood in R^n on which $p^*(\cdot)$ is of class C^1 .

9. Necessary Conditions for Optimality - A Functional Equation

Unless otherwise indicated, the treatment hereafter is restricted to regular optimal trajectories.

Consider a point $y^*(\tau) \in \pi^*(C)$, $\tau < \tau_f^*$. By (ii) of Definition 8.1, $y^*(\tau) \in \text{int } R \times X^*$. Next consider an admissible feedback control $p(\cdot)$ which generates a solution $y(\cdot) : [0,\tau_f] \to R^{n+1}$, with $y(0) = y^*(\tau)$ and trajectory $\pi(\tilde{C})$. Since $y^*(\tau)$ is an interior point of $R \times X^*$, it follows from Theorem 7.1 that

$$\pi(\tilde{C}) \subset A \,/\, \textstyle\sum (C) \cup \sum (C) \qquad (9.1)$$

for sufficiently small τ_f , and

$$\pi^*(C) \subset \textstyle\sum (C) \ . \qquad (9.2)$$

Now recall the function $\Phi(\cdot)$ defined by (6.6). As a consequence of (iii) of Definition 8.1, function $\Phi(\cdot)$ is of class C^2 on a neighborhood of $y^*(\tau)$ in R^{n+1} . Thus, relations (9.1) and (9.2) imply

$$\begin{aligned} &\text{grad } \Phi(y)\, g(x,u) \geqslant 0 \quad \forall\, u \in R^m \\ &\text{grad } \Phi(y)\, g(x,p^*(x)) = 0 \end{aligned} \qquad (9.3)$$

for all $y \in \pi^*(C)$, $y \notin \Theta$, where

$$\text{grad } \Phi(y) = (1, \frac{\partial V^*(x)}{\partial x_1}, \ldots, \frac{\partial V^*(x)}{\partial x_n}) .$$

Conditions (9.3) can be combined into a single functional equation, akin to that of dynamic programming (e.g., see Bellman[14]); namely,

$$\min_u [f_o(x,u) + \sum_{i=1}^{n} \frac{\partial V^*(x)}{\partial x_i} f_i(x,u)]$$
$$= [f_o(x,p^*(x)) + \sum_{i=1}^{n} \frac{\partial V^*(x)}{\partial x_i} f_i(x,p^*(x))] = 0 , \tag{9.4}$$

subject to boundary condition (5.1), $V^*(x) = 0$ for $x \in \theta$. Since an optimal trajectory is presumed to pass through every point of int X, equation (9.4) holds for all $x \in$ int X. Note also that condition (9.4) implies stationarity with respect to u, since u is unconstrained.

10. Necessary Conditions for Optimality - A Minimum Principle

As noted above, $y^*(\tau) \in$ int $R \times X^*$ and hence possesses a neighborhood in R^{n+1} whose points belong to optimal trajectories. Thus, in particular the second of relations (9.3) is an identity in y, and hence may be differentiated with respect to y; this differentiation is valid in view of (iii) and (iv) of Definition 8.1.

Omitting arguments of functions for the sake of brevity, let

$$\frac{d^2\Phi}{dy^2} \triangleq \begin{bmatrix} 0 & 0 & \cdots & 0 \\ 0 & \frac{\partial^2 V^*}{\partial x_1^2} & \cdots & \frac{\partial^2 V^*}{\partial x_1 \partial x_n} \\ \vdots & & & \vdots \\ 0 & \frac{\partial^2 V^*}{\partial x_n \partial x_1} & \cdots & \frac{\partial^2 V^*}{\partial x_n^2} \end{bmatrix}$$

$$\frac{\partial g}{\partial y} \triangleq \begin{bmatrix} 0 & \frac{\partial f_o}{\partial x_1} & \cdots & \frac{\partial f_o}{\partial x_n} \\ \vdots & & & \vdots \\ 0 & \frac{\partial f_n}{\partial x_1} & \cdots & \frac{\partial f_n}{\partial x_n} \end{bmatrix}$$

$$\frac{\partial g}{\partial u} \triangleq \begin{bmatrix} \frac{\partial f_o}{\partial u_1} & \cdots & \frac{\partial f_o}{\partial u_m} \\ \vdots & & \vdots \\ \frac{\partial f_n}{\partial u_1} & \cdots & \frac{\partial f_n}{\partial u_m} \end{bmatrix}$$

$$\frac{dp^*}{dy} \triangleq \begin{bmatrix} 0 & \frac{\partial p_1^*}{\partial x_1} & \cdots & \frac{\partial p_1^*}{\partial x_n} \\ \vdots & & & \vdots \\ 0 & \frac{\partial p_m^*}{\partial x_1} & \cdots & \frac{\partial p_m^*}{\partial x_n} \end{bmatrix}$$

all evaluated at $y = y^*(\tau)$, $u = p^*(x^*(\tau))$.

Differentiation of the second relation (9.3) results in

$$\frac{d^2\Phi}{dy^2}\, g(x,p^*(x)) + \left[\frac{\partial g}{\partial y} + \frac{\partial g}{\partial u}\frac{dp^*}{dy} \right]^T \text{grad}^T\, \Phi(y) = 0 \qquad (10.1)$$

where $y = y^*(\tau)$.

Now introduce the absolutely continuous function

$\lambda(\cdot) : [0,\tau_f^*] \to R^{n+1}$

defined by

$$\lambda(\tau) = \text{grad}^T\, \Phi(y^*(\tau)) \qquad (10.2)$$

for all $\tau \in [0,\tau_f^*)$.

Since

$$\frac{\partial^2 V^*}{\partial x_i \partial x_j} = \frac{\partial^2 V^*}{\partial x_j \partial x_i} \quad \text{at} \quad x = x^*(\tau) \quad , \quad \tau < \tau_f^* \quad ,$$

equation (10.1) imples that $\lambda(\cdot)$ satisfies

$$\dot{\lambda}(\tau) = - \left[\frac{\partial g}{\partial y} + \frac{\partial g}{\partial u}\frac{dp^*}{dy} \right]^T \lambda(\tau) \qquad (10.3)$$

for almost all $\tau \in [0,\tau_f^*]$.

Now consider the function

$H(\cdot) : R^{n+1} \times R^n \times R^m \to R$

defined by

$$H(\lambda,x,u) = \lambda^T\, g(x,u) \; . \qquad (10.4)$$

In terms of this function, relations (9.3) become

$$\min_u H(\lambda(\tau),x^*(\tau),u) = H(\lambda(\tau),x^*(\tau),p^*(x^*(\tau))) = 0 \qquad (10.5)$$

for all $\tau \in [0,\tau_f^*)$. In fact, in view of continuity, relation (10.5) is valid on the closed interval.

A further simplification can be effected by noting that (10.5) implies

$$\lambda^T(\tau)\ \frac{\partial g}{\partial u} = 0$$

whence

$$\lambda^T(\tau)\ \frac{\partial g}{\partial u}\frac{dp^*}{dy} = 0 \quad .$$

Thus, equation (10.3) becomes

$$\dot{\lambda}(\tau) = -\ (\frac{\partial g}{\partial y})^T\ \lambda(\tau)\ . \tag{10.6}$$

Note that the right-hand side of (10.6) is independent of x_o, so that $\lambda_o(\tau) \equiv$ constant; in fact, it follows from (10.2) that

$$\lambda_o(\tau) \equiv 1\ .$$

Hence, leaving aside λ_o , equation (10.6) and the state equation (2.1) constitute a system of order $2n$ in the variables $\lambda_i(\tau)$ and $x_i(\tau)$, $i = 1,2,\ldots,n$.

The initial state is prescribed at $\tau = 0$. If the terminal state is also prescribed - and recall here that $x_n \equiv t$ - then there are $2n$ boundary conditions, and so the problem of solving the system of order $2n$ is well posed; however, it is a boundary value problem, rather than an initial value one. If the prescribed target is not a given point in R^n , additional necessary conditions can be deduced.

11. Transversality

We suppose now that target set θ is a closed set in R^n , whose boundary $\partial\theta$ is a smooth manifold of dimension $n - r$, $r < n$. That

$$\partial\theta \triangleq \{x \in R^n \mid \theta_i(x) = 0 \ , \ i = 1,2,\ldots,r\} \tag{11.1}$$

where the functions $\theta_i(\cdot) : R^n \to R$ are of class C^1, and the vectors $\text{grad}\ \theta_i(x)$, $i = 1,2,\ldots,r$, are linearly independent for $x \in \partial\theta$. If $r = n$, this reduces again to a point target. Thus, suppose $r < n$.

Consider the terminal point $y^*(\tau_f^*)$ of a regular optimal trajectory $\pi^*(C)$. Since $\text{grad}\ \Phi(y)$ is defined in a neighborhood of $y^*(\tau)$, $\tau \in [0,\tau_f^*)$, so is the tangent plane of $\sum(C)$; denote it by $T_\Sigma(y^*(\tau))$. Of course, $\text{grad}\ \Phi(y^*(\tau))$ is normal to $T_\Sigma(y^*(\tau))$.

Since, by (iv) of Definition 8.1, $p^*(\cdot)$ is of class C^1 in a neighborhood of $y^*(\tau_f^*)$, a unique optimal trajectory, $\pi^*(C)$, passes through $y^*(\tau_f^*)$. Also, by (10.2) and the continuity of $\lambda(\cdot)$,

$$\lim_{\substack{\tau\to\tau_f^* \\ \tau<\tau_f^*}} \text{grad}^T\ \Phi(y^*(\tau)) = \lambda(\tau_f^*) \ .$$

Hence, $T_\Sigma(y^*(\tau_f^*))$ is defined.

In view of the linear independence of the $\text{grad}\ \theta_i(x)$, $i = 1,2,\ldots,r$, for $x \in \partial\theta$, boundary $\partial\theta$ possesses a tangent plane, $T_{\partial\theta}(x)$, for $x \in \partial\theta$; hence, so does

$$\sum(C) \cap \partial\Theta = \{y \in R^{n+1} \mid x_o = C \ , \ x \in \partial\theta\}$$

where $\partial\Theta = R \times \partial\theta$. Denote that tangent plane by $T_{\Sigma\cap\partial\Theta}(y)$; of course, it is deduced from $T_{\partial\theta}(x)$ by translation in the x_o-direction. However,

$$T_{\Sigma\cap\partial\Theta}(y^*(\tau_f^*)) \subset T_\Sigma(y^*(\tau_f^*)) \ .$$

But $\lambda(\tau_f^*)$ is normal to $T_\Sigma(y^*(\tau_f^*))$. Hence, $\lambda(\tau_f^*)$ is normal to

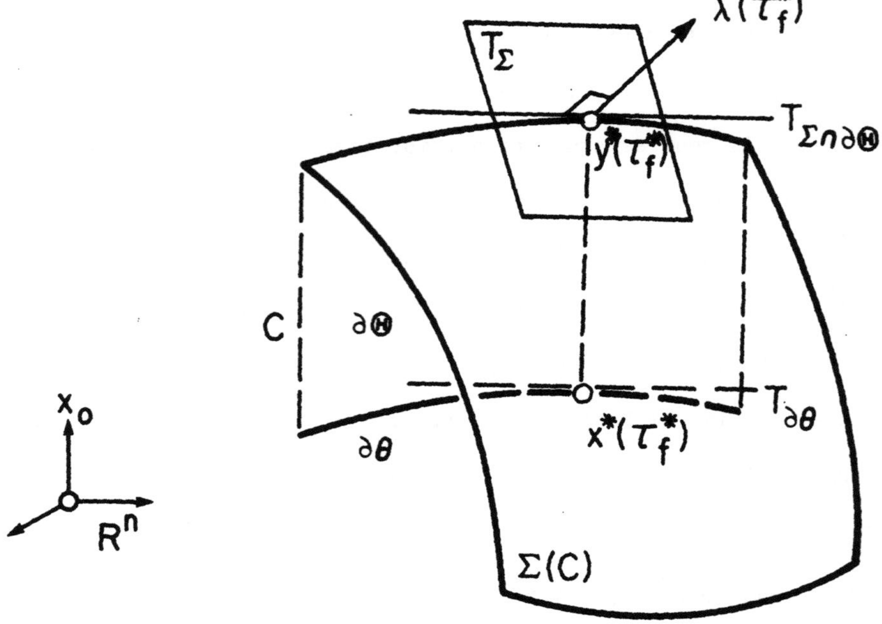

Figure 11.1, Transversality

$T_{\Sigma \cap \partial\Theta}(y^*(\tau_f^*))$; this is illustrated in Figure 11.1. This normality is a <u>transversality condition</u>. It can be expressed as follows. Let $\eta \in R^{n+1}$ be a vector contained in $T_{\Sigma \cap \partial\Theta}(y^*(\tau_f^*))$. Then

$$\lambda^T(\tau_f^*)\, \eta = 0 \tag{11.2}$$

for all such η ; namely, (11.2) holds for all η such that

$$\begin{aligned} &\eta_o = 0 \\ &\sum_{i=1}^{n} \frac{\partial\theta_j(x^*(\tau_f^*))}{\partial x_i}\eta_i = 0 \quad , \quad j = 1,2,\ldots,r \ . \end{aligned} \tag{11.3}$$

Relation (11.2), together with (11.3), yields $n - r$ conditions at $\tau = \tau_f^*$, in addition to the r terminal conditions

$$\theta_i(x^*(\tau_f^*)) = 0 \quad , \quad i = 1,2,\ldots,r \ .$$

Thus, again there is a total of 2n boundary conditions.

12. Two-Person Zero-Sum Games

In the preceding sections we considered dynamical systems controlled by a single agent. Now we present a brief discussion of a situation in which two agents, referred to as <u>players</u>, exert control over the evolution of the state of the system. In particular, in state equation (1.1), the control is now a <u>joint</u> one; namely,

$$u(\tau) = \{u^1(\tau), u^2(\tau)\} \tag{12.1}$$

where

$$u^i(\tau) \in R^{q_i} \quad , \quad q_1 + q_2 = m \ . \tag{12.2}$$

Control $u^i(\tau)$ is selected by player i. As before, we suppose that this choice is made on the basis of a player's knowledge of state $x(\tau)$ by means of a feedback control (player's <u>strategy</u>)

$$p^i(\cdot) : R^n \to R^{q_i} \tag{12.3}$$

so that

$$p(x) = \{p^1(x), p^2(x)\} \ . \tag{12.4}$$

An individual player's feedback control is assumed to be such that a <u>joint</u> feedback control (strategy pair) is admissible according to Definition 2.1; namely, the $p^i(\cdot)$ are such that

$$p(\cdot) = \{p^1(\cdot), p^2(\cdot)\} \in P \ . \tag{12.5}$$

We suppose further that both players desire termination on

prescribed target θ .† Thus, a joint feedback control, or strategy pair, is playable in accord with Definition 3.1. In addition, there is again a performance index defined in (4.1); its value (cost) is now determined by the control choices of both players. Thus, now there is no unique way of defining optimality. In the case of non-cooperation between the players, the concept of equilibrium due to Nash[15] is meaningful. According to this notion, one player desires to minimize while the other one seeks to maximize the cost (4.1); for the sake of definiteness, let player 1 be the minimizer and player 2 be the maximizer. According to Nash's equilibrium concept, each player seeks only to extremize (minimize or maximize) the cost, and he makes his strategy choice under the assumption that such is also the aim of the other player. In the two-player situation under discussion here, this philosophy results in a saddlepoint condition for optimality.

Definition 12.1. A strategy pair $p^*(\cdot) = \{p^{1*}(\cdot), p^{2*}(\cdot)\}$ is optimal on $X \subseteq R^n$ if and only if

i) it is playable at all $x^o \in X$,

and for all $x^o \in X$

ii) $V(x^o, p^{1*}(\cdot), p^2(\cdot), x^1(\cdot)) \leq V(x^o, p^{1*}(\cdot), p^{2*}(\cdot), x^*(\cdot))$
$\leq V(x^o, p^1(\cdot), p^{2*}(\cdot), x^2(\cdot))$

for all terminating plays $\{x^o, p^{1*}(\cdot), p^2(\cdot), x^1(\cdot)\}$, $\{x^o, p^{1*}(\cdot), p^{2*}(\cdot), x^*(\cdot)\}$ and $\{x^o, p^1(\cdot), p^{2*}(\cdot), x^2(\cdot)\}$.

---

†In some two-player games this is not the case; here we do not consider such pursuit-evasion type games.

The conditions that must be met under optimal play can be derived in a manner entirely analogous to that employed earlier for the situation involving a single controller. For this reason, and also because the subject has been discussed elsewhere (e.g., see References 16-20) in some detail, we shall present the results without proof.

The necessary conditions given here are again applicable for regular optimal trajectories. In particular, functional equation (9.4) is now replaced by

$$\begin{aligned}
&\min_{u^1} \Big[f_o(x,u^1,p^{2*}(x)) + \sum_{i=1}^{n} \frac{\partial V^*(x)}{\partial x_i} f_i(x,u^1,p^{2*}(x))\Big] \\
&= \max_{u^2} \Big[f_o(x,p^{1*}(x),u^2) + \sum_{i=1}^{n} \frac{\partial V^*(x)}{\partial x_i} f_i(x,p^{1*}(x),u^2)\Big] \qquad (12.6)\\
&= \Big[f_o(x,p^{1*}(x),p^{2*}(x)) + \sum_{i=1}^{n} \frac{\partial V^*(x)}{\partial x_i} f_i(x,p^{1*}(x),p^{2*}(x))\Big] = 0 \quad ,
\end{aligned}$$

subject to boundary condition $V^*(x) = 0$ for $x \in \theta$.

The minimum principle (10.5) has its counterpart in the minmax principle

$$\begin{aligned}
&\min_{u^1} H(\lambda(\tau),x^*(\tau),u^1,p^{2*}(x^*(\tau))) \\
&= \max_{u^2} H(\lambda(\tau),x^*(\tau),p^{1*}(x^*(\tau)),u^2) \qquad (12.7)\\
&= H(\lambda(\tau),x^*(\tau),p^{1*}(x^*(\tau)),p^{2*}(x^*(\tau))) = 0 \quad ,
\end{aligned}$$

where $\lambda(\cdot)$ is again a solution of equation (10.6).

Also, as before, for a target with smooth boundary, there is the terminal transversality condition (11.2)-(11.3).

13. Example

Here we treat a simple example to illustrate the use of condition (9.4). Consider the system

$$\dot{x}_i(\tau) = u_i(\tau) \quad , \quad i = 1,2,3 \tag{13.1}$$

$$\dot{x}_4(\tau) = 1 \quad (x_4 \equiv t)$$

and performance index

$$\int_0^{\tau_1} m_o\, c^2 \sqrt{1 - u^2(\tau)/c^2}\, d\tau \tag{13.2}$$

where

$$u^2 = \sum_{i=1}^{3} u_i^2$$

and m_o , c are positive constants.

If there is a control $u_i = p_i^*(x)$, $i = 1,2,3$, that minimizes the performance index (13.2), then condition (9.4) must be met, provided the corresponding trajectory is regular. Since the u_i are unconstrained, condition (9.4) is a stationarity condition:

$$\begin{aligned} &\underset{u_1,u_2,u_3}{\text{stat}} \left[m_o\, c^2 \sqrt{1 - u^2/c^2} + \sum_{i=1}^{3} \frac{\partial V^*(x)}{\partial x_i} u_i + \frac{\partial V^*(x)}{\partial t}\right] \\ &= \left[m_o\, c^2 \sqrt{1 - u^2/c^2} + \sum_{i=1}^{3} \frac{\partial V^*(x)}{\partial x_i} u_i + \frac{\partial V^*(x)}{\partial t}\right]_{u_i=p_i^*(x)} = 0 \,. \end{aligned} \tag{13.3}$$

Thus, for $u_i = p_i^*(x)$, the first of (13.3) yields†

$$\frac{\partial V^*(x)}{\partial x_i} = \frac{m_o\, u_i}{\sqrt{1 - u^2/c^2}} \quad , \quad i = 1,2,3, \tag{13.4}$$

† Provided that $u^2 < c^2$

whence, by the second of (13.3),

$$\frac{\partial V^*(x)}{\partial t} = \frac{-m_o\, c^2}{\sqrt{1 - u^2/c^2}} \quad . \tag{13.5}$$

On substituting (13.4)-(13.5) in the second of (13.3), one has

$$\frac{1}{c^2}\left(\frac{\partial V^*(x)}{\partial t}\right)^2 - \sum_{i=1}^{3}\left(\frac{\partial V^*(x)}{\partial x_i}\right)^2 = m_o^2\, c^2 \; . \tag{13.6}$$

14. Concluding Remarks

In the preceding sections we consider dynamical systems whose states evolve under the influence of a single controller or of two controllers (players), respectively. The choice of a control, or of a control pair in the two player case, is presumed to be made with full knowledge of the system's state; that is, a controller selects a control value by means of a mapping from state into his control space. We suppose further that such mappings or feedback controls must belong to a suitable class and must be playable, that is, result in state trajectories which terminate on a prescribed target set.

For such systems, optimality is defined by means of a performance index. In the one controller situation, the value of the performance index is to be minimized. In the two-player case, a saddlepoint value is sought by the two non-cooperating players. Under certain smoothness assumptions, necessary conditions for optimality are presented.

Various extensions and generalizations are possible. For instance, one may wish to restrict the control values to range in given sets; for the two player case, this restriction is treated in References 16-20.

The smoothness (regularity) assumptions utilized here can be relaxed considerably. This is done in References 18-20. The discussion presented here deals only with <u>necessary</u> conditions for optimality of feedback controls. Conditions <u>sufficient</u> to assure optimality of feedback controls may be found in Stalford and Leitmann.[21] It can be shown that condition (9.4) is sufficient for the optimality of $p^*(\cdot)$ on X, provided $p^*(\cdot)$ is playable and (9.4) is satisfied for all $x \in X$. Finally, the two player situation with a single performance criterion may be generalized to the many player one in which each player has his own criterion. Such many player non-cooperative games are treated in Case[22] and Leitmann[23]; sufficient conditions are presented in Stalford and Leitmann.[21]

References

1. Filippov, A. F., Differential equations with many-valued discontinuous right-hand side, *Soviet Mathematics,* Vol. 4, No. 4, 1963.

2. Roxin, E., On generalized dynamical systems defined by contingent equations, *Journal of Differential Equations,* Vol. 1, No. 2, 1965.

3. Hermes, H., Discontinuous vector fields and feedback control, in *Differential Equations and Dynamical Systems,* Hale, J. K. and LaSalle, J. P., Eds., Academic Press, N. Y., 1967.

4. Davy, J. L., Properties of the solution set of a generalized differential equation, *Bulletin of the Australian Mathematical Society,* Vol. 6, 1972.

5. Brunovsky, P., The closed-loop time-optimal control. I: optimality, *SIAM Journal on Control,* Vol. 12, No. 4, 1974.

6. Stalford, H. and Leitmann, G., On the equivalence of two criteria for optimal closed loop control, *Ricerche di Automatica,* Vol. 4, No. 2-3, 1973 .

7. Leitmann, G. and Stalford, H., A note on termination in optimal control problems, *Journal of Optimization Theory and Applications,* Vol. 8, No. 3, 1971.

8. Leitmann, G., A note on optimal open-loop and closed-loop control, *Journal of Dynamical Systems, Measurement, and Control,* Vol. 96, No. 3, 1974.

9. Leitmann, G., Some geometrical aspects of optimal processes, *SIAM Journal on Control,* Vol. 3, 1965.

10. Blaquière, A., Further investigation into the geometry of optimal processes, *SIAM Journal on Control,* Vol. 3, 1965.

11. Leitmann, G., *An Introduction to Optimal Control,* McGraw-Hill Book Co., N. Y., 1966.

12. Blaquière, A. and Leitmann, G., On the geometry of optimal processes, in *Topics in Optimization,* Leitmann, G., Ed., Academic Press, N.Y. 1967.

13. Blaquière, A., Quantitative games: problem statement and examples, new geometric aspects, in *The Theory and Application of Differential Games,* Grote, J. D., Ed., D. Reidel Publishing Co., Dordrecht-Boston, 1975.

14. Bellman, R. E., *Dynamic Programming,* Princeton University Press, Princeton, 1957.

15. Nash, J., Non-cooperative games, *Annals of Mathematics,* Vol. 54, No. 2, 1951.

16. Isaacs, R., *Differential Gamec,* Wiley, N. Y., 1965.

17. Leitmann, G. and Mon, G., Some geometric aspects of differential games, *Journal of the Astronautical Sciences,* Vol. XIV, No. 2, 1967.

18. Leitmann, G. and Mon, G., On a class of differential games, in *Advanced Problems and Methods for Space Flight Optimization,* de Veubeke, B. F., Ed., Pergamon Press, Oxford, 1969.

19. Blaquière, A., Gérard, F. and Leitmann, G., *Quantitative and Qualitative Games,* Academic Press, N. Y., 1969.

20. Blaquière, A. and Leitmann, G., *Jeux Quantitatifs,* Gauthier-Villars, Paris, 1969.

21. Stalford, H. and Leitmann, G., Sufficiency conditions for Nash equilibria in N-person differential games, in *Topics in Differential Games,* Blaquière, A., Ed., North Holland, Amsterdam, 1973.

22. Case, J. H., Toward a theory of many player differential games, *SIAM Journal on Control,* Vol. 7, No. 2, 1969.

23. Leitmann, G., *Cooperative and Non-Cooperative Many Player Differential Games,* Springer Verlag, Vienna, 1974.

WAVE MECHANICS AS A TWO-PLAYER GAME

by

A. BLAQUIERE,

Laboratoire d'Automatique Théorique
Université de Paris 7
Paris, France

1. Introduction.

This article is a complement to an earlier one,[1] in which at least two questions have been left in the shadows. Here we shall focus our attention on them.

The first question is concerned with the validity of a relativistic approach to stochastic processes introduced in.[1] Since such processes need to be defined in a pseudo-Euclidean space, the formulas were obtained as an extension of classical ones which apply to stochastic processes in a properly Euclidean space. In the first part of this paper we shall construct the formulas by another method.

We shall proceed as follows:

In Section 2 we propose a relativistic extension of a method by which Feynman derives the Schrödinger's equation.[2] This method relies on principles which are similar to those of Huygens and Fresnel in optics. The equation of Klein-Gordon for a charged particle in an electro-magnetic field is obtained. There appears the necessity of considering our usual space-time as a 4-dimensional hyperplane in a 5-dimensional space. For the sake of generality, we make no assumptions regarding the way the rest-mass m_o of a particle depends on its position and time, so as to cover the case where it is a constant as well as more complicated possible situations. This point is discussed in Appendix 1.

In Section 3 we make more precise the meaning of the usual time-variable t through a stationarity principle. Connections between our theory and the classical relativistic approximations (e.g. classical Mechanics) are discussed, as well as connections between the equations of Klein-Gordon and Schrödinger. The equation of Schrödinger is viewed as an approximation obtained when the fifth dimension of the space is identified with the fourth.

In Section 4 we construct the formulas of a relativistic stochastic process on which most of the theory developed in Ref.[1] is based. We make more precise the extension to such processes of the special theory of Relativity given in Ref.[1].

In Section 5 we make use of these formulas for extending the second Newton's law to Wave Mechanics, in the case of a charged particle in an electro-magnetic field. The relation thus obtained is also the relativistic extension of a relation given by Nelson[3].

The second question is concerned with a connection between dynamic programming and the theory of differential games, on one hand, and Wave Mechanics on the other hand. Among the prerequisites of the second part of this paper is the fact, discussed in[1], that Wave Mechanics is born of a dynamic programming equation, also called Main equation, obtained by Louis de Broglie in 1923. This equation is situated along the royal road of Hamilton-Jacobi. Now we are ready to extend this point of view along the way laid down by the first part of this article.

In Section 6 we give a complete expression of the stationarity principle discussed in Section 3. Especially we write it in the form of a min-max principle. In Section 7 we establish the connection between this min-max principle and the Huygens-Fresnel-Feynman point of view discussed in Sections 2 and 3.

In Section 8 we show that the min-max principle can be written in the form of sufficiency conditions for optimality of strategies in a two-player zero-sum differential game. Returning to Section 7 we see that the two players have been identified.

In Section 9 we give a complete statement of the two-player zero-sum differential game introduced in the previous sections. The saddle-point condition on which optimality of strategies is based appears to be an extension of the least action principle of Hamilton.

2. A Tentative Generalization of the Feynman's Approach.

First let us introduce the *wave function* ψ of a charged particle in a magnetic field, following a method suggested by the Feynman's approach to Quantum Mechanics.

For the time being let a *point-clock* (or a *point-source*) be a pair (x,θ) where $x = (x_1,x_2,x_3,x_4) \in \mathbf{R}^4$ is its *position* in a 4-dimensional Euclidean space E_4 and $\theta \in [0, 2\pi]$ its *phase-angle*. We shall see later that x_4 is the time to which the small clock is pointing. Otherwise $\underline{x} = (x_1,x_2,x_3)$ is its position at time t in the Galilean frame of reference of an observer $\mathcal{O}$ who makes use of a time t and of a rectangular coordinate system x_1,x_2,x_3. We shall denote by E_5 the 5-dimensional Euclidean space whose points are (x_1,x_2,x_3,x_4,t).

Let a *cloud of point-clocks in* E_4 be a triplet $\Gamma = (\rho(\cdot),\theta(\cdot),\Omega)$, where $\rho(\cdot)\colon x \mapsto \rho = \rho(x)$ is the *density function* of the cloud, defined on E_4 with range in $\mathbf{R}_+$ † ;

† We denote by $\mathbf{R}_+$ the set of non-negative numbers.

$\Omega = \{x : x \in E_4 , \quad \rho(x) > 0\} \subseteq E_4$; and

$\theta(\cdot)$: $x \mapsto \theta = \theta(x)$ is the *phase-angle function* of the cloud, defined on Ω with range in $[0, 2\pi]$.

A *point-clock of the cloud* Γ is a point-clock whose position x belongs to Ω and whose phase-angle is $\theta(x)$. We shall let

$$\tilde{\rho}(x) = \rho(x) \exp (i\theta(x)) \text{ for all } x \in \Omega$$
$$\tilde{\rho}(x) = 0 \quad \text{for all} \quad x \in \text{comp}\ \Omega \text{ in } E_4 ,$$

and, as it concerns the clouds of point-clocks which we shall consider from now on, we shall introduce

Assumption 1. $\tilde{\rho}(x)$ *is of class* C^2 *on* E_4,*and* Ω *is an open subset of* E_4.

We wish to express the fact that, for a given cloud $\Gamma = (\rho(\cdot),\theta(\cdot),\Omega)$, a point-clock of the cloud Γ whose position in Ω is x at the time t of observer $\mathcal{O}$ produces some local "effect" at other points of E_4 and at times greater than t. We shall suppose that there exists a function $q(\cdot) : (x,\Delta x,\Delta t) \mapsto q = q(x,\Delta x,\Delta t)$ defined on $E_4 \times E_4 \times \mathbf{R}^+$ † with range in $\mathbf{C}$ such that, for an arbitrarily given cloud $\Gamma = (\rho(\cdot),\theta(\cdot),\Omega)$, the effect a point-clock of this cloud with position $x \in \Omega$ at the time t of $\mathcal{O}$ produces at point $x + \Delta x \in E_4$, at time $t + \Delta t > t$, is independent of t and is completely characterized by the function $q(\cdot)$.

Now, following the general scheme of the Feynman's approach, we are ready for introducing

Assumption 2. *For all* $x \in E_4$

$$\lim_{\Delta t \to 0^+} \frac{1}{\Delta t} \left(\int q(x,\Delta x,\Delta t)\, d(\Delta x) - 1 \right) = R(x) \tag{1}$$

where $R(x)$ *is defined and continuous on* E_4 *with range in* $\mathbf{C}$.

---

† We denote by $\mathbf{R}^+$ the set of strictly positive numbers.

$$\lim_{\Delta t \to 0^+} \frac{1}{\Delta t} \int \Delta x_\alpha q(x,\Delta x,\Delta t)d(\Delta x) = \Lambda_\alpha(x), \quad \alpha = 1,2,3,4 \tag{2}$$

where $\Lambda_\alpha(x)$ *is defined and of class* C^1 *on* E_4, $\alpha = 1,2,3,4$ *with range in* **R**.

$$\lim_{\Delta t \to 0^+} \frac{1}{\Delta t} \int \Delta x_\alpha \Delta x_\beta q(x,\Delta x,\Delta t)d(\Delta x) = D_{\alpha\beta}(x), \quad \alpha,\beta = 1,2,3,4 \tag{3}$$

where $D_{\alpha\beta}(x) = 0$ for $\alpha \neq \beta$, $D_{\alpha\alpha}(x) = i\varepsilon_\alpha D(x)$, $D(x)$ *is defined and of class* C^2 *on* E_4 *with range in* **R**, *and* ε_α *is a constant*, $\alpha = 1,2,3,4$.

$$\lim_{\Delta t \to 0^+} \frac{1}{\Delta t} \int o(\|\Delta x\|^2)q(x,\Delta x,\Delta t)d(\Delta x) = 0 \tag{4}$$

Here as everywhere in what follows, the integration is to be carried out over the entire space E_4 whenever the range of integration is not specified.

Assumption 3. *For all* $x \in E_4$

$$\lim_{\Delta t \to 0^+} \frac{1}{\Delta t} \left(\int \tilde{\rho}(x-\Delta x)q(x-\Delta x,\Delta x,\Delta t)d(\Delta x)-\tilde{\rho}(x)\right) = 0 \tag{5}$$

Assumption 3 is a *superposition principle* similar to the Huygens-Fresnel principle in optics which states that at point x of E_4, for a given time t of observer $\mathcal{O}$, $\tilde{\rho}(x)$ is obtained to the first order approximation in Δt by summing up the effects due to the point-clocks of the cloud at time $t-\Delta t$, subject to the proviso that a point-clock of the cloud at point $x-\Delta x$ and at time $t-\Delta t$ behaves like a *source* of strength $\rho(x-\Delta x)$ and produces the effect

$$\rho(x-\Delta x)\exp(i\theta(x-\Delta x))q(x-\Delta x,\Delta x,\Delta t)$$

at point x and at time t.

Assumption 4. *For all* $x \in E_4$, *for all* $\Delta x \in E_4$ *and for all* $\Delta t > 0$, $q(x,\Delta x,\Delta t)$ *is twice continuously differentiable with respect to* x_α, $\alpha = 1,2,3,4$.

Then, by expending the function in relation (5) in Taylor series and letting Δt tend to 0^+ one obtains the following partial differential equation

$$- \sum_{\alpha=1}^{4} \frac{\partial}{\partial x_\alpha} (\tilde{\rho}\Lambda_\alpha) + \frac{1}{2} \sum_{\alpha=1}^{4} \frac{\partial^2}{\partial x_\alpha^2} (\tilde{\rho} D_{\alpha\alpha}) + R\,\tilde{\rho} = 0 \tag{6}$$

First, let us consider a special case. Suppose that $x_4 \equiv t$, and that $q(x,\Delta x,\Delta t)$ is given by the formula constructed by Feynman for studying the motion of a non-relativistic charged particle in a magnetic field, namely[2]

$$q(x,\Delta x,\Delta t) = \Bigg[1 - \frac{e}{im_o D_o c} \sum_{\alpha=1}^{3} A_\alpha \Delta x_\alpha - \frac{e}{i2m_o D_o c} \sum_\alpha \sum_\beta \frac{\partial A_\alpha}{\partial x_\beta} \Delta x_\alpha \Delta x_\beta -$$

$$- \frac{e^2}{2m_o^2 D_o^2 c^2} \Big(\sum_{\alpha=1}^{3} A_\alpha \Delta x_\alpha \Big)^2 + \frac{U}{im_o D_o} \Delta t + o(\|\Delta x\|^2) + O(\Delta x)\Delta t \Bigg]$$

$$\gamma \exp \Bigg[- \frac{1}{i2D_o} \sum_{\alpha=1}^{3} \frac{(\Delta x_\alpha)^2}{\Delta t} \Bigg] \tag{7}$$

where m_o is the mass of the particle, D_o is a real coefficient, e is the charge, c is the speed of light, $A_\alpha = A_\alpha(\underline{x},t)$, $\alpha = 1,2,3$, are the components of the vector potential A, and $U = U(\underline{x},t)$ is a scalar potential energy. γ is a normalization factor to insure that $\text{Lim} \int q(x,\Delta x,\Delta t)d(\Delta x)=1$ as $\Delta t \to 0^+$.

Then, as in Ref.[2] p.78, we obtain†

$\Lambda_\alpha(x) = \Lambda_\alpha(\underline{x},t) = -(e/m_o c)A_\alpha(\underline{x},t), \quad \alpha = 1,2,3,$

$\Lambda_4(x) = \Lambda_4(\underline{x},t) = 1, \quad D(x) = D_o, \quad \varepsilon_\alpha = 1, \quad \alpha = 1,2,3, \quad \varepsilon_4 = 0,$

† The calculus is based on properties of the Fresnel integrals and of a class of integrals which are simply related to the latter; see Appendix 2.

$$R(x) = R(\underline{x},t) = -\frac{e}{2m_oc}\sum_{\alpha=1}^{3}\frac{\partial A_\alpha}{\partial x_\alpha} + \frac{e^2}{i2m_o^2D_oc^2}\sum_{\alpha=1}^{3}A_\alpha^2 + \frac{U}{im_oD_o}$$

Substituting in (6), in the case where D_o is independent of x, one gets

$$\frac{\partial\tilde{\rho}}{\partial t} = i\frac{D_o}{2}\nabla^2\tilde{\rho} + \frac{e}{m_oc}\sum_{\alpha=1}^{3}A_\alpha\frac{\partial\tilde{\rho}}{\partial x_\alpha} + \frac{e^2}{i2m_o^2D_oc^2}\left(\sum_{\alpha=1}^{3}A_\alpha^2\right)\tilde{\rho} +$$

$$+ \frac{e}{2m_oc}\left(\sum_{\alpha=1}^{3}\frac{\partial A_\alpha}{\partial x_\alpha}\right)\tilde{\rho} + \frac{U}{im_oD_o}\tilde{\rho}$$

where $\nabla = (\partial/\partial x_1, \partial/\partial x_2, \partial/\partial x_3)$.

This equation may be written

$$\frac{\partial\tilde{\rho}}{\partial t} = i\frac{D_o}{2}\left(\nabla + \frac{e}{im_oD_oc}A\right)\cdot\left(\nabla + \frac{e}{im_oD_oc}A\right)\tilde{\rho} + \frac{U}{im_oD_o}\tilde{\rho} \tag{8}$$

and so, if $D_o = h/2\pi m_o$, we recognize Schrödinger's equation.

We shall see in which sense formula (7) *turns out to be an approximate one*. Let us now suppose that E_4 is an *improperly Euclidean space*, or as we shall say *pseudoeuclidean space*, whose metric has the signature--- + . This inclines us to replacing

$$\sum_{\alpha=1}^{3}\frac{(\Delta x_\alpha)^2}{\Delta t}$$

in the exponential factor of formula (7) by

$$\frac{\sum_{\alpha=1}^{3}(\Delta x_\alpha)^2 - c^2(\Delta x_4)^2}{\Delta t}$$

Accordingly let us replace formula (7) by the formula (9) where $m = m(x)$, $m_o = m_o(x)$, $D = D(x)$ and $\beta = \beta(x)$ are functions of x satisfying the relation (10). We shall determine $\beta(x)$ later.

$$q(x,\Delta x,\Delta t) = \Big[1 - \frac{e}{imDc} \sum_{\alpha=1}^{4} A_\alpha \Delta x_\alpha - \frac{e}{i2mDc} \sum_\alpha \sum_\beta \frac{\partial A_\alpha}{\partial x_\beta} \Delta x_\alpha \Delta x_\beta -$$

$$- \frac{e^2}{2m^2D^2c^2} (\sum_{\alpha=1}^{4} A_\alpha \Delta x_\alpha)^2 + \frac{1}{i2mD} (m_o c^2\sqrt{1-\beta^2})\Delta t + o(\| \Delta x \|^2) + O(\Delta x)\Delta t \Big]$$

$$\gamma \exp \left[- \frac{1}{i2D} \frac{\sum_{\alpha=1}^{3} (\Delta x_\alpha)^2 - c^2(\Delta x_4)^2}{\Delta t} \right] \qquad (9)$$

$$mD = \frac{h}{2\pi}, \qquad 0 \leqslant \beta < 1, \qquad m = \frac{m_o}{\sqrt{1-\beta^2}} \qquad (10)$$

Moreover

$$A_\alpha = A_\alpha(x), \quad \alpha = 1,2,3, \qquad A_4 = - c\Phi(x)$$

where $\Phi = \Phi(x)$ is a scalar potential. Again, γ is a normalization factor to insure that $\mathrm{Lim} \int q(x,\Delta x,\Delta t)d(\Delta x) = 1$ as $\Delta t \to 0^+$. We obtain

$$\Lambda_\alpha(x) = - (e/mc)A_\alpha, \quad \alpha = 1,2,3$$

$$\Lambda_4(x) = - (e/mc^2)\, \Phi$$

$$\varepsilon_\alpha = 1 \ \text{ for } \alpha = 1,2,3 \qquad \varepsilon_4 = - 1/c^2$$

$$R(x) = - \frac{e}{2mc} \sum_{\alpha=1}^{3} \frac{\partial A_\alpha}{\partial x_\alpha} - \frac{e}{2mc^2} \frac{\partial \Phi}{\partial x_4} + \frac{e^2}{i2m^2Dc^2} \sum_{\alpha=1}^{3} A_\alpha^2 -$$

$$- \frac{e^2}{i2m^2Dc^2} \Phi^2 + \frac{1}{i2mD} (m_o c^2\sqrt{1-\beta^2})$$

Substituting in (6) one gets

$$- \Box (\tilde{\rho}D) + \frac{2e}{ic^2} \frac{\partial}{\partial x_4} (\frac{\tilde{\rho}\,\Phi}{m}) + \frac{2e}{ic} \sum_{\alpha=1}^{3} \frac{\partial}{\partial x_\alpha} (\tilde{\rho} \frac{A_\alpha}{m}) - \frac{e}{imc} \sum_{\alpha=1}^{3} \tilde{\rho} \frac{\partial A_\alpha}{\partial x_\alpha} -$$

$$- \frac{e}{imc^2} \tilde{\rho} \frac{\partial \Phi}{\partial x_4} - \frac{1}{mD} \left[m_o c^2\sqrt{1-\beta^2} + \frac{e^2}{mc^2} (\sum_{\alpha=1}^{3} A_\alpha^2 - \Phi^2) \right] \tilde{\rho} = 0 \qquad (11)$$

where $\square = \frac{1}{c^2}\frac{\partial^2}{\partial x_4^2} - \sum_{\alpha=1}^{3}\frac{\partial^2}{\partial x_\alpha^2}$

Then, by letting

$$\psi(x) = D(x)\tilde{\rho}(x) \tag{12}$$

equation (11) can be written as

$$\square\psi + \frac{4\pi i}{h}\left(\sum_{\alpha=1}^{3} G_\alpha \frac{\partial\psi}{\partial x_\alpha} + \frac{e\Phi}{c^2}\frac{\partial\psi}{\partial x_4}\right) + \frac{2\pi i}{h}\left(\sum_{\alpha=1}^{3}\frac{\partial G_\alpha}{\partial x_\alpha} + \frac{e}{c^2}\frac{\partial\Phi}{\partial x_4}\right)\psi +$$

$$+ \frac{4\pi^2}{h^2}\left(\sum_{\alpha=1}^{3} G_\alpha^2 - \frac{e^2}{c^2}\Phi^2 + m_o^2c^2\right)\psi = 0 \tag{13}$$

with $G_\alpha = (e/c)A_\alpha$, $\alpha = 1,2,3.$

Since

$$\left.\frac{\partial\psi(\underline{x},x_4)}{\partial x_4}\right|_{x_4=t} = \frac{\partial\psi(\underline{x},t)}{\partial t}$$

it follows from equation (13) that $\psi(\underline{x},t)$ obeys the Klein-Gordon equation for a charged particle in a magnetic field, obtained simply by replacing in (13) $\psi(x)$ by $\psi(\underline{x},t)$ and the partial derivatives with respect to x_4 by partial derivatives with respect to t .

This result leads us to defining the *space-time of observer O* as the 4-dimensional hyperplane of E_5 defined by† $x_4 = t$.

Let us stress the fact that equation (13) has been obtained by treating x_4 and t as independent variables, and that introducing the condition $x_4 = t$ in it leads to a result which is different from the one we would have obtained by letting $x_4 \equiv t$ at the outset, like in the previous derivation of the Schrödinger's equation.

---

† More generally we could define the space-time of *O* by $x_4 = t +$ constant, by adding an arbitrary constant to the parameter t.

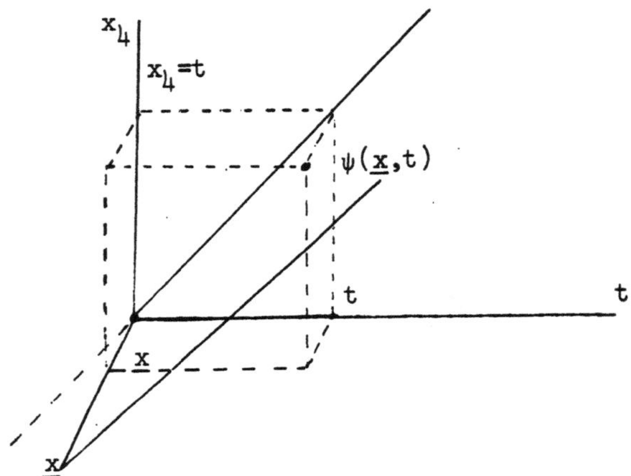

Fig. 1

The way we have introduced the space-time of $\mathcal{O}$ will be enlighten by the discussion of Section 3.

As already mentioned we suppose that the x_4*-component of the position* $(\underline{x},x_4)$ *of a point-clock in* E_4 *is the time to which this small clock is pointing.* This time needs not be the same as the time t of observer $\mathcal{O}$.

3. The Classical Limit and the Definition of Time.

Let us return to the superposition principle on which Assumption 3 is based. It tells us that if $q(x,\Delta x,\Delta t)$ is written in the form

$$q(x,\Delta x,\Delta t) = a(x,\Delta x,\Delta t)\exp(ib(x,\Delta x,\Delta t))$$

where $a(\cdot)$ and $b(\cdot)$ are real functions, then the phase-angle at point $x \in \Omega$, at the time t of observer $\mathcal{O}$, due to a point-clock at position $x - u\Delta t \in \Omega$ and at time $t-\Delta t < t$, with phase-angle $\theta(x-u\Delta t)$, is

$$\varphi(x,u\Delta t,\Delta t) = \theta(x-u\Delta t) + b(x,u\Delta t,\Delta t) + o(u,\Delta t).$$

From the expression (9) of $q(x,\Delta x,\Delta t)$ we can compute $b(x,u\Delta t,\Delta t)$. We find

$$b(x,u\Delta t,\Delta t) = \frac{1}{2D}\left(\sum_{\alpha=1}^{3} u_\alpha^2\Delta t - c^2u_4^2\Delta t\right) - \frac{1}{2mD}\left(m_oc^2\sqrt{1-\beta^2}\right)\Delta t + {}$$
$$+ \frac{e}{mDc}\sum_{\alpha=1}^{4} A_\alpha u_\alpha \Delta t + o(u,\Delta t) \qquad (14)$$

with an additional constant real term due to γ. From the expression of γ given by formula (96) of Appendix 2, it follows that this additional term is $-\pi/2$. We note that it is the same as the one which occurs in optics in connection with the principle of Huygens-Fresnel, and which has been mentioned for the first time by Gouy in the case of isotropic spherical waves. This term plays no role in the following arguments, so that we shall disregard it for simplicity.

Then

$$\varphi(x,u\Delta t,\Delta t) = \theta(x) - \sum_{\alpha=1}^{4} \frac{\partial\theta}{\partial x_\alpha} u_\alpha \Delta t + \frac{\pi m}{h} \left(\sum_{\alpha=1}^{3} u_\alpha^2 \Delta t - c^2 u_4^2 \Delta t \right) -$$

$$- \frac{\pi}{h} (m_o c^2 \sqrt{1-\beta^2}\,) \Delta t + \frac{2\pi}{h} \frac{e}{c} \sum_{\alpha=1}^{4} A_\alpha u_\alpha \Delta t + o(u,\Delta t)$$

Moreover let

$$V(x) = \frac{h}{2\pi} \theta(x)$$

Then

$$\varphi(x,u\Delta t,\Delta t) = \frac{2\pi}{h} (V(x) - S(x,u)\Delta t) + o(u,\Delta t) \tag{15}$$

with

$$S(x,u) = - L(x,u) + \sum_{\alpha=1}^{4} \frac{\partial V}{\partial x_\alpha} u_\alpha = \frac{mc^2}{2} \left(u_4^2 - \frac{1}{c^2} \sum_{\alpha=1}^{3} u_\alpha^2\right) + \frac{m_o c^2}{2} \sqrt{1-\beta^2} -$$

$$- \frac{e}{c} \sum_{\alpha=1}^{4} A_\alpha u_\alpha + \sum_{\alpha=1}^{4} \frac{\partial V}{\partial x_\alpha} u_\alpha \tag{16}$$

where the expression $L(x,u)$ in the right hand side, namely

$$L(x,u) = \frac{m}{2} \left(\sum_{\alpha=1}^{3} u_\alpha^2 - c^2 u_4^2 \right) - \frac{m_o c^2}{2} \sqrt{1-\beta^2} + \frac{e}{c} \sum_{\alpha=1}^{4} A_\alpha u_\alpha$$

is a generalized Lagrangian.

Now *let us write the stationarity condition of the action* $S(x,u)\Delta t$ *with respect to* u_α, $\alpha = 1,2,3,4$. By taking account of (10), these conditions are

$$\frac{\partial V(x)}{\partial x_\alpha} = mu_\alpha + \frac{e}{c} A_\alpha = \frac{m_o u_\alpha}{\sqrt{1-\beta^2}} + \frac{e}{c} A_\alpha , \quad \alpha = 1,2,3 \tag{17}$$

$$\frac{\partial V(x)}{\partial x_4} = - mc^2 u_4 + \frac{e}{c} A_4 = - \frac{m_o c^2 u_4}{\sqrt{1-\beta^2}} - e\,\Phi \tag{18}$$

Let us denote by $u^s(x)$ the vector whose components are given by (17) and (18). Since $u^s(x)$ is independent of t, and since x lies in Ω and Ω is an open subset of E_4, $x-u^s(x)\Delta t$ lies in Ω for Δt sufficiently small. The vector function $u = u^s(x)$, whose components are given by (17) and (18), which makes S(x,u) stationary, defines a *vector field* in Ω. Associated with this vector field is a family of trajectories in Ω defined by the solutions of the set of ordinary differential equations

$$\dot{x}_\alpha = u^s_\alpha(x) \qquad \alpha = 1,2,3,4 \tag{19}$$

If we make the

Assumption 5. $u^s_4(x) \equiv 1$

we see that these trajectoires are the ones which Louis de Broglie calls *guide-trajectories*.

At that point we can parallel also the arguments of Feynman with an additional result concerning the definition of time.

At first sight, from Equation (15) it is not clear how, in the classical limit, some particular u becomes most important. The classical approximation, however, corresponds to the case that the dimensions, masses, time intervals, etc., are so large that the action $S(x,u)\Delta t$ is enormous in relation to $h/2\pi$.

Then the term $(2\pi/h)S(x,u)\Delta t$ in Equation (15) is some very, very large phase-angle. Now if we change $u = (u_1,u_2,u_3,u_4)$, starting from a value $\hat{u}$ which does not make S(x,u) stationary, by a small amount δu, *small on the classical scale,* the change in $S(x,u)\Delta t$ is likewise small on the classical scale, but not when measured in the tiny unit $h/2\pi$. This small change will, generally, make an enormous change in phase, so that the total contribution at point x and at time t of the point-clocks in the neighborhood of a point $x-\hat{u}\Delta t$ at time $t-\Delta t$ will then add to zero. Therefore, no point-clock at time $t-\Delta t$ really needs to be considered if the neighborhing point-clocks give rise to different actions; for the point-clocks in the neighborhood cancel out the contribution at point x and at time t. But for the special point-clock at position $x-u^s(x)\Delta t$ at time $t-\Delta t$, for which $S(x,u^s(x)\Delta t$ is an extremum of $S(x,u)\Delta t$, a small change of u produces, in the first order at least, no change in $S(x,u)\Delta t$. All the contributions at point x and at time t from point-clocks in this region at time $t-\Delta t$ are nearly in phase and do not cancel out. Therefore, only for point-clocks in the vicinity of $x-u^s(x)\Delta t$ at time $t-\Delta t$ can we get important contributions at point x and at time t (Fig.2).

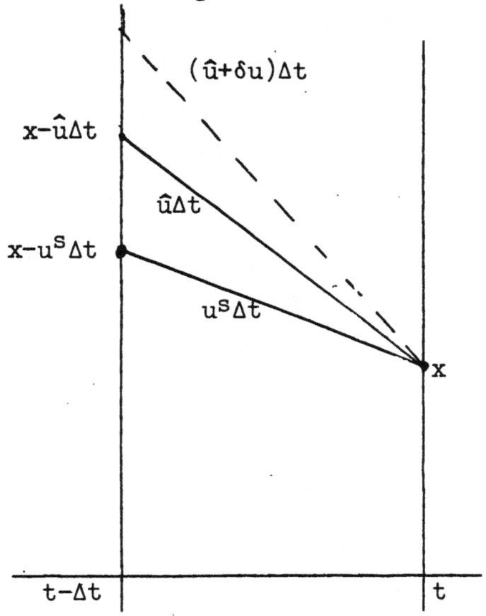

Fig.2

From the remark above and Assumption 5 it follows that only for point-clocks whose x_4-coordinate is in the vicinity of $x_4-\Delta t$ at time $t-\Delta t$ can we get important contributions at point $x = (\underline{x},x_4)$ and at time t. In other words, in the classical limit, the time x_4 to which a small clock is pointing is approximately the same as the time t of observer *O*, with the possible addition of a constant time-lag which can be disregarded.

Another remark concerns the equations of Schrödinger and Klein-Gordon. In the derivation of the Schrödinger's equation, variations of the x_4-variable have been disregarded; that is, we had $\varepsilon_4=0$. However the value of ε_4 derived from (9) turned out to be $\varepsilon_4 = -1/c^2$. So we see how $x_4 \equiv t$ is a non-relativistic approximation: we meet a similar situation as in the classical limit but for another reason. However, in the relativistic equation of Klein-Gordon, this approximation is no longer valid and *two different time variables need be considered*, namely

1 - the time variable t of observer $\mathcal{O}$; and

2 - the time variable x_4 of a point-clock.

One can easily verify that one obtains also the Klein-Gordon equation in a weakly relativistic approximation, with the assumptions $m(x) \equiv m_o(x)$, $\beta(x) \equiv 0$, and ε_4 very small, provided that the two time variables t and x_4 are not identified to one another. Here is the main difference between the equations of Klein-Gordon and Schrödinger.

In the general case, it follows from equation (19), Assumption 5 and the definition of the space-time of observer $\mathcal{O}$ that *a guide-trajectory satisfying the condition* $x_4 = 0$ *at time* $t = 0$ *lies entirely in the space-time of* $\mathcal{O}$.

4. Relativistic Stochastic Processes.

In this paragraph we shall construct relations on which most of the theory developed in Ref.[1] is based.

Among the relations of Ref.[1], we recognize in relations (10) the ones on which we have relied for a tentative extension of the Special Theory of Relativity.[1] We have also defined a *pseudo rest mass* $M_o = M_o(x)$ through the relation

$$\frac{M_o}{\sqrt{1-\underline{u}^2/c^2}} = \frac{m_o}{\sqrt{1-\beta^2}} \qquad u = u^s(x) \qquad \underline{u}^2 = u_1^2 + u_2^2 + u_3^2 \tag{20}$$

Then, by taking account of Assumption 5 relations (17) and (18) can be written

$$\frac{\partial V(x)}{\partial x_\alpha} = \frac{M_o u_\alpha}{\sqrt{1-\underline{u}^2/c^2}} + \frac{e}{c} A_\alpha \qquad \alpha = 1,2,3 \tag{21}$$

$$\frac{\partial V(x)}{\partial x_4} = \frac{-M_o c^2}{\sqrt{1-\underline{u}^2/c^2}} - e\,\Phi \tag{22}$$

from which it follows that

$$\frac{1}{c^2}\left(\frac{\partial V}{\partial x_4} + e\Phi\right)^2 - \sum_{\alpha=1}^{3}\left(\frac{\partial V}{\partial x_\alpha} - \frac{e}{c} A_\alpha\right)^2 = M_o^2 c^2 \tag{23}$$

Now let us compute $\beta(x)$. Let us suppose that $\beta(x)$ does not depend on the electro-magnetic field†, and let $A_\alpha(x) \equiv 0$, $\alpha = 1,2,3,4$, in equation (13). In this special case, equation (13) becomes

$$\Box\,\psi + \frac{4\pi^2}{h^2}\; m_o^2 c^2\; \psi = 0 \tag{24}$$

By letting $D(x)\rho(x) = \exp(-(2\pi/h)W(x))$ and accordingly

$$\psi(x) = D(x)\rho(x)\exp(i\,\frac{2\pi}{h}\,V(x)) = \exp(i\,\frac{2\pi}{h}\,(V(x) + iW(x))) \tag{25}$$

for all $x \in \Omega$, and separating equation (24) into two real equations, one of the equations thus obtained is

$$\frac{1}{c^2}\left(\frac{\partial V}{\partial x_4}\right)^2 - \sum_{\alpha=1}^{3}\left(\frac{\partial V}{\partial x_\alpha}\right)^2 = m_o^2 c^2 + \frac{1}{c^2}\left(\frac{\partial W}{\partial x_4}\right)^2 - \sum_{\alpha=1}^{3}\left(\frac{\partial W}{\partial x_\alpha}\right)^2 - \frac{h}{2\pi}\,\Box\,W \tag{26}$$

By comparing equation (26) and equation (23) in the case where $A_\alpha(x) \equiv 0$, $\alpha = 1,2,3,4$, and taking account of our assumption according to which $\beta(x)$ does not depend on the electro-magnetic field, we obtain

---

† We shall legitimate this assumption later.

$$M_o^2 c^2 = m_o^2 c^2 + \frac{1}{c^2}\left(\frac{\partial W}{\partial x_4}\right)^2 - \sum_{\alpha=1}^{3}\left(\frac{\partial W}{\partial x_\alpha}\right)^2 - \frac{h}{2\pi}\Box W \qquad (27)$$

Let $w = w^s(x)$ be the vector function defined on Ω whose components are given by

$$mw_\alpha = \frac{m_o w_\alpha}{\sqrt{1-\beta^2}} = \frac{\partial W(x)}{\partial x_\alpha} \qquad \alpha = 1,2,3 \qquad (28)$$

$$mw_4 = \frac{m_o w_4}{\sqrt{1-\beta^2}} = -\frac{1}{c^2}\,\frac{\partial W(x)}{\partial x_4} \qquad (29)$$

From equations (20), (27), (28) and (29) we deduce

$$\beta^2 = \sum_{\alpha=1}^{3}\frac{u_\alpha^2}{c^2} + w_4^2 - \sum_{\alpha=1}^{3}\frac{w_\alpha^2}{c^2} - \frac{h}{2\pi m^2 c^2}\Box W \qquad (30)$$

with $u = u^s(x)$, $w = w^s(x)$.

Note that for $W(x) \equiv$ constant we obtain the classical value $\beta^2 = \frac{u^2}{c^2}$.
Other relations are obtained, for $x \in \Omega$, by letting

$$v(x) = u^s(x) - w^s(x) \qquad \overline{v}(x) = u^s(x) + w^s(x) \qquad (31)$$

From relations (28), (29) and (31) we deduce

$$\overline{v}_\alpha(x) = v_\alpha(x) - \frac{2\varepsilon_\alpha}{\rho(x)}\,\frac{\partial(D(x)\rho(x))}{\partial x_\alpha}, \quad \alpha = 1,2,3,4 \qquad (32)$$

where, as before, $\varepsilon_\alpha = 1$, for $\alpha = 1,2,3$, and $\varepsilon_4 = -1/c^2$.

Moreover, as in Ref.[1], let $D^c(x)$ and $\rho^c(x)$ be defined by

$$D^c(x) = (1/2)D(x) \qquad (33)$$

$$D^c(x)\rho^c(x) = kD^2(x)\rho^2(x) \qquad k \equiv \text{constant} \qquad (34)$$

Then we deduce from relation (32)

$$\overline{v}_\alpha(x) = v_\alpha(x) - \frac{2\varepsilon_\alpha}{\rho^c(x)}\,\frac{\partial(D^c(x)\rho^c(x))}{\partial x_\alpha}, \qquad \alpha = 1,2,3,4 \qquad (35)$$

The way relations (10), (33), (34) and (35) appeared in Ref.[1] leads us to defining a *Relativistic Stochastic Process in the pseudo-euclidean space* E_4, for a particle with *rest-mass* $m_o = m_o(x)$ through these formulas. We shall call $D^c = D^c(x)$ the *diffusion coefficient of the process,* $\rho^c = \rho^c(x)$ the *probability density of position* of the particle, $v = v(x)$ and $\overline{v} = \overline{v}(x)$ its *forward and backward velocities,* respectively.

If one separates equation (6) into two real equations, in the case where $q(x,\Delta x,\Delta t)$ is given by relation (9) one obtains

$$\frac{1}{c^2}\left(\left(\frac{\partial V}{\partial x_4} + e\Phi\right)^2 - \left(\frac{\partial W}{\partial x_4}\right)^2\right) - \sum_{\alpha=1}^{3}\left(\left(\frac{\partial V}{\partial x_\alpha} - \frac{e}{c}A_\alpha\right)^2 - \left(\frac{\partial W}{\partial x_\alpha}\right)^2\right) + \frac{h}{2\pi}\Box W = m_o^2 c^2 \tag{36}$$

and

$$\frac{1}{c^2}\frac{\partial W}{\partial x_4}\left(\frac{\partial V}{\partial x_4} + e\Phi\right) - \sum_{\alpha=1}^{3}\frac{\partial W}{\partial x_\alpha}\left(\frac{\partial V}{\partial x_\alpha} - \frac{e}{c}A_\alpha\right) - \frac{h}{4\pi}\left(\frac{1}{c^2}\frac{\partial}{\partial x_4}\left(\frac{\partial V}{\partial x_4} + e\Phi\right) - \sum_{\alpha=1}^{3}\frac{\partial}{\partial x_\alpha}\left(\frac{\partial V}{\partial x_\alpha} - \frac{e}{c}A_\alpha\right)\right) = 0 \tag{37}$$

By making use of relations (17), (18) and (28), (31), (35), equation (37) can be easily written

$$\sum_{\alpha=1}^{4}\frac{\partial(\rho^c u_\alpha)}{\partial x_\alpha} = 0, \qquad \text{with } u = u^s(x), \quad x \in \Omega \tag{38}$$

or equivalently

$$\sum_{\alpha=1}^{4}\frac{\partial(\rho^c v_\alpha)}{\partial x_\alpha} + \Box\,(D^c\rho^c) = 0 \tag{39}$$

Equation (39) is a *generalized Fokker-Planck equation in* E_4 for our relativistic stochastic process.

By taking account of relation (27), equation (23) becomes identical with equation (36), which legitimates our assumption according to which $\beta(x)$ does not depend on the electro-magnetic field.

In addition to the above formulas we shall give the definition of *stochastic forward* and *stochastic backward derivatives* of a real valued function of x, $f(\cdot): x \mapsto f = f(x)$. We shall keep the definitions given in Ref.[1], where they are introduced as extensions of classical formulas which apply to stochastic processes in a properly Euclidean space. According to this extension the stochastic forward and stochastic backward derivatives of f(x) for a relativistic stochastic process with diffusion coefficient $D^c = D^c(x)$ and forward and backward velocities $v = v(x)$ and $\overline{v} = \overline{v}(x)$, respectively, are

$$\left(\frac{df}{dt}\right)_f = \sum_{\alpha=1}^{4} \frac{\partial f(x)}{\partial x_\alpha} v_\alpha - D^c(x)\Box f(x) \tag{40}$$

$$\left(\frac{df}{dt}\right)_b = \sum_{\alpha=1}^{4} \frac{\partial f(x)}{\partial x_\alpha} \overline{v}_\alpha + D^c(x)\Box f(x) \tag{41}$$

respectively.

Clearly these formulas can be extended to the case where $f(\cdot)$ is a vector function. In the following section we shall consider the case where $f(x) \in R^4$.

All the relations of section 4 which hold in E_4 give rise to relations in the space-time of observer $\mathcal{O}$ by simply replacing x by $(\underline{x},t)$ and the partial derivatives with respect to x_4 by partial derivatives with respect to t.

5. An Extension of the Second Newton's Law.

Now let us start from equation (36), and let us make use of the vector identity

$$\operatorname{grad}(m^2\underline{u}^2) = 2m\underline{u}\times(\operatorname{curl} m\underline{u}) + (2m\underline{u}\cdot\nabla)m\underline{u} \tag{42}$$

where $\underline{u} = (u_1, u_2, u_3)$, $\underline{u}^2 = u_1^2 + u_2^2 + u_3^2$ and

$$\text{grad}(\cdot) = \nabla = (\partial/\partial x_1,\ \partial/\partial x_2,\ \partial/\partial x_3)$$

Since

$$m\underline{u} = \text{grad}\ V - \frac{e}{c} A, \qquad A = (A_1, A_2, A_3)$$

we deduce from (42)

$$\text{grad}(m^2\underline{u}^2) = -\ 2m\underline{u} \times (\text{curl}\ \frac{e}{c} A) + (2m\underline{u}\cdot\nabla)m\underline{u}$$

Moreover, from relations (17) and (18) together with Assumption 6 we have, for $\sigma = 1,2,3$

$$\frac{\partial}{\partial x_\sigma}\left(\frac{\partial V}{\partial x_4} + e\Phi\right)^2 = 2\left(\frac{\partial V}{\partial x_4} + e\Phi\right)\left(\frac{\partial^2 V}{\partial x_4 \partial x_\sigma} + e\frac{\partial \Phi}{\partial x_\sigma}\right) =$$

$$= -\ 2mc^2\left(\frac{\partial}{\partial x_4}(m u_\sigma) + \frac{e}{c}\frac{\partial A_\sigma}{\partial x_4} + e\frac{\partial \Phi}{\partial x_\sigma}\right) \tag{43}$$

Let

$$E(x) = -\frac{1}{c}\frac{\partial A(x)}{\partial x_4} - \nabla\Phi(x), \qquad E = (E_1, E_2, E_3) \tag{44}$$

It follows from relations (43) and (44) that

$$\text{grad}\left(\frac{\partial V}{\partial x_4} + e\Phi\right)^2 = -\ 2mc^2\frac{\partial}{\partial x_4}(m\underline{u}) + 2mc^2 e E \tag{45}$$

By similar arguments, from relation (28) and (29) and from the fact that $\underline{w} = (w_1, w_2, w_3)$ is a gradient in R^3, we obtain

$$\text{grad}(m^2\underline{w}^2) = 2m\underline{w} \times (\text{curl}\ m\underline{w}) + (2m\underline{w}\cdot\nabla)m\underline{w} = (2m\underline{w}\cdot\nabla)m\underline{w} \tag{46}$$

where $\underline{w}^2 = w_1^2 + w_2^2 + w_3^2$, and

$$\text{grad}\left(\frac{\partial W}{\partial x_4}\right)^2 = -\ 2mc^2 w_4\frac{\partial}{\partial x_4}(m\underline{w}) \tag{47}$$

Then from equations (36), (42), (45), (46) and (47) we deduce

$$\text{grad}(m_o^2 c^2) = -\ 2m\frac{\partial}{\partial x_4}(m\underline{u}) + 2meE + 2m\underline{u} \times (\text{curl}\ \frac{e}{c} A)$$

$$-(2m\underline{u}\cdot\nabla)m\underline{u} + 2mw_4\frac{\partial}{\partial x_4}(m\underline{w}) + (2m\underline{w}\cdot\nabla)m\underline{w} + \frac{h}{2\pi}\Box\, m\underline{w} \tag{48}$$

Let

$$H(x) = \text{curl } A(x), \qquad H = (H_1, H_2, H_3) \tag{49}$$

Dividing both sides of equation (48) by $2m$ we obtain

$$-\sqrt{1-\beta^2}\,\text{grad}(m_o c^2) + eE + \frac{e}{c}\,\underline{u}\times H =$$

$$u_4 \frac{\partial}{\partial x_4}(m\underline{u}) - w_4 \frac{\partial}{\partial x_4}(m\underline{w}) + (\underline{u}\cdot\nabla)m\underline{u} - (\underline{w}\cdot\nabla)m\underline{w} - \frac{h}{4\pi m}\Box m\underline{w} \tag{50}$$

Let

$$\upsilon = (v_1, v_2, v_3), \qquad \overline{\upsilon} = (\overline{v}_1, \overline{v}_2, \overline{v}_3)$$

From relations (31) we deduce

$$(\underline{u}\cdot\nabla)m\underline{u} - (\underline{w}\cdot\nabla)m\underline{w} = \frac{1}{2}\left((\overline{\upsilon}\cdot\nabla)m\upsilon + (\upsilon\cdot\nabla)m\overline{\upsilon}\right)$$

$$u_4 \frac{\partial}{\partial x_4}(m\underline{u}) - w_4 \frac{\partial}{\partial x_4}(m\underline{w}) = \frac{1}{2}\left(\overline{v}_4 \frac{\partial(m\upsilon)}{\partial x_4} + v_4 \frac{\partial(m\overline{\upsilon})}{\partial x_4}\right)$$

At last equation (50) is written

$$-\sqrt{1-\beta^2}\,\text{grad}(m_o c^2) + eE + \frac{e}{c}\,\underline{u}\times H =$$

$$\frac{1}{2}\Big(\Big(\overline{v}_4 \frac{\partial(m\upsilon)}{\partial x_4} + (\overline{\upsilon}\cdot\nabla)m\upsilon + \frac{h}{4\pi m}\Box m\upsilon\Big) +$$

$$\Big(v_4 \frac{\partial(m\overline{\upsilon})}{\partial x_4} + (\upsilon\cdot\nabla)m\overline{\upsilon} - \frac{h}{4\pi m}\Box m\overline{\upsilon}\Big)\Big)$$

$$= \frac{1}{2}\left(\left(\frac{d(m\upsilon)}{dt}\right)_b + \left(\frac{d(m\overline{\upsilon})}{dt}\right)_f\right) \tag{51}$$

Equation (51) can be considered as an extension of the second Newton's law, with the force

$$F = -\sqrt{1-\beta^2}\,\text{grad}(m_o c^2) + eE + \frac{e}{c}\,\underline{u}\times H$$

acting on the particle. Equation (51) is also the relativistic extension of a formula given by Nelson.[3]

Again all the relations of section 5 which hold in E_4 give rise to relations in the space-time of observer $\mathcal{O}$ by replacing x by $(\underline{x}, t)$ and the partial derivatives with respect to x_4 by partial derivatives with respect to t.

6. Local Stationarity Principles.

In view of relations (36) and (37), the basic equation (6) can be written

$$\frac{1}{c^2}\left(\left(\frac{\partial V}{\partial x_4}+e\Phi\right)+i\frac{\partial W}{\partial x_4}\right)^2-\sum_{\alpha=1}^{3}\left(\left(\frac{\partial V}{\partial x_\alpha}-\frac{e}{c}A_\alpha\right)+i\frac{\partial W}{\partial x_\alpha}\right)^2$$

$$-\frac{ih}{2\pi}\left(\frac{1}{c^2}\frac{\partial}{\partial x_4}\left(\left(\frac{\partial V}{\partial x_4}+e\Phi\right)+i\frac{\partial W}{\partial x_4}\right)-\sum_{\alpha=1}^{3}\frac{\partial}{\partial x_\alpha}\left(\left(\frac{\partial V}{\partial x_\alpha}-\frac{e}{c}A_\alpha\right)+i\frac{\partial W}{\partial x_\alpha}\right)\right)$$

$$=m_o^2c^2 \tag{52}$$

Let

$$v^* = u^s(x)+iw^s(x), \qquad v^* = (v_1^*, v_2^*, v_3^*, v_4^*)$$

$$V^* = V(x)+iW(x) \qquad x\in\Omega$$

In view of relations (17), (18), (28) and (29), equation (52) can also be written

$$\frac{1}{c^2}(mc^2v_4^*)^2-\sum_{\alpha=1}^{3}(mv_\alpha^*)^2-\frac{ih}{2\pi}\left(\Box V^*+\frac{e}{c^2}\left(\frac{\partial\Phi}{\partial x_4}+\sum_{\alpha=1}^{3}c\frac{\partial A_\alpha}{\partial x_\alpha}\right)\right)$$

$$=m_o^2c^2 \tag{53}$$

Here we shall extend a classical relation by introducing

Assumption 6.

$$\sum_{\alpha=1}^{3}\frac{\partial A_\alpha}{\partial x_\alpha}+\frac{1}{c}\frac{\partial\Phi}{\partial x_4}\equiv 0 \tag{54}$$

From relations (53) and (54), and from the definition of D^c, it follows that

$$iD^c\Box V^* = \frac{mc^2}{2}\left((v_4^*)^2-\frac{1}{c^2}\sum_{\alpha=1}^{3}(v_\alpha^*)^2\right)-\frac{m_oc^2}{2}\sqrt{1-\beta^2} \tag{55}$$

Then from (17), (18), (28), (29), (52) and (55) we deduce

$$\sum_{\alpha=1}^{4} \frac{\partial V^*}{\partial x_\alpha} v_\alpha^* + iD^c \Box V^* = \frac{m}{2} \left(\sum_{\alpha=1}^{3} (v_\alpha^*)^2 - c^2 (v_4^*)^2 \right) - \frac{m_o c^2}{2} \sqrt{1-\beta^2} + \frac{e}{c} \sum_{\alpha=1}^{4} A_\alpha v_\alpha^* \tag{56}$$

$$\frac{\partial V^*}{\partial x_\alpha} = m v_\alpha^* + \frac{e}{c} A_\alpha, \quad \alpha = 1,2,3 \qquad \frac{\partial V^*}{\partial x_4} = - mc^2 v_4^* + \frac{e}{c} A_4 \tag{57}$$

By returning to relation (16) we see that the right hand side of relation (56) proceeds from the analytic extension $L(x, u+iw)$ of the Lagrangian $L(x,u)$:

$$L(x,u+iw) = \frac{m}{2} \left(\sum_{\alpha=1}^{3} (u_\alpha + iw_\alpha)^2 - c^2 (u_4 + iw_4)^2 \right) - \frac{m_o c^2}{2} \sqrt{1-\beta^2} + \frac{e}{c} \sum_{\alpha=1}^{4} A_\alpha (u_\alpha + iw_\alpha)$$

By letting

$$H^*(x,u+iw) = L(x,u+iw) - \sum_{\alpha=1}^{4} \frac{\partial V^*(x)}{\partial x_\alpha} (u_\alpha + iw_\alpha) - iD^c(x) \Box V^*(x) \tag{58}$$

the set of equations (56)(57) is equivalent to the set of equations

$$H^*(x,u+iw) = 0 \tag{59}$$

$$\frac{\partial H^*(x,u+iw)}{\partial (u_\alpha + iw_\alpha)} = 0, \qquad \alpha = 1,2,3,4 \tag{60}$$

By comparing relations (16) and (58) we shall be led to an extension of our earlier arguments regarding the stationarity of $S(x,u)$.

Let us separate the real and the imaginary parts of $H^*(x,u+iw)$. One can easily verify that the set of equations (59)(60) is equivalent to (here min max = max min)

$$\min_{(\underline{u},w_4)}\ \max_{(\underline{w},u_4)} \left(L(x,u,w) - \sum_{\alpha=1}^{4} \frac{\partial V(x)}{\partial x_\alpha} u_\alpha + \sum_{\alpha=1}^{4} \frac{\partial W(x)}{\partial x_\alpha} w_\alpha + D^c(x)\Box W(x)\right)$$

$$= L(x,u^s,w^s) - \sum_{\alpha=1}^{4} \frac{\partial V(x)}{\partial x_\alpha} u^s_\alpha + \sum_{\alpha=1}^{4} \frac{\partial W(x)}{\partial x_\alpha} w^s_\alpha + D^c(x)\Box\, W(x) = 0 \qquad (61)$$

$$M(x,u^s,w^s) - \sum_{\alpha=1}^{4} \frac{\partial W(x)}{\partial x_\alpha} u^s_\alpha - \sum_{\alpha=1}^{4} \frac{\partial V(x)}{\partial x_\alpha} w^s_\alpha - D^c(x)\Box\, V(x) = 0 \qquad (62)$$

where

$$\underline{u} = (u_1,u_2,u_3), \quad \underline{w} = (w_1,w_2,w_3), \quad u^s = u^s(x), \quad w^s = w^s(x),\ x \in \Omega$$

and

$$L = -\frac{mc^2}{2}\left(u_4^2 - w_4^2 - \frac{1}{c^2}\sum_{\alpha=1}^{3}(u_\alpha^2 - w_\alpha^2)\right) - \frac{m_o c^2}{2}\sqrt{1-\beta^2} + \frac{e}{c}\sum_{\alpha=1}^{4} A_\alpha u_\alpha \qquad (63)$$

$$M = -\,mc^2\left(u_4 w_4 - \frac{1}{c^2}\sum_{\alpha=1}^{3} u_\alpha w_\alpha\right) + \frac{e}{c}\sum_{\alpha=1}^{4} A_\alpha w_\alpha \qquad (64)$$

Indeed we have

$$\mathrm{L}(x,u+iw) = L(x,u,w) + i\,M(x,u,w)$$

The connection between the stationarity condition embodied in relation (61) and the stationarity condition of section 3 is brought to light through the relation

$$L(x,u,w) - \sum_{\alpha=1}^{4} \frac{\partial V(x)}{\partial x_\alpha} u_\alpha + \sum_{\alpha=1}^{4} \frac{\partial W(x)}{\partial x_\alpha} w_\alpha + D^c(x)\Box\, W(x) =$$

$$= -\,S(x,u) + \Sigma(x,w) \qquad (65)$$

where

$$\Sigma(x,w) = N(x,w) + \sum_{\alpha=1}^{4} \frac{\partial W}{\partial x_\alpha} w_\alpha =$$

$$= \frac{mc^2}{2}\left(w_4^2 - \frac{1}{c^2}\sum_{\alpha=1}^{3} w_\alpha^2\right) + D^c\Box\, W + \sum_{\alpha=1}^{4} \frac{\partial W}{\partial x_\alpha} w_\alpha$$

In other words, *in addition to the fact that* $u^s(x)$ *renders* $S(x,u)$ *stationary,* $w^s(x)$ *renders* $\Sigma(x,w)$ *stationary and moreover*

$$S(x,u^s(x)) - \Sigma(x,w^s(x)) = 0 \tag{66}$$

for all $x \in \Omega$.

As concerns condition (62) one can readily verify that it is equivalent to relation (38), and that $((\underline{u}^s,w_4^s)(\underline{w}^s,u_4^s))$ is also a saddle-point of

$$M(x,u,w) - \sum_{\alpha=1}^{4} \frac{\partial W(x)}{\partial x_\alpha} u_\alpha - \sum_{\alpha=1}^{4} \frac{\partial V(x)}{\partial x_\alpha} w_\alpha - D^c(x)\square\, V(x)$$

However this saddle-point is flat.

In the following we shall turn our attention towards condition (61).

7. Further Discussion of the Stationarity Principles.

Since condition (61) gives a complete expression of the stationarity principle discussed in Section 3, and since this stationarity principle appears now to be twofold, it is natural to extend the arguments of Feynman reported in Section 3, regarding u, to the variable w . A motivation for doing so is the similarity between $S(x,u)$ and $\Sigma(x,w)$.

First note that relation (66) implies that, for $x \in \Omega$,

$$\frac{2\pi}{h} V(x) = \varphi(x,u^s(x)\Delta t,\Delta t) + \frac{2\pi}{h} \Sigma(x,w^s(x))\Delta t + o(\Delta t)$$

where, for Δt sufficiently small

$$\varphi(x,u^s(x)\Delta t,\Delta t) = \frac{2\pi}{h} (V(x) - S(x,u^s(x)\Delta t) + o(\Delta t)$$

is the phase-angle at x *and* t *due to a point-clock at the position* $x-u^s(x)\Delta t \in \Omega$ *at the time* $t-\Delta t$ *and, accordingly*

$$\delta(x,w^s(x)\Delta t,\Delta t) = \frac{2\pi}{h} \Sigma(x,w^s(x))\Delta t + o(\Delta t)$$

is the phase-angle at x *and* t *due to the resulting effect of all the other point-clocks of the cloud at time* $t-\Delta t$†.

We shall give a simple expression of the latter under the following

Assumption 7. *There exists a function* $\mathcal{V}(\cdot): (x,y) \mapsto \mathcal{V} = \mathcal{V}(x,y)$ *defined and continuously differentiable on* $\Omega \times R^4$ *with range in* R*, such that*

$$\mathcal{V}(x,o) = 0, \qquad \text{and}$$

$$\frac{\partial W(x)}{\partial x_\alpha} = - \left. \frac{\partial \mathcal{V}(x,y)}{\partial y_\alpha} \right|_{y=0}$$

for all $x \in \Omega$.

This function can be easily constructed by taking the analytic extension $W(x+iy)$ of $W(x)$; whenever it exists :

$$W(x+iy) = \mathcal{W}(x,y) - i\, \mathcal{V}(x,y)$$

Note that the existence of this analytic extension, for all $x \in \Omega$, presupposes that $W(x)$ is C^∞ on Ω, which seems not to be a too strong assumption for physical purposes.

From the expression of $\Sigma(x,w)$ it follows that, for $x \in \Omega$ and for $\Delta t > 0$ sufficiently small

$$\delta(x,w\Delta t,\Delta t) = \frac{2\pi}{h} \Sigma(x,w)\Delta t + o(\Delta t) =$$

$$= \frac{2\pi}{h} (\mathcal{V}(x-u\Delta t,w\Delta t) + N(x,w)\Delta t) + o(\Delta t)$$

which expression is to be compared with

$$\varphi(x,u\Delta t,\Delta t) = \frac{2\pi}{h} (V(x-u\Delta t) + L(x,u)\Delta t) + o(\Delta t)$$

At last we have

$$\frac{2\pi}{h} V(x) = \varphi(x,u^s(x)\Delta t,\Delta t) + \delta(x,w^s(x)\Delta t,\Delta t)$$

---

† When the additional term $-\pi/2$ mentioned in Section 3 is taken into account in the expression of φ, a term $+\pi/2$ must be added to δ, so that $\varphi + \delta$ is unchanged.

The meaning of the result thus obtained is the following:

For computing the phase-angle $(2\pi/h)V(x)$ *at point* $x \in \Omega$ *at time* t, *the cloud of point-clocks at time* $t-\Delta t$ *for* $\Delta t > 0$ *sufficiently small, can be replaced by two point-sources; that is, the phase-angle at* x *and* t *due to the cloud at time* $t-\Delta t$ *is the sum of the two phase-angles due to these two point-sources, one of which is at the position* $x-u^s(x)\Delta t \in \Omega$ *with phase-angle* $(2\pi/h)V(x-u^s(x)\Delta t)$, *and the other is at the position* $x-u^s(x)\Delta t - w^s(x)\Delta t \in \Omega$ *with phase-angle* $(2\pi/h)\mathcal{V}(x-u^s(x)\Delta t, -w^s(x)\Delta t)$.

The phase-shift due to the first point-source is $(2\pi/h)\,L\,(x,u^s(x))\Delta t + o(\Delta t)$, *and the phase-shift due to the second one is* $(2\pi/h)\,N\,(x,w^s(x))\Delta t + o(\Delta t)$.

The arguments of section 3 can be repeated and extended to both stationarity conditions, the one concerning $u^s(x)$ and the other concerning $w^s(x)$. The above statement will be the crux of our differential game approach to Wave Mechanics.

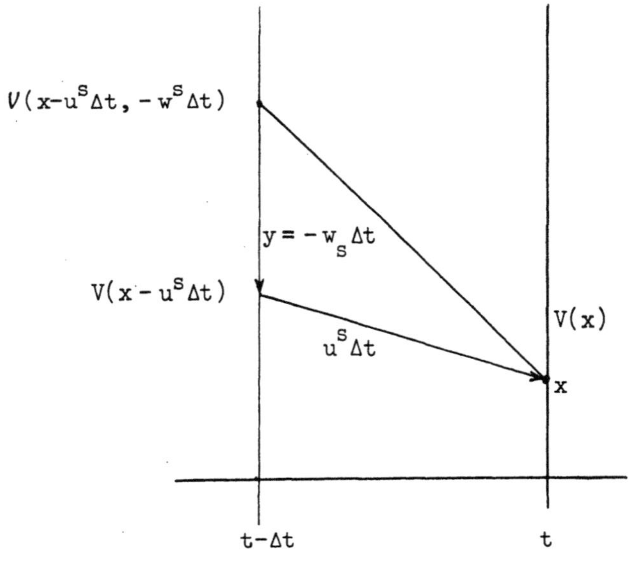

Fig. 3

8. Towards a Differential Game Approach to Wave Mechanics.

Some similarities between condition (61) and necessary conditions for optimal strategies in a two-player zero-sum differential game will motivate a few more developments.

By taking account of relation (55) in relation (58) we have

$$H^*(x,u+iw) = \Lambda^*(x,u+iw) - \sum_{\alpha=1}^{4} \frac{\partial V^*(x)}{\partial x_\alpha}(u_\alpha + iw_\alpha)$$

with

$$\Lambda^*(x,u+iw) = L(x,u+iw) - iD^c(x)\Box V^*(x) =$$

$$= \frac{m}{2}\Big(\sum_{\alpha=1}^{3}(u_\alpha + iw_\alpha)^2 - c^2(u_4 + iw_4)^2\Big) + \frac{e}{c}\sum_{\alpha=1}^{4} A_\alpha(u_\alpha + iw_\alpha) +$$

$$+ \frac{m}{2}\Big(\sum_{\alpha=1}^{3}(u_\alpha^s + iw_\alpha^s)^2 - c^2(u_4^s + iw_4^s)^2\Big)$$

Let

$$\Lambda^*(x,u+iw) = \Lambda(x,u,w) + i\,\Xi(x,u,w)$$

where $\Lambda(\cdot): (x,u,w) \mapsto \Lambda = \Lambda(x,u,w)$ and $\Xi(\cdot) : (x,u,w) \mapsto \Xi = \Xi(x,u,w)$ are real valued functions, and let us extend Assumption 7 by

Assumption 8. *The analytic extensions* $\Lambda^*(x+iy,u+iw)$ *and* $V^*(x+iy)$ *of* $\Lambda^*(x,u+iw)$ *and* $V^*(x)$, *respectively, are defined for all* $(x,y) \in \Omega \times R^4$ *and for all* $(u,w) \in R^4 \times R^4$.

Clearly, Assumption 8 implies Assumption 7.

Let $z = x+iy$ with $x \in \Omega$ and $y \in R^4$, and

$$V^*(z) = \tilde{V}(x,y) + i\tilde{W}(x,y)$$

where $\tilde{V}(\cdot): (x,y) \mapsto \tilde{V} = \tilde{V}(x,y)$, and $\tilde{W}(\cdot): (x,y) \mapsto \tilde{W} = \tilde{W}(x,y)$ are real valued functions which satisfy the Cauchy condition, that is

$$\frac{\partial \tilde{V}}{\partial x_\alpha} = \frac{\partial \tilde{W}}{\partial y_\alpha}, \quad \frac{\partial \tilde{V}}{\partial y_\alpha} = -\frac{\partial \tilde{W}}{\partial x_\alpha}, \quad \alpha = 1,2,3,4$$

Since $\Lambda = L + D^c\Box W$ and $\tilde{V}(x,o) = V(x)$, $\tilde{W}(x,o) = W(x)$ condition (61) can be written (here, again, min max = max min)

$$y = 0 \tag{67}$$

$$\min_{(\underline{u},w_4)}\max_{(\underline{w},u_4)} \left(\Lambda(x,u,w) - \sum_{\alpha=1}^{4}\frac{\partial\tilde{V}(x,y)}{\partial x_\alpha}u_\alpha - \sum_{\alpha=1}^{4}\frac{\partial\tilde{V}(x,y)}{\partial y_\alpha}w_\alpha\right)$$

$$= \Lambda(x,u^s,w^s) - \sum_{\alpha=1}^{4}\frac{\partial\tilde{V}(x,y)}{\partial x_\alpha}u_\alpha^s - \sum_{\alpha=1}^{4}\frac{\partial\tilde{V}(x,y)}{\partial y_\alpha}w_\alpha^s = 0 \tag{68}$$

The set of relations (67)(68) is an extension of the equation of Dynamic Programming (also called Main Equation) introduced by Louis de Broglie in 1923 as one of the starting points of Wave Mechanics. The early theory of de Broglie is reported in Ref.[1]

The form of the condition (68) will motivate an extension of the theory towards the theory of two-players zero-sum differential games.

9. A Differential Game Approach to Wave Mechanics.

At this stage it makes sense to embed our problem in a more general one. Let us consider a pair of points in R^4 whose positions are defined by vectors x and x+y with $x = (x_1,x_2,x_3,x_4) \in R^4$, $y = (y_1,y_2,y_3,y_4) \in R^4$, and let

$$G = \{(x,y): \quad x \in \Omega \ , \quad x+y \in \Omega\}$$

Members (x,y) of G are *states* of the game. The state equations are

$$\begin{aligned} \dot{x} &= \tilde{u}(x,y) \\ \dot{y} &= \tilde{w}(x,y) \end{aligned} \tag{69}$$

where $\tilde{u}(\cdot)$: $(x,y) \mapsto \tilde{u} = \tilde{u}(x,y)$ and $\tilde{w}(\cdot)$: $(x,y) \mapsto \tilde{w} = \tilde{w}(x,y)$ are functions defined on G with ranges in R^4.

We shall suppose that the classes of these functions are such that existence, uniqueness, extension in the future and in the past, and piecing together of the solutions of (69) are insured.

In order to relax the condition (67) we shall consider the analytic extension $\Lambda^*(z, u+iw) = \Lambda^*(x+iy, u+iw)$ of $\Lambda^*(x,u+iw)$, and we shall let

$$\Lambda^*(z,u+iw) = \tilde{\Lambda}(x,y,u,w) + i\,\tilde{\Xi}(x,y,u,w)$$

where $\tilde{\Lambda}(\cdot)$: $(x,y,u,w) \mapsto \tilde{\Lambda} = \tilde{\Lambda}(x,y,u,w)$ and
$\tilde{\Xi}(\cdot)$: $(x,y,u,w) \mapsto \tilde{\Xi} = \tilde{\Xi}(x,y,u,w)$ are real valued functions.

Then we shall suppose that the following condition is satisfied for all $(x,y) \in G$, namely that there exist functions
$\tilde{u}^s(\cdot)$: $(x,y) \mapsto \tilde{u}^s = \tilde{u}^s(x,y)$ and $\tilde{w}^s(\cdot)$: $(x,y) \mapsto \tilde{w}^s = \tilde{w}^s(x,y)$, such that

$$\min_{(\underline{\tilde{u}},\tilde{w}_4)}\ \max_{(\underline{\tilde{w}},\tilde{u}_4)} \left(\tilde{\Lambda}(x,y,\tilde{u},\tilde{w}) - \sum_{\alpha=1}^{4} \frac{\partial\tilde{V}(x,y)}{\partial x_\alpha}\tilde{u}_\alpha - \sum_{\alpha=1}^{4} \frac{\partial\tilde{V}(x,y)}{\partial y_\alpha}\tilde{w}_\alpha\right)$$

$$= \max_{(\underline{\tilde{w}},\tilde{u}_4)}\ \min_{(\underline{\tilde{u}},\tilde{w}_4)} \left(\tilde{\Lambda}(x,y,\tilde{u},\tilde{w}) - \sum_{\alpha=1}^{4} \frac{\partial\tilde{V}(x,y)}{\partial x_\alpha}\tilde{u}_\alpha - \sum_{\alpha=1}^{4} \frac{\partial\tilde{V}(x,y)}{\partial y_\alpha}\tilde{w}_\alpha\right)$$

$$= \tilde{\Lambda}(x,y,\tilde{u}^s,\tilde{w}^s) - \sum_{\alpha=1}^{4} \frac{\partial\tilde{V}(x,y)}{\partial x_\alpha}\tilde{u}^s_\alpha - \sum_{\alpha=1}^{4} \frac{\partial\tilde{V}(x,y)}{\partial y_\alpha}\tilde{w}^s_\alpha = 0 \qquad (70)$$

The conditions (67) (68) are deduced from (70) by letting $y = 0$.

We shall define *strategies* $s(\cdot)$ and $\sigma(\cdot)$ by

$$s(\cdot) = (\tilde{u}_1(\cdot), \tilde{u}_2(\cdot), \tilde{u}_3(\cdot), \tilde{w}_4(\cdot))$$

$$\sigma(\cdot) = (\tilde{w}_1(\cdot), \tilde{w}_2(\cdot), \tilde{w}_3(\cdot), \tilde{u}_4(\cdot))$$

and a *trajectory* generated by a strategy pair $(s(\cdot),\sigma(\cdot))$ is an ordered set

$$\pi^{ij} = \{(x,y):\ x = x(t),\ y = y(t),\quad t \in [t_i, t_j]\}$$

where $x(\cdot)$: $t \mapsto x = x(t)$ and $y(\cdot)$: $t \mapsto y = y(t)$ are solutions of (69) on some interval of time $[t_i,t_j]$. The ordering is insured by the parametrization in t.

Let there be given an *initial set* $\Theta \subset G$.

A trajectory π^{ij} such that $(x(t_i),y(t_i)) \in \Theta$, $(x(t),y(t)) \notin \Theta$ for

$t > t_i$ and $(x(t_j),y(t_j)) = (x^j,y^j)$ is termed an admissible trajectory relative to (x^j,y^j), and the corresponding strategy pair $(s(\cdot),\sigma(\cdot))$ is also called admissible[†].

We shall denote by $I(x^j,y^j)$ the set of all admissible strategy pairs relative to (x^j,y^j).

With an admissible trajectory relative to (x^j,y^j), generated by $(s(\cdot),\sigma(\cdot)) \in I(x^j,y^j)$, given by $(x(t),y(t))$ on interval $[t_i,t_j]$, we shall associate a pay-off. For that purpose let there be given a function $V_o^*(\cdot)$: $z \mapsto V_o^*(z)$ defined on the set Θ^*

$$\Theta^* = \{z = (x+iy): \quad (x,y) \in \Theta^*\}$$

with range in C. Let

$$V_o^*(z) = \tilde{V}_o(x,y) + i\tilde{W}_o(x,y)$$

where $\tilde{V}_o(\cdot)$: $(x,y) \mapsto \tilde{V}_o = \tilde{V}_o(x,y)$ and $\tilde{W}_o(\cdot)$: $(x,y) \mapsto \tilde{W}_o = \tilde{W}_o(x,y)$ are real valued functions defined on Θ, which satisfy the Cauchy conditions.

Now the pay-off is

$$J(x^j,y^j,s(\cdot),\sigma(\cdot)) = \int_{t_i}^{t_j} \tilde{\Lambda}(x(t),y(t),\tilde{u}(x(t),y(t)),\tilde{w}(x(t),y(t)))dt + \tilde{V}_o(x(t_i),y(t_i)) \quad (71)$$

At last, at a point (x^j,y^j) such that $(s^s(\cdot),\sigma^s(\cdot)) \in I(x^j,y^j)$ where

$$s^s(\cdot) = (\tilde{u}_1^s(\cdot),\ \tilde{u}_2^s(\cdot),\ \tilde{u}_3^s(\cdot),\ \tilde{w}_4^s(\cdot))$$

$$\sigma^s(\cdot) = (\tilde{w}_1^s(\cdot),\ \tilde{w}_2^s(\cdot),\ \tilde{w}_3^s(\cdot),\ \tilde{u}_4^s(\cdot))$$

one can readily deduce from condition (70) that

---

† More usually, in the theory of games Θ is a prescribed terminal set, also called target set, and the concept of admissible trajectory is replaced by the one of terminating trajectory.

$$J(x^j,y^j,s^s(\cdot),\sigma(\cdot)) \leqslant J(x^j,y^j,s^s(\cdot),\sigma^s(\cdot))$$
$$\leqslant J(x^j,y^j,s(\cdot),\sigma^s(\cdot)) \quad (72)$$

for all $(s^s(\cdot),\sigma(\cdot)) \in I(x^j,y^j)$ and for all $(s(\cdot),\sigma^s(\cdot)) \in I(x^j,y^j)$ and that

$$J(x^j,y^j,s^s(\cdot),\sigma^s(\cdot)) = \tilde{V}(x^j,y^j) \quad (73)$$

Hence if we think of two players J_1 and J_2, the children of the Maxwell's devil, which make use of the strategies $s(\cdot)$ and $\sigma(\cdot)$, respectively, such that J_1 strives to minimize whereas J_2 strives to maximize the pay-off along a trajectory emanating from Θ and ending at state (x^j,y^j), the game we have defined is *zero-sum*. $\tilde{V}(x^j,y^j)$ is the *value* of the game at state (x^j,y^j).

The saddle-point condition (72) appears to be an extension of the least action principle of Hamilton.

Acknowledgments

I am indebted to Prof. A. Marzollo of the University of Trieste, Deputy Secretary General of CISM, Udine, and sometime Visiting Professor at the University of Paris 7, for stimulating discussions and for his valuable criticism of this paper.

References

1. Blaquière, A.; *J. Opt. Th. Appl.*, 27, n°1, January 1979
2. Feynman, R.P., and Hibbs, A.R., *Quantum Mechanics and Path Integrals*, McGraw-Hill, 1965
3. Nelson, E., *Phys. Rev.*, 150, n°4, 1079, 1966
4. Blaquière, A., *Comptes rendus*, 288, Série A, 319, 1970
5. Feyman, R.P., Theory of Electromagnetic Interaction in Selected papers on Quantum Electrodynamics edited by J. Schwinger, 1950.

Appendix 1: Classical Mechanics Revisited.

Let us define classical mechanics as a mechanics in which $\rho^c(x)$ *is bounded and*

$$\rho^c(x)D^c(x) \equiv \text{constant}, \textit{ or equivalently } \rho(x)D(x) \equiv \text{constant} \tag{74}$$

for all x *in* Ω.

From a physical point of view, that amounts to assuming that, in general

$$\rho^c(x)D^c(x) \equiv \text{constant} + f(x, \tfrac{h}{m}), \qquad x \in \Omega$$

where the function $f(\cdot)$: $(x,\varepsilon) \mapsto f(x,\varepsilon)$ is such that $f(x,\varepsilon) \to 0$ as $\varepsilon \to 0$ for all x in Ω, so that $f(x, \frac{h}{m})$ can be neglected at the classical scale.

From the definition of W(x) (Eq.(25)) it follows that relation (74) implies that

$$W(x) \equiv \text{constant} \tag{75}$$

Then, from relations (31) and (35) we deduce

$$v_\alpha(x) = \bar{v}_\alpha(x) = u^s_\alpha(x) \qquad \alpha = 1,2,3,4 \tag{76}$$

from relations (28)(29) and (30) we have

$$\beta^2 = \sum_{\alpha=1}^{3} \frac{u_\alpha^2}{c^2}, \qquad u = u^s(x) \tag{77}$$

and from relation (36) we have

$$\frac{1}{c^2}\left(\frac{\partial V}{\partial t} + e\Phi\right)^2 - \sum_{\alpha=1}^{3}\left(\frac{\partial V}{\partial x_\alpha} - \frac{e}{c}A_\alpha\right)^2 = m_o^2 c^2 \tag{78}$$

in the space-time of observer O.

Relation (78) must be complemented by

$$\frac{\partial V(x)}{\partial x_\alpha} = mu_\alpha + \frac{e}{c}A_\alpha = \frac{m_o u_\alpha}{\sqrt{1-\beta^2}} + \frac{e}{c}A_\alpha, \quad \alpha = 1,2,3 \tag{79}$$

$$\frac{\partial V(x)}{\partial t} = -mc^2 + \frac{e}{c}A_4 = -\frac{m_o c^2}{\sqrt{1-\beta^2}} - e\,\Phi \tag{80}$$

deduced from (17)(18) and Assumption 5, for $x = (\underline{x},t)$, that is, for x in the space-time of observer $\mathcal{O}$.

Equation (78) is the Hamilton-Jacobi equation of classical relativistic Mechanics.

Further, from relation (37) and Assumption 6 we deduce

$$\Box V(x) = 0 \tag{81}$$

It follows from (17) (18) and Assumption 6 that relation (81) is equivalent to

$$\sum_{\alpha=1}^{4} \frac{\partial}{\partial x_\alpha} (mu_\alpha) = 0, \qquad u = u^s(x) \tag{82}$$

and since it follows from (74) and $D^c = h/4\pi m$ that

$$\frac{\rho^c(x)}{m(x)} \equiv \text{constant (and equivalently } \frac{\rho(x)}{m(x)} \equiv \text{ constant)} \tag{83}$$

relation (82) is equivalent to

$$\sum_{\alpha=1}^{4} \frac{\partial(\rho^c u_\alpha)}{\partial x_\alpha} = 0, \qquad u = u^s(x) \tag{84}$$

which is just relation (38).

From relation (84) we deduce

$$\frac{\partial \rho^c}{\partial t} = - \sum_{\alpha=1}^{3} \frac{\partial(\rho^c u_\alpha)}{\partial x_\alpha}, \qquad u = u^s(\underline{x},t) \tag{85}$$

in the space-time of observer $\mathcal{O}$. This is the continuity equation of Hydrodynamics.

At last relation (51) becomes

$$- \sqrt{1-\beta^2}\, \text{grad}(m_o c^2) + eE + \frac{e}{c}\, \underline{u} \times H = \frac{d}{dt}(mu) \tag{86}$$

with

$$m = \frac{m_o}{\sqrt{1-\beta^2}} \tag{87}$$

Relations (77)(78)(79)(80)(86)(87) are well known relations of classical relativistic Mechanics, *in which area the rest-mass* m_o *of the par-*

ticle is supposed to be a constant. This restriction does not affect our theory since we have made no assumption regarding the function $m_o(x)$.

However if $m_o(x) \equiv$ constant, then relation (86) accounts for electro-magnetic forces only, so that if there is no electro-magnetic field or if the particle is not charged the force acting on it is identically zero.

From relation (86) we see that we can account for forces other than electro-magnetic ones, in the case where these forces derive from a potential function, such as gravitational forces at the non-relativistic approximation for instance, in two ways.

The first way is to suppose that the rest-mass m_o depends on x and t. If we let†

$$m_o c^2 = U_o(\underline{x},t) + \overline{m}_o c^2$$

where $U_o(\cdot) : (\underline{x},t) \mapsto U_o = U_o(\underline{x},t)$ is a function of $(\underline{x},t)$ and $\overline{m}_o$ is a constant, relation (86) becomes

$$-\sqrt{1-\beta^2}\ \text{grad}\ U_o + eE + \frac{e}{c}\,\underline{u} \times H = \frac{d}{dt}(mu)$$

with m given by (87).

Another way which is *not* equivalent to the latter is to suppose that m_o is a constant and that there exists a vector $B = B(x)$, with

$$B_\alpha = B_\alpha(x) = 0, \qquad \alpha = 1,2,3 \qquad\qquad B_4 = -cU_g(x)$$

so that $eA(x)$ is replaced by

$$eA(x) + B(x)$$

in formula (9), with the proviso that the force other than electro-magnetic ones acting on the particle is given by

$$F_\alpha(x) = -\frac{\partial U_g(x)}{\partial x_\alpha}, \qquad \alpha = 1, 2, 3$$

---

† Note that, at the non-relativistic approximation $m_o = \overline{m}_o + (U_o/c^2)$ is replaced by $m_o \simeq \overline{m}_o \equiv$ constant, whereas grad $m_o c^2 =$ grad $U_o(\underline{x},t)$.

Then relation (86) becomes

$$- \text{grad } U_g + eE + \frac{e}{c} \underline{u} \times H = \frac{d}{dt} (mu)$$

Again m is given by (87). An important difference with the first assumption is that, in that case, $m_o \equiv$ constant.

Now let us return to formula (9) and, in order to make the junction with formula (7), let $\Delta x_4 \equiv \Delta t$ in it. Then we see that formula (9) becomes identical with formula (7), even if β is not supposed to be "very small"†, provided that

$$U(\underline{x},t) = \frac{1}{2} \left(m_o c^2 \sqrt{1-\beta^2} + \frac{m_o c^2}{\sqrt{1-\beta^2}}\right) + U_g(\underline{x},t) + e\,\Phi(\underline{x},t) \tag{88}$$

Relation (88) can be written

$$U(\underline{x},t) = m_o c^2 + U_g(\underline{x},t) + e\,\Phi(\underline{x},t) + o(\beta) \tag{89}$$

At last when β is sufficiently small so that o(β) can be neglected, (89) is approximated by

$$U(\underline{x},t) = m_o c^2 + U_g(\underline{x},t) + e\,\Phi(\underline{x},t) \tag{90}$$

Indeed, we can account for the two cases treated above by writing

$$U(\underline{x},t) = \overline{m}_o c^2 + U_o(\underline{x},t) + U_g(\underline{x},t) + e\,\Phi(\underline{x},t)$$

with $U_g(\underline{x},t) \equiv 0$ in the first case and $U_o(\underline{x},t) \equiv 0$ in the second one.

---

† Since $m_o D_o = mD = h/2\pi$

Appendix 2.

Difficulties due to the fact that integrals of the form

$$\int_{-\alpha}^{+\alpha} x^{2k} \exp(ix^2)dx, \qquad k \text{ integer} > 0$$

have no limit as $\alpha \to +\infty$ are avoided by letting

$$\int_{-\infty}^{+\infty} x^{2k} \exp(ix^2)dx = 2 \int_{0}^{\infty} x^{2k} \exp(ix^2)dx =$$

$$= \lim_{\lambda \to 0^+} 2 \int_{0}^{\infty} x^{2k} \exp((-\lambda+i)x^2)dx$$

Let $x^2 = \tau$, $\lambda - i = s$, then we have

$$\int_{0}^{\infty} 2x^{2k} \exp((-\lambda+i)x^2)dx = \int_{0}^{\infty} \tau^{(k-\frac{1}{2})}\exp(-s\tau)d\tau = \frac{\Gamma(k+\frac{1}{2})}{s^{(k+\frac{1}{2})}}$$

which is the Laplace transform of $\tau^{(k-\frac{1}{2})}$; that is†

$$\int_{-\infty}^{+\infty} x^{2k} \exp(ix^2)dx = \frac{\Gamma(k+\frac{1}{2})}{(-i)^{(k+\frac{1}{2})}} \tag{91}$$

Note that this formula also holds for $k = 0$, in which case

$$\int_{-\infty}^{+\infty} \exp(ix^2)dx = \frac{\sqrt{\pi}}{\sqrt{-i}} = \sqrt{i\pi} \tag{92}$$

Let γ be defined by the normalization condition

$$\gamma \int_{-\infty}^{+\infty} \exp\left[\frac{i}{2D_o} \frac{(\Delta x)^2}{\Delta t}\right] d(\Delta x) = 1 \tag{93}$$

---

† Remember that the Laplace transform of t^K, $\mathrm{Re}(K) > -1$, is $\frac{\Gamma(K+1)}{s^{(K+1)}}$

From (92) and (93) we deduce

$$\gamma = 1/\sqrt{2i\pi D_o \Delta t} \tag{94}$$

Then from (91) and (94) we deduce

$$\gamma \int_{-\infty}^{+\infty} (\Delta x)^{2k} \exp\left[\frac{i}{2D_o} \frac{(\Delta x)^2}{\Delta t}\right] d(\Delta x) = \frac{(2iD_o)^k}{\sqrt{\pi}} \Gamma\left(k+\frac{1}{2}\right)(\Delta t)^k \tag{95}$$

A few concluding comments will be in order. First, the way we have defined

$$\int_{-\infty}^{+\infty} x^{2k} \exp(ix^2)dx$$

amounts to replacing $q(x,\Delta x,\Delta t)$ in formulas (7) and (9) by $q(x,\Delta x,\Delta t)\exp(-\lambda(\Delta x)^2)$, then to letting $\lambda \to 0^+$ at the end of the calculus.

Secondly, the relations (92) and (95) for $k = 1,2,\ldots$ will be applied to the Δx_α, $\alpha = 1,2,3$ or $\alpha = 1,2,3,4$, separately. It follows that the normalization factor γ in formula (9) for instance will not be given by (94); instead it will be the product of four normalization factors, namely

$$\gamma = \frac{1}{(\sqrt{2i\pi D\Delta t})^3} \; \frac{1}{\sqrt{-2i\,\pi \frac{D}{c^2}\,\Delta t}} = -\frac{ic}{(2\pi D\Delta t)^2} \tag{96}$$

Thirdly, the fact that $D = D(x)$ in formula (9) and that, according to Appendix 1, $D_o = D_o(\underline{x},t)$ in formula (7), is unimportant for the present calculus since the integrations are carried over the Δx_α, $\alpha = 1,2,3,4$ or $\alpha = 1,2,3$, for fixed x or $(\underline{x},t)$.

At last

$$\int_{-\infty}^{+\infty} x^{(2k+1)} \exp(ix^2)dx = 0, \qquad k = 0,1,2,\ldots$$

STOCHASTIC QUANTISATION IN A RIEMANNIAN MANIFOLD

Etim Etim
Laboratori Nazionali dell'INFN, Frascati
Italy

ABSTRACT.

There is an ambiguity in quantising a classical system in a curved background. It concerns the arbitrariness of the coefficient multiplying the curvature scalar in the Hamiltonian. This problem has somehow been confused, in the method of stochastic quantisation, with the difficulty of defining forward and backward derivatives of tensor fields in a Riemannian space. Two arguments are offered to show that the ambiguity is not avoided in stochastic methods. In the first we compare the kernel of the Feynman path integral with that of the Markovian semi-group transformations and in the second we randomise the metric itself keeping terms up to second order in Wiener differentials.

The Riccati equation has a revealing application in non-relativistic quantum mechanics. This insight is due to Nelson(1), but its genesis goes back to the early days of quantum mechanics(2). To begin with consider the equation

$$\frac{du(x)}{dx} = a(x)\, u^2 + b(x)\, u + c(x) \tag{1}$$

It is linearized by means of the substitution(3)

$$u(x) = -\frac{1}{a(x)} \frac{d}{dx} \ln \varphi(x) \tag{2}$$

to

$$-\frac{1}{a(x)} \frac{d^2\varphi(x)}{dx^2} + \left(\frac{a'(x)}{a^2(x)} + \frac{b(x)}{a(x)} \right) \frac{d\varphi(x)}{dx} - c(x)\, \varphi(x) = 0 \tag{3}$$

It is therefore equivalent to the one-dimensional time-independent Schrödinger equation

$$-\frac{\hbar^2}{2m} \frac{d^2\varphi(x)}{dx^2} + V(x)\, \varphi(x) = 0 \tag{4}$$

if $b(x)=0$ and $a(x)$ is a constant or $a(x)=a(x_0)\exp(-\int_{x_0}^{x} b(z)dz)$.

This relationship extends to systems of more than one degree of freedom. In a space of dimension n, $u(x)$ is a vector $u_\mu(x)$ and eqs.(1) and (2) generalise to

$$\frac{\hbar}{2}\, \partial_\mu u_\mu(x) + \frac{m}{2}\, u^2(x) = V(x) \tag{1'}$$

$$u_\mu(x) = \frac{\hbar}{m}\, \partial_\mu \ln \varphi(x) \tag{2'}$$

in a flat space.

Taking the gradient of both sides of eq.(1') and making use of (2') one gets ($\Delta \equiv \partial_\mu^2$)

$$(u_\nu \partial_\nu + \frac{\hbar}{2m} \nabla) u_\mu = \frac{1}{m} \partial_\mu V(x) \tag{5}$$

The main idea of the stochastic approach to quantum mechanics is that the operator

$$A = b_\nu(x)\partial_\nu + \frac{\hbar}{2m} \nabla \tag{6}$$

is the infinitesimal generator of the stochastic process defined by the Langevin equation

$$dx_\mu (t) = b_\mu(x,t)\, dt + dW_\mu (t) \tag{7}$$

It is thus possible, by means of this identification, to relate quantum mechanics to classical probability theory. What is achieved is a theory in which the random classical trajectories are hidden variables. This method of quantisation therefore puts in differential form the basic ideas of Feynman's path integral(4). Both methods of quantisation agree in flat space but seem puzzingly to differ in a curved background.

I will assume that the ambiguity associated with quantisation in a curved space is known(6,7). It concerns the arbitrariness of the numerical factor multiplying the scalar curvature in the Hamiltonian. This ambiguity has somehow been confused, in the method of stochastic quantisation with the difficulty of finding appropriate generalisations of forward and backward derivatives of tensor fields(5). The latter problem is kinematical. That of the presence of the scalar curvature term in the Hamiltonian, on the other hand, is dynamical, for if the scalar curvature is not a constant the energy spectrum will be modified. The presence of a scalar curvature term in the Hamiltonian obtained through stochastic quantisation can be ascertained in two ways. First by comparing the kernel in the Feynman path integral with the kernel of the semigroup operator associated with the process (7). Both kernels must have the same normalisation if the transformations induced by them are to be compared. Secondly by randomising the metric, the same type of reasoning which leads to the identification of Planck's constant in eq.(7) also legitimates the presence of the curvature scalar in the Hamiltonian.

Consider the system with Lagrangian

$$L(\dot{x}(t),x(t)) = \frac{m}{2} g_{\mu\nu}(x)\dot{x}^{\mu}\dot{x}^{\nu} - V(x);\mu,\nu = 1,2\ldots n \tag{8}$$

Since V(x) is a sure function stochastic elements of motion about the trajectories of $L(\dot{x},x)$ are defined relative to the quadratic form

$$Lo\ (\dot{x},x) = \frac{m}{2}\ g_{\mu\nu}(x)\dot{x}^{\mu}\ x^{\nu} \tag{9}$$

and its geodesic

$$\frac{d^2\ x^{\mu}(t)}{dt^2} + \Gamma^{\mu}_{\lambda\tau}\ \frac{dx^{\lambda}}{dt}\ \frac{dx^{\tau}}{dt} = 0 \tag{10}$$

The basic assumption which gives rise to eq.(7) is that variations $|\ \Delta x^{\mu}\ |$ about these trajectories are of order $(\Delta t)^{1/2}$.

Expanding $x^{\mu}(t+\Delta t)$ in powers of Δt and making use of eq.(10) we have

$$\begin{aligned} x^{\mu}(t+\Delta t) &= x^{\mu}(t) + \dot{x}^{\mu}(t)\Delta t + \frac{\ddot{x}^{\mu}}{2!}(t)(\Delta t)^2 + \frac{1}{3!}\dddot{x}^{\mu}(t)(\Delta t)^3 + \ldots \\ &= x^{\mu}(t) + \dot{x}^{\mu}(t)\Delta t - \frac{1}{2}\Gamma^{\mu}_{\nu\lambda}\ \dot{x}^{\nu}\dot{x}^{\lambda}(\Delta t)^2 \\ &\quad - \frac{1}{3!}(\partial_{\sigma}\Gamma^{\mu}_{\nu\lambda} - 2\Gamma^{\mu}_{\rho\nu}\ \Gamma^{\rho}_{\lambda\sigma})\ \dot{x}^{\mu}\ \dot{x}^{\lambda}\ \dot{x}^{\sigma}(\Delta t)^3 + \ldots. \end{aligned} \tag{11}$$

Introducing the assumption $|\Delta x^{\mu}(t)| \lesssim (\Delta t)^{1/2}$ in (11) one defines therefrom the stochastic process

$$\begin{aligned} dw^{\mu}(t) &= d\xi^{\mu}(t) - \frac{1}{2}\Gamma^{\mu}_{\nu\lambda} d\xi^{\nu}(t) d\xi^{\lambda}(t) - \frac{1}{3!}(\partial_{\sigma}\ \Gamma^{\mu}_{\nu\lambda} - 2\Gamma^{\mu}_{\rho\nu}\Gamma_{\lambda\sigma})\cdot \\ &\quad \cdot d\xi^{\nu}(t)\ d\xi^{\lambda}(t) d\xi^{\sigma}(t) \end{aligned} \tag{12}$$

The distribution of the $d\xi^\mu(t)$ is the Gaussian

$$
\begin{aligned}
P_o(x(t)),\Delta\xi,\Delta t) &= N \exp\left[-\frac{1}{\hbar}\int^{t+\Delta t} d\tau\, H(x(\tau),c(\tau))\right] \\
&= N \exp\left(-\frac{V(x)\Delta t}{\hbar}\right) \exp\left(-\frac{m}{2\hbar\Delta t} g_{\mu\nu}(x)\Delta\xi^\mu\Delta\xi^\nu\right)
\end{aligned}
\tag{13}
$$

where H is the Hamiltonian corresponding to L and N is a normalization factor. From eqs.(12) and (13) we have the conditional expectations

$$
\begin{aligned}
E(dw^\mu(t)\,|\,x(t)) &= -\frac{\hbar}{2m} g^{\lambda\nu}\Gamma^\mu_{\lambda\nu}\,\Delta t \\
E(dw^\mu(t)dw^\nu(t)|x(t)) &= \frac{\hbar}{m} g^{\mu\nu}\,\Delta t
\end{aligned}
\tag{14}
$$

If the Wick rotated distribution

$$
\begin{aligned}
P_o(x(t);\Delta\ ,i\Delta t)) &= N \exp \frac{i}{\hbar}\int_t^{t+\Delta t} d\tau L(x(\tau),\dot{x}(\tau)) \\
&= N \exp\left(-\frac{i}{\hbar} V(x)\Delta t\right)\exp\left(\frac{im}{2\hbar\Delta t} g_{\mu\nu}\Delta\xi^\mu\Delta\xi^\nu\right)
\end{aligned}
\tag{15}
$$

is used as the kernel in the Feynman path integral one gets a Schrödinger equation without a scalar curvature term. However $d\xi^\mu(t)$ is <u>not</u> the stochastic process associated with trajectories of the Lagrangian L. It agrees with $dw^\mu(t)$ only in a flat space where the connection coefficients $\Gamma^\mu_{\lambda\nu}$ are identically zero.

Solving eq.(12) by iteration for $d\xi^\mu(t)$ we have

$$
\begin{aligned}
d\xi^\mu(t) &= dw^\mu(t) + \frac{1}{2}\Gamma^\mu_{\lambda\nu}\, dw^\nu\, dw^\lambda \\
&+ \frac{1}{3!}\left(\partial_\sigma \Gamma^\mu_{\nu\lambda} + \Gamma^\mu_{\rho\sigma}\right) dw^\nu dw^\lambda dw^\sigma + \ldots
\end{aligned}
\tag{16}
$$

Now substitute for $d\xi^\mu$ in $P_o(x;\Delta\xi,\Delta t)$ and expand in a series for quasi-moments of $dw^\mu(t)$ i.e.

$$P_o(x;\Delta\xi,\Delta t) \equiv P\,(x);dw,\Delta t) =$$

$$= P_o(x;dw,\Delta t).\Big[1 - \frac{m}{2\hbar\Delta t}\; g_{\mu\nu}\,\Gamma^{\mu}_{\lambda\tau}\; dw^{\nu}dw^{\lambda}dw^{\sigma}$$

$$- \frac{m}{8\hbar\Delta t}\; g_{\mu\nu}\,\Gamma^{\mu}{}_{\lambda\tau}\; dw^{\lambda}dw^{\tau}dw^{\rho}dw^{\sigma} -$$

$$- \frac{m}{6\hbar\Delta t}\; g_{\mu\nu}\,\Gamma^{\mu}_{\lambda\tau}\,\Gamma^{\nu}_{\rho\sigma}\; dw^{\nu}dw^{\lambda}dw^{\tau}dw^{\sigma}$$

$$- \frac{m^2}{8\hbar^2(\Delta t)^2}\; g_{\mu\nu}(\partial_{\mu\nu}g_{\lambda\tau}\,\Gamma^{\mu}_{\rho\sigma}\,\Gamma^{\lambda}_{\alpha\beta}\; dw^{\nu}dw^{\tau}dw^{\rho}dw^{\sigma}dw^{\alpha}dw^{\beta} +$$

$$+ \;.....\Big] \tag{17}$$

The Wick rotated distribution $P(x, dw, i\Delta t)$ as a kernel in the Feynman path integral with integration over $dw\ (t)$ gives rise to the scalar curvature term $\hbar^2 R/6m$ in the Hamiltonian. This comes about through the change in the normalization of $P_o(x,dw,\Delta t)$ induced by the additional terms in eq.(17). In fact we have

$$\int d^n(\Delta w)P(x;\Delta w,\Delta t)\,\Big[\sqrt{g} + \partial_{\mu}\sqrt{g}\;\Delta w^{\mu} + \frac{1}{2}\,\partial^2_{\mu\nu}\;\sqrt{g}\;\Delta w^{\mu}dw^{\nu} + ...\Big]$$

$$= N\,\exp\,(-\frac{V\Delta t}{\hbar})(1 - \frac{\hbar\Delta t\;R}{6m})(2\pi h\Delta t)^{n/2} \tag{18}$$

$$\simeq N\,\exp\Big[-\frac{1}{\hbar}\int_{\Delta t}^{t+\Delta t}(V + \frac{\hbar^2 R}{6m}\Big]\,(2\pi\hbar\Delta t)^{n/2}$$

Comparing this with eq.(15) one sees that the effective potential of the motion perturbed by $dw^{\mu}(t)$ is $V + \hbar^2 R/6m$. The corresponding Riccati equation therefore reads

$$\frac{\hbar}{2}\,\nabla_{\mu}\;u^{\mu}(x) + \frac{m}{2}\;u^{\mu}u_{\mu} = V + \frac{\hbar^2 R}{6m} \tag{19}$$

The same result is obtained by randomising the metric i.e.

$$dg^{\mu\nu} = \frac{m}{\hbar(n+1)}(u^{\mu}(x)d\xi^{\nu}(t) + u^{\nu}(t)d\xi^{\mu}(t)) + \frac{\lambda R(x)}{n}d\xi^{\mu}(t)d\xi^{\nu}(t) + ... \tag{20}$$

$\hbar$ and $R(x)$ appear in eq.(20) on dimensional grounds; λ is a pure number. Using eq.(12) we have the conditional expectations

$$D_\mu g^{\mu\nu}(x) = \lim_{\Delta t\to 0} \frac{1}{\Delta t} E(d\xi_\mu dg^{\mu\nu}|x(t)) = u_\mu\, g^{\mu\nu} \tag{21}$$

$$Dg^{\mu\nu} = \lim_{\Delta t\to 0} \frac{1}{\Delta t} E(dg^{\mu\nu}|x(t)) = \frac{\lambda\hbar}{mn} R\, g^{\mu\nu} \tag{22}$$

From eq.(20) we find that

$$\frac{\lambda\hbar}{m} R = \hbar\, g_{\mu\nu} Dg^{\mu\nu} = \hbar \lim_{\Delta t\to 0} E\left(\frac{d(\ln g(x))}{\Delta t}|x\ (t)\right) \tag{23}$$

$$g(x) = \det\ (g_{\mu\nu})$$

may be interpreted as the (vacuum) energy fluctuation due to perturbation of the matric. To order $\hbar$ the operator D_μ is an extended covariant derivative.

$$D_\mu = \frac{\hbar}{m} \nabla_\mu + u_\mu \tag{24}$$

Eq.(19) therefore expresses the energy relation

$$\left(\frac{m}{2} D_\mu D_\nu - \frac{\lambda\hbar}{m} R_{\mu\nu}\right) g^{\mu\nu} = V \tag{25}$$

The constant λ cannot be determined from eq.(18) only. We therefore conclude that just as $dw^\mu(t)$ must be expanded up to second order in $d\xi^\mu(t)$ in order to see the connection coefficient, $dg^{\mu\nu}$ must be expanded to second order in $d\xi^\mu(t)$ in order to see the curvature scalar.

Taking the gradient of both sides of eq.(25) allows to identify

$$A_{\mu\nu} = \left(b^\alpha\, \nabla_\alpha + \frac{\hbar}{2m}\ g^{\alpha\beta}\, \nabla_\alpha\, \nabla_\beta\right) g_{\mu\nu} - \frac{\hbar}{2m} R_{\mu\nu} \tag{26}$$

as the generator of the process (7) in a curved space.

Although $A_{\mu\nu}$ contains the curvature tensor it is independent of λ. This circumstance is responsible for the conclusion that stochastic quantisation leads to no ambiguity in the Hamiltonian.

REFERENCES

(1) E. Nelson, Phys. Rev. 150, 1079 (1965).

(2) L. De Broglie; Compt. Rend. 183, 447 (1926).

(3) E.L. Ince, Ordinary Differential Equations; Dover Publications Inc. N.Y. (1956), p.23.

(4) R.P. Feynman; Rev. Mod. Phys. 20. 327 (1948).

(5) D. Dohrn and F. Guerra; Lett. Nuovo Cimento 22, 121 (1978) and references quoted therein.

(6) B.S. De Witt; Rev. Mod. Phys. 29, 277 (1957).

(7) K.S. Cheng; Jour. Math. Phys. 13, 1723 (1972).

KINEMATICS AND DYNAMICS OF ITÔ PROCESSES

Laura M. Morato

Istituto di Elettrotecnica ed Elettronica, Università di Padova and Laboratorio per Ricerche di Dinamica dei Sistemi e di Bioingegneria, C.N.R., Padova, C.P.1075, Padova

In this paper we deal with an extension of the Nelson's stochastic mechanics of a quantum particle, Nelson.[1] The basic Nelson's hypothesis is that the "position variable" $q(t) \in R^3$ must be interpreted as a continuous Markov process described by the stochastic differential equation

$$dq(t) = b(q(t),t)dt + (h/2\pi m)^{\frac{1}{2}} dw \tag{1}$$

with initial condition $q(0)=q_0$, h being the Planck's constant, m the mass of the particle and w(t) the standard Brownian motion. The drift b(q(t), t) has to be determined, the crucial constraint being that the Newton law F=ma be satisfied by the mean acceleration a(q(t),t), defined by

$$a(q(t),t) = [(DD^* + D^*D)/2]\, q(t) \tag{2}$$

where D and D^* are the mean forward and backward derivatives of q(t), namely

$$Dq(t) = \lim_{\Delta \downarrow 0} E^{q(t)} \{(q(t+\Delta) - q(t))/\Delta\}$$
$$D^{*}q(t) = \lim_{\Delta \downarrow 0} E^{q(t)} \{(q(t)-q(t-\Delta))/\Delta\} \qquad (3)$$

Denoting by $\varrho(q,t)$ the probability density of $q(t)$ and by v the current velocity field defined by $v(q(t),t) \triangleq [(D+D^{*})/2\ q(t)]$, the two equations

$$\frac{\partial \varrho}{\partial t} = - \operatorname{div} \varrho v \qquad (4)$$

$$ma = F \qquad (5)$$

completely describe the process $q(t)$.

If force F can be derived from the gradient of the potential $V(q)$, then the wave function $\psi(q,t) \triangleq \varrho(q,t)^{1/2} e^{iS(q,t)}$ (where $S(q,t)$ is defined by $v = \frac{h}{m}$ grad S) satisfies the Schrödinger equation

$$i \frac{\partial \psi}{\partial t} = (- \frac{\hbar}{2m} \nabla^2 + \frac{1}{\hbar} V)\psi \qquad (6)$$

If F is the Lorentz force produced by the electromagnetic field described by the four-vector potential $(\mathcal{A}, \varphi)$, and if phase S is defined by $v = \frac{h}{m}$ (grad $S - \mathcal{A}$) one has

$$i \frac{\partial \psi}{\partial t} = (\frac{\hbar}{2m} (i\nabla + \mathcal{A})^2 + \frac{1}{\hbar} \varphi)\psi \qquad (7)$$

Note that in this second case $S(q,t)$ is an arbitrary differentiable real valued function of q and t.

Let us now consider a n-dimensional continuous Markov process described by the Itô stochastic differential equation

$$dx(t) = b(x(t),t)dt + G(x(t),t)dw \qquad (8)$$

with initial condition $x(0)=x_0$. Here b and G are respectively n-vector and nxm matrix-valued functions of x and t and w is the m-dimensional standard Brownian motion.

It has been shown in Morato[4] that the kinematics proposed by Nelson for the class of processes described by (1) can be naturally extended to Itô processes.

This result is obtained by the explicit constructing of the backward representation of x(t), under regularity conditions and for all $x \in \mathcal{X}$, where $\mathcal{X}$ denotes the set where the probability density $\varrho(x,t)$ of x(t) at time t does not vanish. We report here the final kinematical formulae of the current velocity v and of the mean acceleration a defined as before. One has

$$v = b - \frac{1}{2}\,(GG^T \mathrm{grad}\,\ln\varrho + (\nabla^T GG^T)^T) \tag{9}$$

$$a = -(\tfrac{1}{2}(GG^T\nabla)\nabla^T u + (u^T\nabla)u) + \frac{\partial v}{\partial t} + (v^T\nabla)v) \tag{10}$$

where u is defined by $u \triangleq [(D-D^*)/2]\, x(t)$ and may be computed as follows

$$u = \frac{1}{2}\,(GG^T \mathrm{grad}\,\ln\varrho + (\nabla^T GG^T)^T) \tag{11}$$

One may prove the following Theorem

Theorem 1 - Let $x(t) \in \mathcal{X}$ be an Itô process described by (8) with initial condition $x(0)=x_o$ and probability density $\varrho(x,t)$ which does not vanish on $\mathcal{X} \times [0,\infty]$. Assume that x(t) admits a current velocity field and a mean acceleration field expressed by (9)-(11). Assume also that there exists some number $c>0$ such that $y^T GG^T y \geq c y^T y$ for all $y, x \in \mathcal{X}$ and $t \in [0,\infty)$. Let S and S' be any real-valued differentiable function of x and t. Then there exists a class of (n+1) vector valued functions $(\mathcal{A}_S, \varphi_S)$ of x and t that is invariant under the gauge transformation

$$\begin{aligned} \mathcal{A}_{S'} &= \mathcal{A}_S + \mathrm{grad}(S'-S) \\ \varphi_{S'} &= \varphi_S - \frac{\partial(S'-S)}{\partial t} \end{aligned} \tag{12}$$

such that the operator

$$\mathcal{H}_S = \frac{1}{2}\,(i\nabla + \mathcal{A}_S)^T\, GG^T (i\nabla + \mathcal{A}_S) + \varphi_S \tag{13}$$

determines the time evolution of the wave function $\psi_S = \varrho^{1/2} e^{iS}$ by the Schrödinger-type equation

$$i \frac{\partial \psi_S}{\partial t} = \mathscr{H}_S \psi_S \qquad (14)$$

Proof. By assumption we can define the n-vector valued function of x and t $\mathscr{A}_S(x,t)$ $x \in X$, $t \in [0,\infty)$, by the equality

$$\mathscr{A}_S = -(GG^T)^{-1} v + \text{grad } S \qquad (15)$$

By introducing (15) in the definition of $\mathscr{H}_S$ and by expliciting (14) for $\psi_S = \varrho^{1/2} e^{iS}$, one gets, by separating the real and the imaginary parts

$$\frac{\partial \varrho}{\partial t} = - \text{div } v\varrho \qquad (16)$$

$$- \frac{1}{2}\left(\frac{1}{2} \nabla^T GG^T \text{grad } \ln \varrho + \frac{1}{4} \text{grad } \ln \varrho^{\ T} GG^T \text{grad } \ln \varrho \right) + \frac{1}{2} v^T (GG^T)^{-1} v = - \left(\varphi_S + \frac{\partial S}{\partial t}\right) \qquad (17)$$

By assumption v is given by (9) and then (16) represents the correct continuity equation of x(t).

The searched class is uniquely defined by (15) and (17). Note that, as (4) and (5) completely describe the process q(t) if F is given, here if φ_S is given and G is fixed, (16) and (17) completely describe the more general process x(t).

The theorem shows that the time evolution operator (13) is not typical of the quantum description of a particle in an external electromagnetic field, but it seems to represent a property of a wide class of continuous Markov processes.

What can we say about the analogue of the Newton law F=ma in this case? A partial answer is contained in the following Proposition.

Proposition 1. Let x(t) be a stochastic process satisfying the same hypothesis of Theorem 1 and let G not depend on x. Then the mean acceleration a can be written in the quasi-Lorentz form

$$a = GG^T(E + v \wedge H) + (\frac{\partial}{\partial t} GG^T)(GG^T)^{-1}v \tag{18}$$

where E and H are vector fields defined by

$$E = -\operatorname{grad} \mathscr{A}_S - \frac{\partial S}{\partial t} \tag{19}$$

$$H = \operatorname{curl} \varphi_S \tag{20}$$

$(\mathscr{A}_S, \varphi_S)$ being defined as in (15) and (17).

Proof. The assertion follows by taking gradients of both sides of (17) and then comparing, for G independent of x, with the explicit expression of a given by (9) and (10). One gets

$$a - \frac{\partial v}{\partial t} - (v^T\nabla)v = -GG^T \operatorname{grad}(\varphi_S + \frac{\partial S}{\partial t} + \frac{1}{2} v^T(GG^T)^{-1}v) \tag{21}$$

Then (18) follows by recalling (15) and by applying the identity

$$\frac{1}{2} \operatorname{grad} y^TQy = (y^TQ\nabla)y + Qy \wedge \operatorname{curl} y \tag{22}$$

for any $y \in R^n$ and any nxn symmetric and constant matrix Q

From proposition 1, it is appearent that if there exists a differentiable real-value function S(x,t) of x and t such that $v=GG^T \operatorname{grad} S$ and G is a constant matrix then there exists a "potential" V(x,t) such that $a=-GG^T \operatorname{grad} V$. Note that in this case it is not necessary to assume that GG^T can be inverted.

We can conclude that, if the diffusion matrix G does not depend on x, the continuous Markov process defined by (8) is completely described by the analogue of (4) and (5), where F/m may be in general expressed through a (n+1)-vector potential.

In this case the Schrödinger-type equation (14) may be seen as a compact way of writing (4) and (5).

To complete the analysis we now stress the crucial fact that distinguishes the stochastic processes which are associated, by the Nelson's theory, with the usual quantum description of a physical system from the more general ones considered in this paper. The point is that the potentials which describe the stochastic force in Nelson's theory are, by construction, the "classical" ones. There is no reason that this has to be a general rule. To clarify the point let us consider the simple example of a one-dimensional gaussian process x(t) with mean s(t) and constant variance, described by the stochastic differential equation

$$dx(t) = - k(x(t)-s(t)-\dot{s}(t)/k)dt + kgdw \tag{23}$$

with initial condition $x(0)=x_0 \sim N(s_0,k^2g/2)$. Here k and g are positive constants and s(.) is a fixed trajectory which is assumed to be solution of the equation:

$$\ddot{s} = - \frac{\partial U(s,t)}{\partial s} \tag{24}$$

for the initial conditions $s(0)=s_0$, $\dot{s}(0)=\dot{s}_0$. Clearly the limit process for g going to zero is the fixed trajectory s(.). By applying (9)-(11) one gets

$$a(x,t) = - \frac{\partial V(x,t)}{\partial x} \tag{25}$$

where V(x,t) is defined, up to an arbitrary function of t, by

$$V(x,t) = \frac{1}{2} k^2 x^2 + (\ddot{s}(t) + k^2 s(t))x \tag{26}$$

It is appearent that one has, in general, $V(.,t) \neq V(.,t)$. This particular case, which is the quantum case, occurs if (23) is the stochastic representation of a quantum state of a one-dimensional oscillator. One has, in this case $\ddot{s}(t)=-k^2 s(t)$ and then the stochastic potential V(.) is the classical one.

REFERENCES

1. Nelson, E., Derivation of the Schrödinger equation from Newtonian mechanics, *Phys. Rev.*, 150, n.4, 1966.
2. Nelson, E., *Dynamical theories of Brownian motion*, Princeton University Press, 1967.
3. Dankel Jr., T.G., Mechanics on manifold and incorporation of spin into Nelson's stochastic mechanics, *Arch. Rational Mech. Ann.*, 37, 1970.
4. Morato, L.M., Notes on stochastic mechanics, Int.Rep. LADSEB-CNR, 1978.

QUANTUM MARKOV PROCESSES

LUIGI ACCARDI
Istituto di Matematica
Università di Milano - Italy

O.) Quantum mechanics is both a new mechanics and a new probability theory. The probabilities obtained as an output of quantum theoretical computations are interpreted by the physicists exactly in the same way as the probabilities computed with more traditional tools (namely, in most cases, as expected relative frequencies).
The existence of two, quite different mathematical formalisms both aimed at the computation of the same type of quantities (probabilities, expectation values, ...) naturally raises some questions about their mutual relations.

1.) Does the difference of the two mathematical formalisms reflect a difference in physical assumptions? Can we consider these formalisms as necessary consequences of physically plausible requirements?

2.) To what extent the quantum mechanical formalism is necessary? Or, stated otherwise, is it true that all the conclusions of quantum theory can be deduced within the framework of classical probability theory?
The opinions concerning the answer to question (2) are widely different. On one side, the proposals of various models of stochastic mechanics, electrodynamics ... seem to indicate the existence of a reductionist program implying that the quantum mechanical models are just classical models

appropriately perturbed by some classical random noise. On the opposite side Bell's inequality is invoked to conclude that there are some probabilistic assertions of quantum theory which cannot be deduced by whatsoever "physically acceptable" classical probabilistic model.

Concerning question (1), many serious attempts have been made but, at the moment, a really convincing solution is lacking. As a preliminary step towards the solution of this question we can consider the following problem:

3.) Is there a general theory which includes both the classical and the quantum probabilistic formalism? (just as the theory of differentiable manifolds includes the euclidean and the classical non euclidean geometries).

That is, one first looks for an unification of the two theories at a formal mathematical level, then within this unified theory we investigate the physical requirements which determine the choice of one of the different and inequivalent models (just as different assumptions on the curvature of a riemannian manifold will isolate different classes of mathematical models). The first attempt to answer this question goes back to I. von Neumann who, in 1935, formulated a set of axioms defining a general probabilistic theory such that both classical probability and quantum theory where different mathematical models of these axioms. This problematic lies at the roots of one of the most fruitfull lines of research of XX-th century's mathematics: the general theory of operator algebras. We cannot even mention here the salient points of this development. One of the achievements of this research was the establishment of a general algebraic theory of measure and integration which allows, in particular, to deal with the non commuting observables arising in quantum theory. However, in order to include in this scheme a general algebraic formulation of probability theory, some further steps were necessary. First of all one had to find the quantum theoretical counterpart of the concepts of *conditional expectation* and of *conditional density* (this necessity was felt by von Neumann himself who dealt with this question in a long, unpublished manuscript of 1937).

A second, necessary, step was the quantum theoretical formulation of the concept of *stochastic process*. And, finally, the quantum counterpart of the most important stochastic processes (Markov, Gaussian, ...) had to be formulated. There is a large literature dealing with these problems which, for reasons of space, we cannot review here. We will limit ourselves to describe a new proposal for the solution of these problems, differing from the proceeding ones in the main ideas and in the fact that it satisfies the following requirements:

i) mathematical coherence and generality
ii) to prescribe positive expectation values to positive observables
iii) to include the classical theory (i.e. to reduce to it when all the observables commute).

The aim of this exposition is to explain the main ideas of the definitions and the constructions involved by illustrating them with the simplest possible examples. For this reason, as well as for the limited space, the exposition will be mainly heuristic and, sometimes, schematic and incomplete.

1.) The Conditional Density Matrix

We shall discuss only systems with a finite number of states. *Algebraic probability theory* is the study of the couples $\{A , \mu\}$ where A is an algebra of complex matrices (containing the identity matrix) and μ is a *state* on A (i.e. a linear functional, positive on positive matrices and such that $\mu(1) = 1$).

Example 1. $A = \{$ all diagonal $n \times n$, $\mathbb{C}$-matrices$\} = D$
Every state on A has the form

$$\mu(a) = \sum_{i=1} p_i a_i \; ; \quad a = \mathrm{diag}(a_1 , \dots , a_n)$$

where $p_i \geq 0$ and $\Sigma p_i = 1$. Conversely any matrix $W = \mathrm{diag}(p_1, \dots, p_n)$, with p_i as above, defines a state on A through the formula

$$\mu(a) = \tau_0 (W a)$$

where τ_0 denotes the trace on $n \times n$ matrices. Thus, in our example, the couple $\{A , \mu\}$ is equivalent to the probability space $\{S , \bar{\mu}\}$ where $S = \{1, \dots , n\}$ and $\mu(i) = p_i$, $i \in S$.

Example 2. $A = \{$ all $n \times n$, $\mathbb{C}$-matrices$\} = M$
In this case it is known that a linear functional μ on A is a state if and only if

it has the form

$$\mu(a) = \tau_o(W a)$$

where $W \geqslant 0$ and $\tau_o(W) = 1$.
Any $n \times n$ matrix W with the above listed property is called a *density matrix*. Since example (2) includes example (1), the concept of density matrix generalizes the classical concept of probability density.
In order to generalize the concept of *conditional probability density*, let us first of all formulate it in algebraic way, in the language of example (2).
Let $S = \{1, \dots, n\}$ be a space with n points; let $\mu : (i, j) \in S \times S \to p(i, j) \geqslant 0$, $\sum_{i,j \in S} p(i, j) = 1$, be a probability measure on $S \times S$; and let $\mu_o : i \in S \to p_i^o$ be the marginal probability defined by: $p_i^o = \sum_{j \in S} p(i, j)$. The conditional probability $p(j \mid i)$ of the couple $(i, j) \in S \times S$ given its first component is

$$p(j \mid i) = \frac{p(i, j)}{p_i^o} \tag{1}$$

The conditional expectation of a random variable $a : (i, j) \in S \times S \to a(i, j) \in \mathbb{C}$ is defined by

$$E(a)(i) = \sum_{j \in S} a(i, j)\, p(j \mid i) \tag{2}$$

and satisfies the important relation

$$\mu(a) = \mu_o(E(a)) \tag{3}$$

Let now e_{ij} denote the $n \times n$ matrix all whose elements are O but the (i, j)-th one which is equal to 1. According to example (1) the probability space $\{S \times S, \mu\}$ is equivalent to the couple $\{A, \mu\}$ where A is the algebra of diagonal $n^2 \times n^2$ matrices, i.e. $D \otimes D$, and μ is the state on $A = D \otimes D$ given by

$$\mu(a) = \tau(W a) \quad ; \quad a \in D \otimes D$$

$$W = \sum_{i,j} p(i,j) \; e_{ii} \otimes e_{jj}$$

$\tau (= \tau_o \otimes \tau_o)$ is the trace on $M \otimes M$. Similarly the probability space $\{S, \mu_o\}$ is equivalent to

$$\{D, \mu_o\} \; ; \; \mu_o(a_o) = \tau_o(W_o \, a_o) \; ; \; a_o \in D \; ;$$

$$W_o = \Sigma \; p_i^o \; e_{ii}$$

Denoting now $E^o : M \otimes M \rightarrow M$, the partial trace:

$$E^o : a = \sum_{i,j} a_{ij,hk} \; e_{ij} \otimes e_{hk} \rightarrow \sum_{ij} [\Sigma \; a_{ij,hh}] \, e_{ij}$$

we have the equalities

$$W_o = E^o(W) \tag{4}$$

$$E(a) = \sum_i E(a)(i) \, e_{ii} = E^o(\underline{W}_o^{-1} \, W \, a)$$

where $\underline{W}_o = W_o \otimes 1$, or, symmetrising:

$$E(a) = E^o(\underline{W}_o^{-1/2} \, W_o^{1/2} \cdot a \cdot W^{1/2} \, \underline{W}_o^{-1/2}) \tag{5}$$

the quantity

$$K = W^{1/2} \, \underline{W}_o^{-1/2} \tag{6}$$

denotes, in the classical case, the diagonal matrix

$$K = \sum_{ij} p(j \mid i)^{1/2} \, e_{ii} \otimes e_{jj}$$

and from the relations

(7) $$\mu(a) = \tau(W\,a)$$

(8) $$E(a) = E^{o}(K^{*}\,a\,K)$$

we see that just as W is a density with respect to the trace, K can be considered as a density with respect to the partial trace. Remark that, because of (4), one has

(9) $$E^{o}(K^{*}K) = 1 \quad (= \text{identity } n \times n \text{ matrix})$$

In the above formulas a, W, W_o are diagonal matrices. Dropping this assumption, all formulae from (3) to (9) still hold. This shows that the quantity $K = W^{1/2} W_o^{-1/2}$ is defined also in a quantum context and plays the role of the (square root of the) conditional density matrix. More generally, we will call a *conditional density matrix* any matrix $K \in M \otimes M$ satisfying (9).

Remark 1. Throughout the discussion we have assumed the invertibility of the operator W_o.

Remark 2. In the non-commutative case, the appearance of the square root in (6) can be justified in terms of Tomita's theory.

2.) Quantum Markov Chains

We now use the concept of conditional density matrix to construct a quantum analogue of the Markov chains with a finite number of states. Let N be the set of the natural integers, considered as a 1-dimensional lattice. To each point n of N we associate a 2-state system σ_n with state space $S = \{+1, -1\}$ – say, a spin system. If the σ_n interact through a nearest neighbour potential, the equilibrium situation of the infinite system $\sigma \equiv \{\sigma_n\,;\, n \in N\}$ – considered as a classical system – is described by a Markov chain $\mu \equiv \{\mu_o, P\}$, where μ_o is a state on the algebra D (= diagonal 2×2 matrices) and $P = (P_{ij})\ 1 \leqslant i, j \leqslant 2$ is a stochastic matrix (i.e. $p_{ij} \geqslant 0$; $\sum_{j=1}^{2} p_{ij} = 1$). To each point n of the lattice, we associate the corresponding algebra of observables $D_n \cong D$. The statistical description of the system is determined by the assignment of all the correlation functions $\mu(a_o \cdot \ldots \cdot a_n)$, $a_j \in D_j$. For a classical Markov chain these correlations are expressed in terms of μ_o and P as follows: let

$$K = \sum_{ij} p_{ij}\, e_{ii} \otimes e_{jj}$$

and denote $\mathcal{E}(x) = E^o(K^* \times K)$; $x \in D \otimes D$. Then

$$\mu(a_o \cdot a_1 \cdot \ldots \cdot a_n) = \mu_o(\mathcal{E}(a_o \otimes \mathcal{E}(a_1 \otimes \ldots \otimes \mathcal{E}(a_n \otimes 1)) \qquad (10)$$

The change from the classical to the quantum description is accomplished by considering, instead of D, the algebra M of all 2×2 complex matrices. Again, the statistical description of the whole quantum system is determined by the assignment of the correlation functions. It is easy to verify that for any quantum conditional density matrix the expressions (10) define an agreeing family of correlation functions, hence a state of the infinite quantum system.

Remark The consideration of the Ising and the Heinsenberg model shows that the *conditional density matrix is completely determined by the interaction hamiltonian.*

3.) Quantum Markov Processes: Continuous Parameter

The construction of N. 2 can be generalized in case of continuous parameter. In this case, instead of N we consider R^+ (= real numbers ≥ 0); to each $t \in R^+$ we attach the algebra $B_t \cong B(\mathcal{H})$ = algebra of all bounded operators on the complex separable Hilbert space $\mathcal{H}$ – the algebra of "observables at time t"; and again we consider the statistical description of the system completely determined by all the correlation functions $\mu(a_o \cdot a_{t_1} \cdot \ldots \cdot a_{t_n})$ where $a_t \in B_t$. In this case, instead of the operator $\mathcal{E} : D_n \otimes D_{n+1} \to D_n$, we will have a 2-parameter family of operators $\mathcal{E}_{t,s} : B_s \otimes B_t \to B_s$; and we look for correlation functions of the form:

$$\mu(a_o\, a_{t_1} \ldots a_{t_n}) = \mu_o(\mathcal{E}_{t_1,o}(a_o \otimes \mathcal{E}_{t_2,t_1}(a_{t_1} \otimes \ldots \otimes \mathcal{E}_{t,t_n}(a_{t_n} \otimes 1) \ldots) \qquad (11)$$

where $0 < t_1 < \ldots < t_n < t$.

If we want that the right hand side of (11) defines an *agreeing* family of correlation functions, then there will be some compatibility equations among the $\mathcal{E}_{t,s}$, of the type:

(12) $$\mathcal{E}_{t,s}(1 \otimes 1) = 1$$

(13) $$\mathcal{E}_{s,r}(a_r \otimes \mathcal{E}_{t,s}(1 \otimes a_t)) = \mathcal{E}_{t,r}(a_r \otimes a_t)\ ;\ r < s < t$$

Moreover if we want positive correlations for positive observables, the $\mathcal{E}_{t,s}$ have to be taken completely positive. Now define, for each $t > s$, the map

$$Z(t, s) : a \in B(\mathcal{H}) \cong B_t \rightarrow \mathcal{E}_{t,s}(1 \otimes a) \in B(\mathcal{H}) \cong B_s$$

this map is completely positive, maps 1 into 1, and satisfies the Chapman-Kolmogorov equation:

(14) $$Z(s, r) \cdot Z(t, s) = Z(t, r)\ ;\ r < s < t$$

i.e. it is a quantum dynamical (or quantum markovian) evolution.

Remark. In the classical case (i.e. each $B_t \equiv$ a commutative algebra) the $\mathcal{E}_{t,s}$ (transition expectations) are completely determined by the markovian evolution $Z(t, s)$ through the relation:

$$\mathcal{E}_{t,s}(a_s \otimes a_t) = a_s \cdot Z(t,s)(a_t)$$

where the dot denotes multiplication.
It is clear that such a relation cannot take place in the quantum case since in this case the right hand side might be nonpositive even if $a_s \otimes a_t$ is positive.

Thus, in general, in the quantum case, *differently from the classical case the evolution $Z(t, s)$ does not determine the transition expectations.* The following theorem shows that Heisenberg evolution are a lucky exception:

Theorem 1. Given a (initial) state μ_0 on $B(\mathcal{H})$ and an Heisenberg evolution

$$Z(t, s)(a) = e^{i(t-s)H} a e^{-i(t-s)H}$$

then there exists a ***unique*** quantum Markov process $\{\mu_o, \mathcal{E}_{t,s}\}$ whose canonical evolution is the given one.

Theorem 2. Quantum Markov processes associated to usual quantum systems are characterized by the property that their canonical evolution is reversible.

Remark. In the classical case there is a $1-1$ correspondence between deterministic systems and classical Markov processes whose canonical evolution is reversible and whose initial state is pure.
Moreover, as it happens with classical Markov processes with reversible evolution and pure initial state, the correlation functions of the quantum Markov processes corresponding to usual quantum systems factorise, i.e.

$$\mu(a_o\, a_{t_1} \dots a_{t_n}) = \mu_o(a_o)\, \mu_{t_1}(a_{t_1}) \dots \mu_{t_n}(a_{t_n})$$

Theorem 3. If the map $s, t \mapsto \mathcal{E}_{t,s}$ is "regular enough" then $\mathcal{E}_{t,s}$ has the form

$$\mathcal{E}_{t,s}(a \otimes b) = T_s(a \otimes Z(t, s)(b))$$

where $T_s : B_s \otimes B_t \to B_s$ is a completely positive linear map.
The operator T_s should be interpreted as the operator of "instantaneous coupling" at time s. In the classical case it is just the multiplication ($T_s(a \otimes b) = a \cdot b$).
Remark now that, for a given evolution $Z(t, s)$ the compatibility equations for the $\mathcal{E}_{t,s}$ become equations in the unknowns T_s. Let us introduce the condition of "evolution covariance"

$$T_s(Z(t, s) \otimes Z(t, s)) = Z(t, s) \cdot T_t$$

(i.e. if one first couples two observables at time t and then brings back the result at time s, the result is the same as first bringing back both observables at time s, and then coupling the result). With this condition the unknowns are reduced to the single T_0 (coupling at time 0).

Theorem 4. Let Z(t, s) be reversible (i.e. Heinsenberg). And assume that

$$T_0(B_0 \otimes B_0) = M_0 \text{ is an algebra .}$$

Then there are only two classes of solutions of the compatibility equations (in the unknown T_0). The former corresponds to the quantum Markov processes described in Theorem (2) – i.e. usual quantum systems.

The latter gives rise to a class of quantum Markov processes which correspond to a generalization of von Neumann's theory of the quantum measurement process.

Moreover also for this class the joint correlations are explicitly computed, and it is found that in the classical case they reduce to the usual correlation functions of ergodic theory and classical statistical mechanics.

BIBLIOGRAPHY

[1] L. Accardi, Non commutative Markov chains, Proceedings, School in Math. Phys., Camerino 1974.

[2] L. Accardi, Non relativistic quantum mechanics as a noncommutative Markov process, Advances in Math. 20 (1976), 32 9-366.

[3] L. Accardi, Non commutative Markov chains with a preassigned evolution: an application to the quantum theory of measurement, Advances in Math. 1978.

[4] L. Accardi, Quantum Markov processes, Proceedings, Symp. "Mathematical problems in the quantum theory of irreversible processes", Arco Felice, 1978.

[5] L. Accardi, On the quantum Feynman-Kec formula, Rendiconti Seminario Matematico e Fisico, Università di Milano, 1978.

[6] L. Accardi, On the non-commutative Markov property. Functional Anal. and its appl. 1 (1975), 1-6.

STOCHASTIC ELECTRODYNAMICS
GENERAL CONSIDERATIONS

by

Simon DINER
Institut de Biologie Physico-chimique
Laboratoire de Chimie Quantique
13 rue Pierre et Marie Curie.75005 PARIS
FRANCE.

Stochastic Electrodynamics is a tentative to account for the peculiarities of quantum physics from a completely classical point of view. It has the advantage over many other attempts on that way to be physically well defined and to lead to equations which are not impossible to solve, so that it is possible to check the theory against experiment or other theories, as quantum mechanics.

The epistemological challenge to contemporary physics is the difficulty of coexistence, if not of passage between quantum mechanics and the whole body of classical theories. Such a situation is one of crisis, because the main goal of scientific activity seems to be the unification of different theories, or at least the building of a dictionary of correspondance between the concepts and laws of different theories. Contemporary theoretical physics has not yet been able to achieve a satisfactory unification, despite some efforts in this direction (1-4). Most of the physicists think that quantum ideas are going to prevail and be the nucleus of the future unification of physics ; they think that after the mechanical view of nature and the electromagnetic view of nature (5) there will be a quantum view of nature. Such a trend becomes apparent when fundamental concepts of classical theories are considered as less relevant as corresponding concepts in quantum theory. A good example of such a situation is given by the tendancy to present statistical physics from a quantum point of view (e.g. Kittel (6)), considering that "a correct definition of entropy is only possible in the framework of Q.M., whereas in classical mechanics entropy can only be introduced in a somewhat limited and artificial manner" (7).

Another example is given by the definition of a "macroscopic level of quantum mechanics" (8) and by the current speaking about quantum behaviour on a macroscopic scale for lasers, supra-conductors and superfluids (9).

From the historical point of view it is interesting to stress that quantum theory arose in the very moment where the existence of the proof of atomic reality would have seemed to lead to a definitive triumph of some mechanical picture of the world. After fifty years of progressive domination of the quantum ideology, classical theories develop again, showing their power for the description of reality. "The increasing importance of dynamical macro-systems in physics, chemistry and biology has moved classical physics out of the shadow of quantum theory into the actuality of modern research" (10). In fact " a primary atomism" where the discontinuity of matter and radiation are irreducible a priori explanations, is progressively replaced by an "organic atomism" which gives a generative description of discontinuities from continuous and inorganized systems. The successes of the qualitative theory of dynamical systems (including ergodic theory), the field of non-linearity (including soliton theory) and the modern development of phase transition theory show in an obvious way that "classical complexity" has not exhausted its explanatory power and reveal continuously unexpected possibilities.

Stochastic electrodynamics is precisely a tentative to see if such an unexpected possibility of understanding quantum behavior would not lie in a formulation of classical electrodynamics taking in account the existence of a universally present random electromagnetic field.

To say truth, the challenges to which such a tentative is going to be faced are strong and numerous. It will have to wind around the many "impossibility theorems" which seem to prevent from a smooth passage between classical theories and quantum mechanics. But on the other side we shall see that such a classical theory provides so many answers to "mysterious facts" of quantum theory, that it is temptating to consider it as more than a simple mirage in the desert of ignorance.

First, I shall rapidly review the main points of difficulties in a free passage beteween classical concepts and theories and quantum theory. In a second part I shall briefly present the main successes of Stochastic Electrodynamics. P. Claverie will develop the mathematical formalism of that theory and show how, despite it's first success, it does not seem in it's present state to be able to answer to the more fundamental question of microphysics. Nevertheless such a theory is very useful to point out clearly the main apparently irreducible concepts and facts of quantum mechanics, and we hope will help in some way the further developments of microphysics.

I. - CONNECTIONS AND INCOMPATIBILITIES BETWEEN QUANTUM THEORY AND CLASSICAL THEORIES.

1. - Quantum theory and classical mechanics

The umbilical cord between classical mechanics and quantum theory is quantization. Rules of quantization allow to give a quantum description of a classical system. But in the quantum theory the umbilical cord is broken, and classical concepts remain metaphorical. Axiomatic quantum theory (II), tries to avoid this initial recourse on classical description, attempting the creation of autonomous quantum concepts. But then one is faced with a fundamentally unsolved problem : the classical limit of quantum theory.

The situation has not drastically changed from the time Bohm has written at the end of his famous book on "Quantum Theory" : (12, p. 625)

"This means that without an appeal to a classical level quantum theory would have no meaning. We conclude then that quantum theory presupposes the classical level and the general correctnesse of classical concepts in describing this level ; it does not deduce classical concepts as limiting cases of quantum concepts".

From the classical mechanical point of view Q.M. has strange features : it is built from the hamiltonian theory (without dissipation) and for all systems harmonic oscillator model is at work, especially as concerne light absorption and emission, without the possibility of identifying actual physical oscillations.

2. - Quantum theory and statistical mechanics

Quantum theory cannot be formulated as a classical statistical mechanical theory, because no probability distribution in phase-space F (q , p) can be defined, such as the expectation of observables calculated in the classical manner should be equal to those obtained using the operator formalism of Q.M. The formalism of Q.M. is strictly speaking incompatible with any phase-space picture of microphysics (13).

But on the other hand it has been proven (14) that for high enough temperature the quantum partition function of quantum statistics converges towards the classical one, implying an analogous convergence for the various thermodynamical functions.

The use of Wigner function (15) allows to make a formal analogy between quantum and classical statistical mechanics, but does'nt help much in clarifying the relationships between these theories.

Another problem of interest is the theory of linear response to an external perturbation of the system. Both in the quantum mechanical and the classical cases, for a system in thermal equilibrium (Gibbs ensemble), there is a relation between the detailed behaviour at equilibrium (fluctuations) and the specifical phenomena taking place out of equilibrium (transport coefficients, dissipation of energy). More precisely one has to calculate simultaneously the two-time correlation function and the response function of the system. For systems with detailed balance (as hamiltorian systems) these two functions are linearly related to one another by a fluctuation-dissipation theorem (FDT). (16) (17)

But in general for a stochastic process neither detailed balance nor a FDT does exist. There are however processes for which a DFT does indeed hold, but the theorem has a different form depending on the form of the stationary probability.

For a stochastic process with detailed balance, and a Maxwell. Boltzman equilibrium stationary probability, leading to the classical limit of the quantum FDT, a law of dissipation given by the Lorentz radiation reaction corresponds to a law of fluctuation given by the Rayleigh. Jeans law (spectrum $\sim \omega^2$). This is another way of presenting the classical Einstein. Hopf result of 1910.

The amazing result is that one obtained by Callen and Welton (18) who have shown that to the Lorentz radiative reaction, the quantum FDT associates a randomly fluctuating field with an energy density given by the Planck radiation law including the zero point contribution. The quantum FDT involves in fact a condition of detailed balance which is very strong. It is radiation balance condition : at every frequency the absorbed power equilibrates the emitted power. This is the fundamental assumption which leaded Einstein to the Planck law in his celebrated demonstration of 1917. But such a condition is never true for a classical system, except the harmonic oscillator and equivalent systems.

So that the Kirchoff law, which is a consequence of the second principle of thermodynamics, seems incompatible with a simple mechanical model of matter.

3. - Quantum theory and probability theory

Quantum theory shows the difficulty of defining clear cut events in the microphysical world, satisfying laws of boolean algebras. Classical probability theory is not at work in quantum theory, which is a probability theory of it's own. This is clearly reflected in the central role played by amplitude of probability (wave function) in quantum mechanics, and in the fact that the superposition principle differs from the law of addition of independant probabilities, not only by the use of amplitude of probability instead of probability but by considering the same outcome of different type of trials instead of different outcomes of the same trial. The peculiarities of the quantum theory of measurement add a new non standard feature which makes even more difficult any tentative to account for quantum mechanics in the framework of the classical theory of stochastic processes. Recently, due to the development of quantum optics, a theory of quantum stochastic processes has been built (19) (20).

II. - WHY TRY TO DEVELOP STOCHASTIC ELECTRODYNAMICS ?

1. - Qualitative features and successes of stochastic electrodynamics (S.E.D)

S.E.D consider that the vacuum electromagnetic field which appears in quantum electrodynamics (Q.E.D) is an actual field which can be described classicaly. S.E.D considers the influence of this random field on classical electrodynamical systems of localized point charges.

The main characteristic of this random field is it's spectral density which is taken equal to

$$\mathcal{S}(\omega) = \frac{2}{3}\frac{\hbar}{c^3}|\omega|^3$$

This is the unique spectrum of random electromagnetic radiation which is :

Lorentz invariant (21) (22)

invariant under an adiabatic compression of a reflecting cavity (23) (demonstration of the Wien law).

The main successes of S.E.D are (24) :

a quantum picture for the harmonic oscillator (with some specific differences between SED and quantum theory) (25).

a Planck law for the spectrum of the blackbody radiation, when the vacuum field is introduced in the original analysis of Einstein and Hopf based on harmonic dipole oscillators (22) (23)

Qualitative explanation of fact which are considered as typically quantal : tunnel effect, beginning of the photoelectric emission immediately after the lightbeam is switched on, Lamb shift, spontaneous emission, fluctuations in the electromagnetic field and photon counting statistics, Van der Waals forces and diamagnetic behaviour.

2. - A typical success of S.E.D : derivation of the black-body radiation spectrum

Let us follow the derivation due to Theimer (26), which examplifies the methods and results of S.E.D

It is assumed that the fluctuating radiation energy density in a blanckbody cavity is the sum of two stochastically independant terms

$$E(\omega,T) = E_0(\omega) + E_T(\omega,T)$$

E_0 is a zero point energy density with the mean value

$$\overline{E_0(\omega)} = \frac{\omega^2}{\pi^2 c^3}\,\frac{1}{2}\hbar\omega = \frac{3}{4\pi^2}\,\mathcal{S}(\omega)$$

$E_T(\omega,T)$ is a temperature dependant energy density, which follows the laws of classical statistical mechanics.

We make the fundamental assumption that a principle of "radiation balance" is valid, so that termal equilibrium is established separately for each frequency or each mode. Now on we consider each mode separately.

The fluctuations of E_T satisfy :

$$\delta E_T = \overline{E_T^2} - \overline{E_T}^2 = kT \frac{\partial \overline{E_T}}{\partial T}$$

According to a famous calculus guessed by Einstein and made by Lorentz, the fluctuations of the electromagnetic field occurring because of local interference of waves give a fluctuation in energy equal to the square of the mean value, so that

$$\delta E_0 = \overline{E_0}^2$$
$$\delta E = \overline{E}^2$$

As a consequence of the independance between the fluctuations of the zero field and the temperature dependant field

$$\delta E_T = \delta E - \delta E_0$$

$$\delta E = \overline{E}^2 = \overline{E_0}^2 + \overline{E_T}^2 + 2\, \overline{E_0} \cdot \overline{E_T}$$

and

$$\delta E_T = \delta E - \overline{E_0}^2 = \overline{E_T}^2 + 2\, \overline{E_0} \cdot \overline{E_T}$$

which is the formula for the fluctuations of energy obtained by Einstein for a black body radiation field (the second term is responsible for the introduction of the concept of photon). Then

$$\overline{E_T} + 2\, \overline{E_0} \cdot \overline{E_T} = k\, T^2 \frac{\partial \overline{E_T}}{\partial T}$$

which on integration gives

$$\overline{E_T} = \frac{2\, E_0}{e^{2\overline{E_0}/kT} - I} = \frac{\hbar\omega}{e^{\hbar\omega/kT} - I}$$

that is the Planck law.

3. - Discussion and conclusions

The last result was obtained without any apparent quantum

hypotheses, but under the fundamental assumption of "radiation balance". Such an assumption is not in general valid for classical electrodynamical systems, except for the harmonic oscillator. This is the reason of the success of Boyer's demonstration of Planck law according to Einstein. Hopf analysis. But when Boyer replaces the harmonic oscillator by an anharmonic one, he only obtains the Rayleigh Jeans law (27) (28).

This example shows that S.E.D in it's present state does not seem to incorporate a correct description of particles. The vacuum electromagnetic field seems to be an important physical phenomena, responsible of fundamental physical effects. But it appears that at this level of sophistication particles cannot be described merely as points acted on by the fields. Particles and fields belong to the same physical reality and cannot be described in a stochastic way as an ordinary brownian motion. The organic connection (to be described in the future) between particles and fields must be responsible for the peculiarities of quantum physics. Stochastic electrodynamics is a modest step toward a deeper understanding of microphysics ; it has the merit to show one aspect of the interdependance of "the physical objects" through the active influence of the vacuum electromagnetic field.

R E F E R E N C E S

I. LEVY-LEBLOND,J.M.,Int.J.Quant.Chem.,Suppl. I,4I5,I977.

2. FROHLICH,H.,Riv.Nuovo Cimento,3,490,I973

3. PRIMAS,H.,J.Math.Biol.,4,28I,I977.

4. PRIMAS,H. and GANS,W.,in "Materie-Geist-Leben:zum Probleme der Reduktion der Wissenschaften",ed.by Kanitschneider,B., Dunker und Humblot, Berlin, P I5,I979

5. MAC CORMACH,R.,Isis,6I,459,I970.
6. KITTEL,C.,Thermal Physics,Wiley,I969.
7. WEHRL,A.,Rev.Mod.Phys.,50,22I,I978.
8. GEORGE,C.,PRIGOGINE,I. and ROSENFELD,L., Detkgl.Danske Videnskabernes Selskab,mat-fys Meddelelser,38,I2,I972.
9. ROGOVIN,D.and SCULLY,M., Physics Reports,25 C (n°3),I75,I976.
I0. ENZ,C.P.,Physica,89A,I,I977.
II. GUDDER,S.P.,in "The uncertainty principle and foundations of quantum mechanics",ed.by Price,W.C. and Chissik,S.S., Wiley,p ,I977.
I2. BOHM,D.,Quantum Theory,Prentice Hall,I95I.
I3. COHEN,L.,Philosophy of Science,33,3I7,I966.
I4. KIRKWOOD,J.G.,Phys.Rev.,44,3I,I933.
I5. DE GROOT,S.R., La transformation de Weyl et la fonction de Wigner:une forme alternative de la mécanique quantique, Montréal,I974.
I6. ZUBAREV,D.N.,Non equilibrium statistical thermodynamics,Plenum Publishing Corporation,I974.
I7. HANGGI,P.,Helv.Phys.Acta,5I,202,I978.
I8. CALLEN,H.B. and WELTON,T.A.,Phys.Rev.,83,34,I95I.
I9. HELSTROM,C.W.,Quantum detection and estimation theory, Academic Press,I976.
20. DAVIES,E.B.,Quantum theory of open systems,Academic Press,I976.
2I. MARSHALL,T.W.,Proc.Cambridge Philos.Soc.,6I,537,I965.
22. BOYER,T.H.,Phys.Rev.,I82,I374,I969.
23. BOYER,T.H.,Phys.Rev.,I86,I304,I969.
24. MILONNI,P.W.,Phys.Reports,25,I,I976.
25. MARSHALL,T.W.,Proc.Roy.Soc.,A276,475,I963.
26. THEIMER,O.,Phys.Rev.D,4,I597,I97I.
27. BOYER,T.H.,Phys.Rev.,AI8,I228,I978.
28. BOYER,T.H.,Phys.Rev.,DI3,2382,I976.

STOCHASTIC ELECTRODYNAMICS :

METHODS AND RESULTS

Pierre CLAVERIE

Laboratoire de Chimie Quantique
(Université de Paris VI)
Institut de Biologie Physico-Chimique
13, rue Pierre et Marie Curie
75005 Paris (FRANCE)

I. INTRODUCTION : The basic stochastic differential equation (Braffort-Marshall).

Stochastic Electrodynamics (SED) is a classical theory of particles and fields. The difference with respect to usual Electrodynamics is the assumption of a universal stochastic electromagnetic field ("background" field or "zero-point" field), which could be conceived as a classical counterpart to the vacuum field of Quantum Electrodynamics (QED). Thus this stochastic field (uniform and isotropic) has zero mean and spectral density [1-4].

$$\int_{-\infty}^{+\infty} e^{-i\omega t}\,\mathcal{B}_E(t)\,dt = \mathcal{S}_E(\omega) = \frac{2\hbar}{3c^3}\,|\omega|^3 \qquad (1)$$

(the energy density being $\rho(\omega) = (3/4\pi^2)\mathcal{S}_E(\omega) = (\hbar/2\pi^2c^3)\omega^3$, with $\omega > 0$). The stochastic magnetic field $\vec{B}$ also has zero mean and the same spectral density $\mathcal{S}_B(\omega) = \mathcal{S}_E(\omega)$, but we shall not use it in the present work.

Consequently, the motion of a charged particle (mass m, charge e) in an external field of force $\vec{F}(\vec{r})$ is determined in SED by a Lorentz-Dirac type equation where, in addition to the external force $\vec{F}(\vec{r})$ and the usual radiation damping force $(2e^2/3c^3)\dddot{\vec{r}} = m\tau\dddot{\vec{r}}$ (where $\tau = 2e^2/3m^3$), we have a stochastic electromagnetic force $\vec{f} = e[\vec{E} + (\dot{\vec{r}}/c) \times \vec{B}]$. We therefore have the basic stochastic differential equation (called "Braffort-Marshall" equation) [1,2,4,5,6]:

$$m\ddot{\vec{r}} = \vec{F}(\vec{r}) + m\tau\dddot{\vec{r}} + \vec{f}(\vec{r},t) \tag{2}$$

Under non-relativistic conditions ($|\dot{\vec{r}}| \ll c$) it is possible to simplify the stochastic force $\vec{f}(\vec{r},t)$ by neglecting the magnetic term $(\dot{\vec{r}}/c) \times \vec{B}$ and the $\vec{r}$-dependence of $\vec{E}(\vec{r},t)$, and, afterwards, the equation is modified in such a way as to eliminate the run-away solutions [1,2,6], and we finally get :

$$m\ddot{\vec{r}} = \vec{F}(\vec{r}) + \tau(\vec{\nabla}\vec{F}).\dot{\vec{r}} + e\,\vec{E}^{(\tau)}(t) \tag{3}$$

where the modified stochastic field

$$E^{(\tau)}(t) = \int_0^{+\infty} \vec{E}(t+\tau s)\, e^{-s}\, ds \tag{4}$$

still has zero mean, and a modified spectral density (for each component)

$$\mathcal{S}_E^{(\tau)}(\omega) = \mathcal{S}_E(\omega)/(1+\tau^2\omega^2) = \frac{2\hbar}{3c^3}\,\frac{|\omega|^3}{1+\tau^2\omega^2} \tag{5}$$

But even so we still have a stochastic differential equation which exhibits, in general, two difficulties : it is non-linear (since $\vec{F}(\vec{r})$ is in general a non-linear function of $\vec{r}$), and the stochastic term $\vec{f}(t) = e\,\vec{E}(t)$ is <u>not</u> a white noise ($|\omega|^3$ spectrum instead of a constant spectrum). Thus, non standard methods must be used in order to treat this problem, and this feature makes SED a very interesting problem in mathematical physics, independently of the results obtained in the field of microphysics.

For future development, it is appropriate to switch from this 2nd-order differential equation in configuration space to a 1st-order differential system in some configuration-momentum space (phase space). The procedure is not strictly unique, and two main possibilities will be indicated [1, section 3.C]:

1) The "mechanical" momentum $\vec{p} = m\dot{\vec{r}}$. Then eq. (3) becomes :

$$\frac{d\vec{r}}{dt} = \vec{p}/m \tag{6a}$$

$$\frac{d\vec{p}}{dt} = \vec{F}(\vec{r}) + \frac{\tau}{m} (\vec{\nabla}\vec{F})\, \vec{p} + e\, \vec{E}^{(\tau)}(t) \tag{6b}$$

2) The "electro-kinetic" momentum [2] p^{ek}. We define :

$$\frac{d\vec{p}^{ek}}{dt} = \vec{F}\,[\,\vec{r}(t)\,] \qquad \text{or} \qquad \vec{p}^{ek}(t) = \int^{t} \vec{F}\,[\,\vec{r}(t')]\, dt' \tag{7}$$

In terms of the variables $\vec{r}$ and $\vec{p}^{ek}$, eq. (3) becomes :

$$m\, \frac{d\vec{r}}{dt} = p^{ek} + \tau\, \vec{F}(\vec{r}) + e\, \mathcal{A}^{(\tau)}(t) \tag{8a}$$

$$\frac{d\vec{p}^{ek}}{dt} = F(\vec{r}) \tag{8b}$$

eq. (8b) is nothing but the definition (7), while eq. (8a) is obtained by integrating eq. (3) once with respect to time t [Santos, private communication] . Thus

$$\mathcal{A}^{(\tau)}(t) = \int^{t} \vec{E}^{(\tau)}(t')\, dt' \tag{9a}$$

is a new stochastic process with zero mean and spectrum :

$$\mathcal{S}_{\mathcal{A}}^{(\tau)}(\omega) = \frac{\mathcal{S}_E^{(\tau)}(\omega)}{\omega^2} = \frac{\mathcal{S}_E(\omega)}{\omega^2(1+\tau^2\omega^2)} = \frac{2\kappa}{3c^3}\, \frac{|\omega|}{1+\tau^2\omega^2} \tag{9b}$$

Further discussion concerning the two kinds of phase-space representation may be found in the section 3.C of [1] . It may be mentioned that the "reduced" Fokker-Planck equation, which will be derived according to Haken's procedure in section III below, and which corresponds to the formal limit $\tau \to 0$, turns out to be the same whatever the phase-space representation chosen.

One of the main problems in SED is to determine whether a charged particle in a given external field of force will undergo some "stationary" stochastic motion (generating a stationary phase-space density defined through a time-averaging procedure), as a consequence of some statistical balancing between the energy loss due to radiation damping and the energy gain from the stochastic electromagnetic field.

It is possible to give a qualitative analysis of such an energy balance for the harmonic oscillator and for the circular orbits of the Kepler problem (Hydrogen atom) [3, p. 800][7, p. 442]. The restriction to circular orbits appears essential, because quite different results will emerge from a complete treatment.

In the special case where the "deterministic" force $\vec{F}(\vec{r})$ is linear with respect to $\vec{r}$ (free particle, constant field of force, harmonic oscillator), the Braffort-Marshall equation becomes linear, and it may solved easily by applying the Fourier transform method, as is done in the theory of the Brownian motion [8, section 10.B]. Although not being fully identical with those of Quantum Theory, the results are rather satisfactory : the position and momentum probability distributions for the stationary state, and consequently the average energy, are the same as those found by Quantum Theory for the ground state of the harmonic oscillator [1,2,3,4,9]; the response function is also the same in both theories [1, section 8 B]. The main differences lie in the existence of a non-zero dispersion of the energy for the stationary state, and in the lack of physical excited states in the framework of SED [4,9].

II. APPROXIMATE FOKKER-PLANCK TYPE EQUATIONS IN PHASE SPACE.

In order to deal with the general case, we shall derive approximate Fokker-Planck type equations, which govern the evolution of the probability density (the stationary density being of special interest). The lack of uniqueness of such equations is due to the fact that the solution $\{\vec{r}(t), \vec{p}(t)\}$ of the Braffort-Marshall equation (3) is not a Markov process (here $\vec{p}$ denotes the momentum m $\dot{\vec{r}}$), since the stochastic force is not a white noise : accordingly, there is a generalized Fokker-Planck equation of a rather complicate form, and several approximations of the usual type (second-order partial derivatives) are possible. A basic feature underlying these approximations is the fact that the damping and stochastic forces are small with respect to the "deterministic" hamiltonian, which makes it possible to use perturbation methods. The essential idea underlying the use of these methods is to switch to an "interaction representation", namely to introduce as new variables a set of "constants of motion" corresponding to the unperturbed problem (essentially two choices are possible for the unperturbed problem : we take the deterministic force only, or both the deterministic and damping forces).

Two main lines of approximations will be described. For definiteness, they will be denoted as Lax-5 and Lax-6, respectively, according to the sections 5 and 6 of Lax's important paper [10]:

(a) in the Lax-5 procedure, the exact (non Markovian) process is first approximated by a genuine Markov process, the Fokker-Planck equation of which is then derived, and used as an approximate evolution equation for the probability density of the exact process. This type of approach was initiated, independently from Lax, by Stratonovich [11, and references therein], and was further elaborated in a mathematically more rigorous way by Khas'minskii [12], Papanicolaou, Hersh and other authors [13-16].

(b) in the Lax-6 procedure, the exact generalized Fokker-Planck equation is obtained first, by using the stochastic Liouville equation [17, 18, 19,6], and applying the projection operator (smoothing) method [20, 21,22,19], or cumulant expansions techniques [23,19]. This equation looks extremely cumbersome : it contains an infinite series of terms, each one involving an integration with respect to past time and derivatives of higher and higher order with respect to the phase-space variables [6,19]. Fortunately, by keeping terms up to order τ only, it is possible to get a second-order partial derivative equation of the familiar Fokker-Planck type [10, section 6]; [19, section III. 17]. Nevertheless, it must be emphasized that the matrix of the coefficients of the second-order terms may not be positive definite [1, Section 4], hence we use the name "Fokker-Planck type" or "pseudo Fokker-Planck" equation.

Besides these references, we may quote recent works by Tropper [24] and Hänggi [25] which also have some relevance concerning the treatment of non Markovian processes and the possibility of their approximation through Fokker-Planck type evolution equations.

Let us now describe the two main kinds of approximate Fokker-Planck equations. Following Lax's notations [10, section 5], [1, section 4.B], we consider a general system of first-order stochastic differential equations

$$\frac{da_i}{dt} = B_i(\vec{a},t) + G_i(\vec{a},t;\ \vec{f}(t)) \quad (i=1,\dots,n) \tag{10}$$

where a_i $(i=1,\dots,n)$ denote the components of the phase-space variable $\vec{a}$, B_i the components of the "unperturbed force", G_i the components of the perturbing force, and $\vec{f}(t) = \vec{f}(t;\varpi)$ is the stochastic driving term, ϖ denoting the basic random variable labeling the various samples of this term. For the sake of brevity, we usually do not write ϖ explicitly, but it must always be understood to be present whenever we write f(t), and ensemble averages, denoted with brackets < >, are taken with respect to this "sample variable" ϖ. By contrast, time averaging will be denoted with an upper bar with the indication of the time interval when needed.

Let us make two preliminary remarks :

(1) in SED, we have an important simplification when dealing with the motion of point particles : the stochastic term $\vec{G}$ reduces to the force $\vec{f}(t)$ without any explicit dependence with respect to the phase-space point $\vec{a}$. This property would no more hold for a rigid rotator, because here the stochastic force is involved in a vector product with

the dipole moment, hence an explicit angular dependence [1, section 8.C].

(2) there is some possibility of choice in the splitting of the total force (which appears in the right-hand sides of eqs. (10) or (12)) into an unperturbed part $\vec{B}$ and a perturbation part $\vec{G}$: we may indeed either include the damping term in the unperturbed part, or conversely include it into the perturbation part. If we consider, for definiteness, the system (10), these two choices would be, respectively :

$$\vec{B} = \begin{bmatrix} p/m \\ \\ \vec{F}(\vec{r}) + \frac{\tau}{m}(\vec{\nabla}\vec{F})\vec{p} \end{bmatrix} \qquad \vec{G} = \begin{bmatrix} 0 \\ \\ e\,\vec{E}^{(\tau)}(t) \end{bmatrix} \tag{11a}$$

and

$$\vec{B}^{(o)} = \begin{bmatrix} \vec{p}/m \\ \\ \vec{F}(\vec{r}) \end{bmatrix} \qquad \vec{G} = \begin{bmatrix} o \\ \\ \frac{\tau}{m}(\vec{\nabla}\vec{F})\vec{p} + e\,\vec{E}^{(\tau)}(t) \end{bmatrix} \tag{11b}$$

Both choices may be used, and they lead, as they must, to identical results in the formal limit $\tau \to 0$ (with which we shall be mainly concerned). In Lax's paper [10], the assumption $<\vec{G}> = 0$ corresponds to our first choice : thus, if the second choice is made, Lax's formulae cannot be used directly, and the treatment must be generalized in order to include the possibility $<\vec{G}> \neq 0$. On the contrary, Khas'minskii formulae [12] are completely general and may be used for $<\vec{G}> \neq 0$ as well. We shall see in section III below that, whatever the choice made, the "reduced" Fokker-Planck equation obtained through the Haken procedure is actually the same.

We now describe the two kinds of Fokker-Planck type equations, using for definiteness the phase-space representation (6) (it would be a trivial matter to write the corresponding equations for the second representation (8)). Our general notation for the Fokker-Planck equation will be :

$$\frac{\partial W}{\partial t} = - \sum_i \frac{\partial}{\partial a_i} (A_i W) + \frac{1}{2} \sum_i \sum_j \frac{\partial}{\partial a_i} \frac{\partial}{\partial a_j} (D_{ij} W) \tag{12}$$

A_i and D_{ij} denote the "drift" and "diffusion" coefficients, respectively, and W denotes the probability density. We shall sometimes use $D'_{ij} = \frac{1}{2} D_{ij}$.

(A) Lax 5-Khas'minskii equation

We start from Lax's equations (5.33) and (5.50,51) in ref. [10] :

$$A_i(\vec{a},t) = B_i(\vec{a},t) + \frac{1}{\Delta t}\int_t^{t+\Delta t} \sum_k \frac{\partial a_i^{[s]}}{\partial a_k} \langle G_k[s] \rangle \, ds \; +$$

$$+ \frac{1}{\Delta t}\int_t^{t+\Delta t} ds \int_t^s ds' \sum_l \left\langle \frac{\partial}{\partial a_l^{[s']}} \left[\sum_k \frac{\partial a_i^{[s]}}{\partial a_k} G_k[s] \right] G_l[s'] \right\rangle \tag{13a}$$

$$D_{ij}(\vec{a},t) = \frac{1}{\Delta t}\int_t^{t+\Delta t} ds \int_t^{t+\Delta t} ds' \sum_k \sum_l \frac{\partial a_i^{[s]}}{\partial a_k} \frac{\partial a_j^{[s']}}{\partial a_l} \langle G_k[s] \, G_l[s'] \rangle \tag{13b}$$

where

$$\vec{a}^{[s]} = \vec{a}(s,t) \qquad \vec{a}^{[s']} = \vec{a}(s',t) \tag{14}$$

denote the phase space points which would be obtained at time s and s' respectively according to the (deterministic) unperturbed motion with initial condition $\vec{a}$ at time t. Since these points will depend only of the time intervals $\theta = s-t$ and $\theta' = s'-t$ in the case of a time-independent unperturbed force, we shall also use in this case the notation:

$$\vec{a}^{[s]} = \vec{a}^{(s-t)} = \vec{a}^{\theta} \quad \text{and} \quad \vec{a}^{[s']} = \vec{a}^{(s'-t)} = \vec{a}^{\theta'} \tag{15}$$

The meaning of the partial derivatives such as $\frac{\partial a_i^{[s]}}{\partial a_k}$ is the following : the unperturbed (deterministic evolution generates, for every value of s and t, a mapping $\vec{S}_{(s,t)}$ of the phase space onto itself :

$$\vec{S}_{(s,t)}[\vec{a}] = \vec{a}^{(s,t)} \tag{16}$$

Then $\frac{\partial a_i}{\partial a_k^{[s]}}$ are the elements of the Jacobian matrix of the inverse mapping $\vec{S}^{-1}_{(s,t)}$, which by definition fulfills the relation $S^{-1}_{(s,t)}[\vec{a}^{(s,t)}] = \vec{a}$. Finally, we also use in (13) the shortened notation :

$$\vec{G}[s] = \vec{G}\left[\vec{a}^{(s,t)}, s;\ f(s)\right] \quad \vec{G}[s'] = \vec{G}\left[\vec{a}^{(s',t)}, s';\ f(s')\right] \tag{17}$$

It must be emphasized that the second term in the right-hand side of (13a) is lacking in Lax's formula (5.50,5.51), because Lax supposed from the outset $< \vec{G} > = 0$.

If $\vec{G}[s]$ is a stationary process, we denote

$$< G_k[s]\ G_l[s'] > = \mathcal{B}_{kl}(s-s') \tag{18}$$

the correlation functions, and we may elaborate further the expressions (13) by introducing the new variables $u = s-s'$, $v = (s-t)+(s'-t)$, in terms of which we have :

$$s = t + (v+u)/2 \qquad s' = t + (v-u)/2 \tag{19}$$

We use the fact that we may choose an intermediate time scale Δt large with respect to the correlation time t_c of the stochastic force and with respect to the deterministic time scale t_d (e.g. the period if the deterministic motion is periodic), but simultaneously small with respect to the "dissipative time scale" of order $1/\tau$ (see [10] section 5). This is possible in the case of SED since t_c and t_d are of the order of 1 atomic unit, while $1/\tau \simeq 10^6$, thus $\Delta t \simeq 10^3$ a.u. would be a convenient intermediate time scale.

We finally get the formulae :

$$A_i(\vec{a},t) = B_i(\vec{a},t) + \overset{t \text{ --- } s \text{ --- } +\infty}{\overline{\sum_k \frac{\partial a_i}{\partial a_k^{[s]}} < G_k[s]>}} +$$

$$+ \int_0^{+\infty} du \, \overset{o \text{ --- } v \text{ --- } +\infty}{\overline{\sum_k \sum_l < G_k[s] \; G[s']> \frac{\partial}{\partial a_l^{[s']}} \left(\frac{\partial a_i}{\partial a_k^{[s]}} \right)}} +$$

$$+ \int_0^{+\infty} du \, \overset{o \text{ --- } v \text{ --- } +\infty}{\overline{\sum_k \sum_l < \frac{\partial G_k[s]}{\partial a_l^{[s']}} G[s']> \frac{\partial a_i}{\partial a_k^{[s]}}}} \tag{20a}$$

$$D_{ij}(\vec{a},t) = \int_{-\infty}^{+\infty} du \, \overset{o \text{ --- } v \text{ --- } +\infty}{\overline{\sum_k \sum_l < G_k[s] \; G_l[s']> \frac{\partial a_i}{\partial a_k^{[s]}} \frac{\partial a_j}{\partial a_l^{[s']}}}} \tag{20b}$$

which will be valid alternatives to the previous formulae (13) provided the integrals and time-averages extended to infinity are actually convergent.

In the application to SED (eqs. 6 or 8), if we make the choice (11a) for $\vec{G}$, it will depend only on t, and we shall have a triple simplification :

(a) $< G(t) > = 0$ (b) $\frac{\partial G_k[s]}{\partial a_l^{[s']}} = 0$ (c) $< G_k[s] \; G_l[s']> =$

$\mathcal{B}_{kl}(s-s') = \mathcal{B}_{kl}(u)$ does not depend on v, thus the general formula (20) reduce to :

$$A_i(\vec{a},t) = B_i(\vec{a},t) + \int_0^{+\infty} du \sum_k \sum_l \mathcal{B}_{kl}(u) \, \overset{o \text{ --- } v \text{ --- } +\infty}{\overline{\frac{\partial}{\partial a_l^{[s']}} \left(\frac{\partial a_i}{\partial a_k^{[s]}} \right)}} \tag{21a}$$

$$D_{ij}(\vec{a},t) = \int_{-\infty}^{+\infty} du \sum_k \sum_l {}_{kl}(u) \, \overset{o \text{ --- } v \text{ --- } +\infty}{\overline{\frac{\partial a_i}{\partial a_k^{[s]}} \; \frac{\partial a_j}{\partial a_l^{[s']}}}} \tag{21b}$$

Considering now the choice (11b) for $\vec{G}$:

$$\vec{G} = \vec{f}^{damp}(\vec{a},t) + \vec{f}(t) \tag{22}$$

$\vec{f}^{damp}$ is proportional to τ, and keeping only terms up to order τ, we get from (13) the formulae

$$A_i(\vec{a},t) = B_i^{(o)}(\vec{a},t) + \overline{\sum_k \frac{\partial a_i}{\partial a_k^{[s]}} f_k^{damp}[s]}^{\,t\,\text{———}\,s\,\text{———}\,+\infty}$$

$$+ \int_0^{+\infty} du \sum_k \sum_l \mathcal{B}_{kl}(u) \overline{\frac{\partial}{\partial a_l^{[s']}} \left(\frac{\partial a_i}{\partial a_k^{[s]}} \right)}^{\,o\,\text{———}\,v\,\text{———}\,+\infty} \tag{23a}$$

$$D_{ij}(\vec{a},t) = \int_{-\infty}^{+\infty} du \sum_k \sum_l \mathcal{B}_{kl}(u) \overline{\frac{\partial a_i}{\partial a_k^{[s]}} \frac{\partial a_j}{\partial a_l^{[s']}}}^{\,o\,\text{———}\,v\,\text{———}\,+\infty} \tag{23b}$$

Taking into account the relation

$$\vec{B}(\vec{a},t) = \vec{B}^{(o)}(\vec{a},t) + \vec{f}^{damp}(\vec{a},t) \tag{24}$$

we see that the essential difference between (21) and (23) is the replacement of $\vec{f}^{damp}(\vec{a},t)$ in (21a) by the time average (along the undamped unperturbed orbit) appearing as the second term of (23a).

It is worth emphasizing that, despite their differences, all approximate equations obtained in the framework of the Lax 5-Khas'minskii procedure are genuine Fokker-Planck equations, since the matrix of their diffusion coefficients is non-negative, as a consequence of the way in which these coefficients are obtained : they correspond to a genuine Markov process, which is an approximation to the exact (non Markovian) process. We shall see that the situation is rather different in the second family (Lax 6 - Van Kampen) of approximate equation, that we shall now consider.

(B) Lax-6 - Van Kampen equations

Here we shall give much less details since this approach was developped in a previous work [1, sections 4.B, 5,6], to which we refer for more complete information. The complete generalized Fokker-Planck equation was derived (for the case of SED) by de la Peña-Auerbach and Cetto [6, section II], and the Fokker-Planck type equation of usual form (second-order partial derivatives) may be found in the papers by Lax [10, section 6] and Van Kampen [19, section III.19]. As explained in previous work [1, section 4.A.2], there is at first sight a difference between the equation (6.36) of Lax and the equation (19.7) of Van Kampen : more precisely, the last term of Van Kampen's equation is missing in Lax's equation. Now, as mentioned earlier [1, section 4.2.A], [26, section 3, footnote on p.13], this extra term derived by Van Kampen precisely cancels with the third term that Lax obtained in his formula (5.36) for the drift coefficient (derived from his eq. (6.32)).

As a result, the correct formula for the drift coefficients are obtained by keeping only the first two terms of Lax's formula (6.35), while the diffusion coefficients are correctly given by Lax's formula (6.34) (following Van Kampen, we replace t_o by $-\infty$) :

$$A_i(\vec{a},t) = B_i(\vec{a},t) + \int_{-\infty}^{t} ds \sum_k \left\langle G_k[s] \frac{\partial G_i[t]}{\partial a_k[s]} \right\rangle \tag{25a}$$

$$D_{ij}(\vec{a},t) = 2\int_{-\infty}^{t} ds \sum_k \left\langle G_k[s] \; G_i[t] \right\rangle \frac{\partial a_j}{\partial a_k[s]} \tag{25b}$$

(the factor 2 in (25b) is not present in Lax's formula (6.34) because Lax includes 1/2 in the definition of his diffusion coefficient).

These formulae hold for the case $\langle \vec{G} \rangle = 0$; we shall not consider here the generalization to the case $\langle \vec{G} \rangle \neq 0$.

There is some similarity between the formulae (25) and the Lax 5-Khas'minskii formulae (20), a similarity which is better seen by introducing the variable u = t-s in (25), thereby converting the integrals into the form $\int_0^{+\infty}$ du. But these formulae are nevertheless definitely different, one of the most striking differences being the occurence of a double sum $\sum_k \sum_l$ in (20) instead of a simple sum $\sum_k$ in (25). As a consequence, if the stochastic vector $\vec{G}$ has some component G_i identically zero, the corresponding i-th line of the diffusion matrix D_{ij} (j=1,..., n) will also be zero according to (25), while this is not the case, in general, according to (20). In the case of SED, with $\langle\vec{G}\rangle = 0$ and $\vec{G}$ independent of a, we get the, simplified formulae (which should be compared with (21) :

$$A_i(\vec{a},t) = B_i(\vec{a},t) \tag{26a}$$

$$D_{ij}(\vec{a},t) = 2\int_0^{+\infty} du \sum_k \mathcal{B}_{ik}(u) \frac{\partial a_j}{\partial a_k}\bigg|_{-u} \tag{26b}$$

(C) Comparison between Lax-5 and Lax-6 approximations

When applied, for example to the harmonic oscillator, the Lax-5 formulae give a diffusion matrix of the form : $D_{qq} \neq 0$, $D_{pp} \neq 0$, $D_{qp} = D_{pq} = 0$, while the Lax-6 formulae give [1, section 6] : $D_{qq} = 0$, $D_{qp} = 0$, $D_{pq} \neq 0$, $D_{pp} \neq 0$, which is clearly different from the previous one. Even after symmetrization, i.e. by using $D'_{qp} = D'_{pq} = D_{pq}/2$, the Lax-6 matrix is not positive definite.

Having made clear the existence of differences between the Lax-5-Khas'minskii and Lax-6-Van Kampen approaches, we would like now th understand better the connection between them, and in particular their agreement as concerns the long time behaviour and the stationary solution.

A possible way, followed by Marshall [27, part II] consists in using new phase-space variables which are all constants of the unperturbed (deterministic) motion : then, it may be shown that the Lax 5-Khas'minskii coefficients may be derived from the Lax 6-Van Kampen ones by performing a suitable time-averaging over some "intermediate time-scale" Δt such as previously defined in section A

Another way will be presented in the next section III (a more detailed treatment will appear in [31]): it consists in applying the Haken reduction procedure to both kinds of Fokker-Planck type equations, and it will be seen that the same reduced Fokker-Planck equation is obtained, which proves the consistency between the Lax_5-Khas'minskii and Lax_6-Van Kampen approximations.

(D) Applications of the Fokker-Planck (type) equations.

Before leaving the topic of these approximate Fokker-Planck equations in phase-space, we still consider some related developments (see [1] for further detail). These equations involve a "Liouville part" (the same in all cases) and a damping-diffusion part (which changes from one equation to the other). By applying the inverse Wigner transform, these equations may be transformed into equations of evolution for a function $\rho^S(x,x';t)$ resembling to the quantum "density matrix" $\rho^Q(x,x';t)$, but definite differences persist with respect to the quantum Von Neuman equation (which governs the evolution of ρ^Q [1, section 5].

From the stationary probability density of a given system, it is possible to derive the absorption coefficient (with respect to an external electromagnetic field) by using Kubo's linear response theory [32]. It is not possible to develop this aspect of SED here. Some preliminary results were presented in [1, section 8].

III. DERIVATION OF THE "REDUCED" FOKKER-PLANCK EQUATION THROUGH HAKEN'S PROCEDURE.

In order to study more complicated systems, it is interesting to write down a "reduced Fokker-Planck equation" in terms of new coordinates which are "relevant constants of motion". Indeed, since the deterministic orbits are only slowly perturbed by the damping and stochastic forces, it is physically intuitive that it will be possible to describe the stochastic motion as a "diffusion of the orbits" rather than a diffusion of the phase points themselves. Thus, for one-dimensional systems, a single constant of motion, characterizing the orbit (such as energy or action), is sufficient. For multiperiodic systems, action variables [33,34] are appropriate : see for example the 3-dimensional Kepler problem (Hydrogen atom) in section IV.

The reduced Fokker-Planck equation in terms of the relevant constants of motion may be obtained in two main ways :

(1) We apply directly the Lax 5 method to this reduced set of new variables, i.e. we calculate by perturbation method the variations of these "constants of motion" under the effect of the damping and stochastic forces, and by averaging these variations we get the drift and diffusion coefficients. The perturbation method may be applied directly to the equations of motion [27, 28,30] or, following the main idea of Kubo's response theory [32], it may be applied to the Liouville equation of the problem [35].

(2) We apply to the previously obtained Fokker-Planck type equations in phase-space a "reduction procedure" devised by Haken [29]. It is this second possibility that we shall briefly describe here (see ref. [31] for a more complete account).

We recall the main features of this reduction procedure as it is described in the section XI.C2 of ref. [29]. The Fokker-Planck equation is written

$$\frac{\partial W}{\partial t} = (\mathcal{L}_0 + \mathcal{L}_1)\, W \tag{27}$$

where $\mathcal{L}_0$ is the "Liouville part" of the evolution operator corresponding to the unperturbed deterministic evolution <u>without</u> damping. For example, in the case of SED :

$$\mathcal{L}_0 = - \frac{\partial}{\partial q}[\, F(q)\, W] - \frac{\partial}{\partial p}[\frac{p}{m} W\,] \tag{28}$$

while $\mathcal{L}_1$ involves the contribution of the damping and stochastic forces to the drift and diffusion coefficients, and goes to zero with τ. The reduced set of "constants of the unperturbed motion" is denoted :

$$h_1, h_2, \ldots, h\ , \ldots h_M \quad (M < N) \tag{29}$$

The operator $\mathcal{L}_1$ is changed from its usual form

$$\mathcal{L}_1 W = - \sum_i \frac{\partial}{a_i}[\, A_i^{(1)} W] + \frac{1}{2}\sum_i \sum_j \frac{\partial}{a_i}\ \frac{\partial}{a_j}[\, D_{ij}\ W] \tag{30}$$

to the "current form"

$$\mathcal{L}_1 W = \sum_i \frac{\partial}{\partial a_i}[\, l_i W\,] + \sum_i \sum_j \frac{\partial}{\partial a_i}\left[D'_{ij}\, \frac{\partial W}{\partial a_j}\right] \tag{31}$$

where

$$l_i = - A_i^{(1)} + \sum_j \frac{\partial}{\partial a_j}\left(\frac{D_{ij}}{2}\right) \qquad D'_{ij} = \frac{D_{ij}}{2} \tag{32}$$

Then, for a probability density depending only on the reduced set of constants of motion, it is possible to derive the time-dependent reduced Fokker-Planck equation (in [29] Haken derives the stationary equation only) :

$$C\, \frac{\partial W}{\partial t} = \sum_\mu \frac{\partial}{\partial h_\mu}[\, C^\mu W] + \sum_\mu \sum_\nu \frac{\partial}{\partial h_\mu}\left[\, C^{\mu\nu}\, \frac{\partial W}{\partial h_\nu}\right] \tag{33}$$

where the coefficients C, C^μ and $C^{\mu\nu}$ are given by :

$$C = \mathbb{T}[\,1] \qquad C^\mu = \sum_j C^\mu_j \qquad C^{\mu\nu} = \sum_i \sum_j C^{\mu\nu}_{ij} \tag{34a}$$

$$C^\mu_j = \mathbb{T}\left[\, l_j\, \frac{\partial h_\mu}{\partial a_j}\right] \quad C^{\mu\nu}_{ij} = \mathbb{T}\left[\, D'_{ij}\, \frac{\partial h_\mu}{\partial a_i}\, \frac{\partial h_\nu}{\partial a_j}\right] \tag{34b}$$

where the $\mathbb{T}$ averaging operation is defined for an arbitrary phase-space function $\varphi(\vec{a})$ as :

$$\mathbb{T}\, [\varphi(\vec{a})] = \int \prod_{\lambda=1}^{M} \delta[\, h_\lambda - h_\lambda(\vec{a})]\ \varphi(\vec{a})\ d\vec{a} \tag{35}$$

$\mathbb{T}[\varphi]$ is a function of the M values $\{h_1,\ldots,h_M\}$, which must be carefully distinguished from the phase-space functions $h_\lambda(a)$, which denote the expressions of the relevant constants of motion in terms of the original phase-space variables $\{a_1,\ldots,a_N\}$. In the eqs. (34a), $\mathbb{T}[\,1]$ denotes the averaging applied to the constant function equal to 1.

The meaning of the $\mathbb{T}$ operation is clear : it consists in taking the average over the phase-space points which correspond to prescribed values $\{h\}$ of the constants of motion, i.e. over the set of all orbits which are characterized by these values. Since these orbits are actually dynamically equivalent, the averaging may essentially be reduced to a time-averaging over any one of these orbits.

We shall now apply this reduction procedure to both Lax 5-

Khas'minskii and Lax 6-Van Kampen approximate Fokker-Planck type equations, we shall see that the same reduced Fokker-Planck equation is obtained in all cases, and some comments will be made concerning this interesting uniqueness property. For the sake of simplicity, we consider here the pecular case where the stochastic force involves no explicit dependence on the phase point (as is usually the case in SED). The general case is treated in ref. [31].

(A) Lax 5-Khas'minskii equation

Let us start from formulae (21), by skipping out the deterministic (undamped) part of the drift coefficient, according to the definition of $\mathcal{L}_1$ above. First of all, we must convert these coefficients to those of the current form, namely we have essentially to evaluate, according to (32), the "supplementary drift" $l_i^{suppl} = \frac{1}{2}\sum_j \frac{\partial}{\partial a_j}(D_{ij})$. For that purpose, it is more efficient to introduce non symmetric coefficients D^+_{ij} ($\neq D^+_{ji}$) :

$$D^+_{ij} = \int_0^{+\infty} du \sum_k \sum_l \mathcal{B}_{kl}(u)\ \overline{\frac{\partial a_i}{\partial a_k^{[s]}}\ \frac{\partial a_j}{\partial a_l^{[s']}}}^{\,0\ldots v\ldots +\infty} \tag{36}$$

which fulfill the relation $D_{ij} = D_{ji} = D^+_{ij} + D^+_{ji}$. Now, using the identity (3.33) of ref. [10], we get very simple expressions for the coefficients of the "current form" equation : $l_i = -B_i^{damp}$ and $D'_{ij} = D^+_{ij}$. Then, applying Haken's reduction formulae (34,35) to these coefficients, we get :

$$C^\mu = -\ [\sum_i \frac{\partial h_\mu}{\partial a_i} B_i^{damp}\ ] \tag{37a}$$

$$C^{\mu\nu} = \int_0^{+\infty} du \sum_k \sum_l \mathcal{B}_{kl}(u)\ [\frac{\partial h_\mu}{\partial a_k}\ \frac{\partial h_\nu}{\partial a_l^{-u}}\ ] \tag{37b}$$

In (37b), the time averaging over v which was present in (36) has disappeared because the $\mathbb{T}$ operation essentially involves a time averaging over the deterministic orbits, as mentioned above, which makes the v-averaging unnecessary.

If we start from the formulae (23), corresponding to choice (11b) instead of (11a), we finally recover exactly the same formulae (37).

(B) Lax 6-Van Kampen equation

We now start from formulae (26), and, as before, we keep only the damping part B_i^{damp} of the drift coefficient (26a). In order to switch to the "current form", we evaluate the "supplementary drift" of (32), and using again the identity (3.33) of [10], we find a zero value for it. The "current form" coefficients are therefore $l_i = -B_i^{damp}$ and $D'_{ij} = D_{ij}/2$, where D_{ij} is given by (26b). Since l_i is the same as the one previously obtained from the Lax_5 equation, the reduced drift coefficient C^{μ} will obviously be given by the same formula (37a), and the evaluation of $C^{\mu\nu}$ according to the reduction formulae (34b) also leads to the same result (37b) as obtained previously. This shows the uniqueness of the reduced Fokker-Planck equation. Note that this is a genuine Fokker-Planck equation (non negative diffusion matrix), because it is derived from the Lax_5 equation through a transformation (34,35) which keeps this property unchanged.

For one-dimensional systems (with 2-dimensional phase space (q,p), the reduced Fokker-Planck equation becomes 1-dimensional (energy E, or action J, may be used as reduced variable), and the treatment therefore becomes quite simple. This method has been applied in SED to several non linear problems : anharmonic oscillator [36a, 37], rigid dipolar rotator in a constant electric field (Stark effect) [36b].

IV. THE HYDROGEN ATOM PROBLEM.

We denote the Coulomb potential $V(r) = -\kappa/r$ (for the H atom, $\kappa = e^2$). Owing to the isotropy of this potential, it is possible to use a reduced description in terms of two variables only, chosen among the "constants of motion" which define the size and shape of the orbit : energy E, total angular momentum M, eccentricity $\varepsilon = [1 + (2E\kappa M^2 / m^2)]^{1/2}$, or $\eta = \sqrt{1-\varepsilon^2} = [-2EM^2 m\kappa^2]^{1/2}$, Delaunay action variable $L = [mx^2/(-2E)]^{1/2} = M/\eta$. The reduced Fokker-Planck equation may be obtained (a) by direct calculation of the drift and diffusion coefficients for the action variables L, M [30] , or (b) by application of Haken's reduction procedure [31]. I give here the results for the variables E, M (see [31] for details) :

$$8\pi^2 MT \frac{\partial W}{\partial t} = \frac{\partial}{\partial E}[C^E W] + \frac{\partial}{\partial M}[C^M W] + \frac{\partial}{\partial E}[C^{EE}\frac{\partial W}{\partial E}] + \frac{\partial}{\partial E}[C^{EM}\frac{\partial W}{\partial M}]$$

$$+ \frac{\partial}{\partial M}[C^{ME}\frac{\partial W}{\partial E}] + \frac{\partial}{\partial M}[C^{MM}\frac{\partial W}{\partial M}] \qquad (38)$$

where $T = 2\pi\kappa\,[m/|2E|^3]^{1/2}$ denotes the period of the orbit with energy E, and

$$C^E = 16\pi^3\tau m\kappa^2 \left(\frac{E}{M^2} + \frac{3}{2}\frac{m\kappa^2}{M^4}\right) \qquad C^M = 16\pi^3\tau \frac{m\kappa^2}{M} \qquad (39a)$$

$$C^{EE} = 16\pi^3\tau\kappa M \frac{|2E|^4}{m\kappa^2}\Phi_3(\varepsilon) \qquad C^{MM} = 16\pi^3\tau\kappa M|2E|\,\Phi_1(\varepsilon)$$

$$C^{EM} = 16\pi^3\tau\kappa M \frac{|2E|^{5/2}}{(m\kappa^2)^{1/2}}\Phi_2(\varepsilon) \qquad (39b)$$

where the functions $\Phi_r(\varepsilon)$ may be expressed as series in terms of Bessel functions (so-called Kapteyn series, see [38, chap. XVII]) :

$$\Phi_r(\varepsilon) = \frac{1}{2}\sum_{n=-\infty}^{+\infty} \mathrm{sg}(n)\, n^r \left[\frac{\eta}{\varepsilon} J_n(n\varepsilon) + J_n'(n\varepsilon)\right]^2 \qquad (40)$$

where sg(n) denotes the sign of n, and $\eta = \sqrt{1-\varepsilon^2}$. Alternatively, the functions Φ_r may be expressed under integral form, which makes it possible to study the behaviour of the Φ_r near $\eta = 0$, where they diverge as powers of $1/\eta$, but also involve smaller terms containing Log [39,30]. Let us rewrite eq. (38) with vector notations :

$$8\pi^2 MT \frac{\partial W}{\partial t} = \mathrm{div}\,[W\vec{C} + \bar{\bar{C}}\,\overrightarrow{\mathrm{grad}}\,W] \qquad (41)$$

where $$\vec{C} = \begin{bmatrix} C^E \\ C^M \end{bmatrix} \qquad \bar{\bar{C}} = \begin{bmatrix} C^{EE} & C^{EM} \\ C^{ME} & C^{MM} \end{bmatrix} \qquad (42)$$

A so-called "Detailed Balance" [28,29 section XI.D] stationary solution W_o would exist if it would be possible to have the probability current identically zero :

$$W_o\vec{C} + \bar{\bar{C}}\,\overrightarrow{\mathrm{grad}}\,W_o = 0 \qquad (43)$$

i.e. W_o would be given by

$$\overrightarrow{\mathrm{grad}}\,\mathrm{Log}\,W_o = -(\bar{\bar{C}})^{-1}\vec{C} \qquad (44)$$

Hence the Detailed Balance condition : such a solution will exist if and only if the vector field $(\bar{\bar{C}})^{-1}\,\vec{C}$ is a gradient field (hence the name "potential condition" used by Haken [29, section XI.D]). The detailed knowledge of the behaviour of the coefficients near $\varepsilon = 0$ and $\varepsilon = 1$ enables us to study the Detailed Balance condition, which turns out to be satisfied <u>at these boundary values only</u>. This situation suggests that a solution calculated according to the Detailed Balance recipe could give a convenient approximation to the exact stationary solution in the neighbourhood of these boundaries. Now, a Detailed Balance approximate solution is

$$W_{app}(M,\eta) = \text{Const. } \exp\left[\, -b(\eta)\; M/\mathcal{M} \,\right] \tag{45}$$

with

$$b(1) = 2 \quad \text{and} \quad b(0) = \frac{9\pi}{10\sqrt{3}} = 1.636... \tag{46}$$

and, when taking into account the volume element (due to the reduction from full phase space to the reduced $\{M,\eta\}$space), namely M^2/η^2 dMd , we see that the complete probability density $W_{app}(M,\eta)\; M^2/\eta^2$ has a divergent integral, due to it behaviour near $\eta = 0$:

$$\int_0^{+\infty} dM \int_0^1 d\eta\; M^2/\eta^2\; W_{app}(M,\eta) = +\infty \tag{47}$$

It may be seen in the same way that the corresponding average value of the energy is zero. In order words, W_{app} corresponds to a diffusion of the electron towards infinity (self-ionization). But these results are a consequence of the behaviour of $W_{app}(M,\eta)$ precisely near one of the boundaries (η=0) where the exact solution may be expected to be very close to W_{app} : consequently, the possibility that the exact stationary density W exhibits the same kind of divergence must be seriously considered. We therefore developed [40] a method for getting this exact stationary solution under the form :

$$-\text{Log } W(M,\eta) = g_0(\eta) + \left(\frac{M}{\mathcal{M}}\right) g_1(\eta) + ... + \left(\frac{M}{\mathcal{M}}\right)^k g_k(\eta) + ... \tag{48}$$

which, for large M, may be conveniently converted into a Padé approximant form. Upon inserting this expression (48) into the Fokker-Planck equation, we get a sequence of second-order differential equations (which are linear except the first one for $g_o(\eta)$), from which the $g_k(\eta)$ may be determined recurrently. Some problems are presently investigated concerning the choice of the boundary conditions (we may impose the vanishing of the probability current along some boundary, for example, but the question remains whether different conditions could be accepted). But the results obtained up to now suggest that, whatever reasonable choice is made, the behaviour of an exact stationary W would be qualitatively similar to that of W_{app}, namely the integral of W over the whole reduced phase-space would be divergent, with all the ensuing physical consequences already described.

V. CONCLUSION

Considering the results obtained for the hydrogen atom and for the anharmonic oscillator [36,37], and other results (no treated here) concerning the lack of "radiation balance" at every frequency for general (multi-) periodic systems (for each the Fourier analysis of the deterministic motion does not reduces to a single frequency, as is the case for the harmonic oscillator and rigid rotator), we cannot avoid the conclusion that the discrepancies between SED, on one hand, and Quantum Theory and experiment, on the other hand, appear very serious as soon as we deal with non-linear systems (or, more precisely, systems with harmonics present in their deterministic motion). These results support the view that, quite generally, it really appears non trivial, in the framework of a classical stochastic theory, to account for such features as sharp energy levels, sharp transition frequencies (spectral lines) which are considered as well established facts in microphysics (actually, experimental evidence, if it is more or less indirect as concerns the levels themselves, is rather direct as concerns the spectral lines). Thus, it seems preferable, at the present time, to adopt a very cautious opinion concerning the prospects of purely classical stochastic models in microphysics.

REFERENCES

[1] P. CLAVERIE and S. DINER, Intern. J. Quant. Chem. 12, Suppl.1, 41 (1977).

[2] E. SANTOS, Nuovo Cimento (série 11) 19B, 57 (1974).

[3] T.H. BOYER, Phys. Rev. D11, 790 (1975).

[4] T.W. MARSHALL, Proc. Roy. Soc. (London) A276, 475 (1963).

[5] L. De La PENA-AUERBACH and A.M. CETTO, Rev. Mex. Fis. 25, 1 (1976).

[6] L. De La PENA-AUERBACH and A.M. CETTO, J. Math. Phys. 18, 1612 (1977).

[7] P. CLAVERIE and S. DINER, p. 395 in "Localization and Delocalization in Quantum Chemistry", ed. by O. Chalvet, R. Daudel, S. Diner and J.P. Malrieu (Reidel, Dordrecht, 1976).

[8] Ming Chen WANG and G.E. UHLENBECK, Rev. Mod. Phys. 17, 323 (1945). Reprinted in "Selected Papers on Noise and Stochastic Processes", ed. by N. Wax (Dover, New York, 1954).

[9] L. De La PENA and A.M. CETTO, J. Math. Phys. 20, 469 (1979).

[10] M. LAX, Rev. Mod. Phys. 38, 541 (1966).

[11] R.L. STRATONOVICH, chap. 1, in "Conditional Markov Processes and their Application to the Theory of Optimal Control" (American Elsevier, New York, 1968) (Russian original : Moscow University Press, Moscow, 1966).

[12] R.Ž. KHAS'MINSKII, Theory Prob. Appl. 11, 390 (1966).

[13] G.C. PAPANICOLAOU and J.B. KELLER, SIAM J. Appl. Math. 21, 287 (1971).

[14] G.C. PAPANICOLAOU and R. HERSH, Indiana Univ. Math. J. 21, 815 (1972).

[15] R. COGBURN and R. HERSH, Indiana Univ. Math. J. 22, 1067 (1973).

[16] G.C. PAPANICOLAOU, p.209 in "Modern Modelling of Continuum Phenomena", Lectures in Applied Mathematics, vol. 16, Ed. by R.C. Di Prima (American Mathematical Society, Providence, Rhode Island, 1977).

[17] R. KUBO, J. Math. Phys. 4, 174 (1963).

[18] A. BRISSAUD and U. FRISCH, J. Math. Phys. 15, 524 (1974).

[19] N.G. VAN KAMPEN, Phys. Repts. (Phys. Letters C) 24C, 171 (1976).

[20] R. ZWANZIG (a) J. Chem. Phys. 33, 1338 (1964); (b) Physica, 30, 1109 (1964).

[21] U. FRISCH (a) Ann. Astrophys. 29, 645 (1966) and 30, 565 (1967) (b) p. 75 in "Probabilistic Methods in Applied Mathematics", ed. by A.T. Bharucha-Reid (Academic Press, New York, 1968).

[22] R.H. TERWIEL, Physica, 74, 248 (1974).

[23] N.G. VAN KAMPEN, Physica 74, 215 and 239 (1974).

[24] M.M. TROPPER, J. Stat. Phys. 17, 491 (1977).

[25] P. HÄNGGI, Zeitschr. Phys. B31, 407 (1978).

[26] P. CLAVERIE and S. DINER "Some Remarks about the LAX Approximation in Stochastic Electrodynamics". Technical Report (1977).

[27] T.W. MARSHALL, "Brownian motion and quasi-Markov processes", parts I and II (submitted to Physica)

[28] T.W. MARSHALL and P. CLAVERIE, Brownian motion and quasi-Markov processes, part III (submitted to Physica).

[29] H. HAKEN, Rev. Mod. Phys. 47, 67 (1975). (See especially section XI.C.2.).

[30] T.W. MARSHALL and P. CLAVERIE "Stochastic Electrodynamics of non-linear system. I. Particle in a central field of force" J. Math. Phys. (in press).

[31] P. CLAVERIE, L. De La PENA-AUERBACH and S. DINER, "Stochastic Electrodynamics of non-linear systems. II. Derivation of a reduced Fokker-Planck equation in terms of relevant constants of motion" (to be published).

[32] R. KUBO, J. Phys. Soc. Japan, 12, 570 (1957).

[33] H. GOLDSTEIN, "Classical Mechanics", Addison-Wesley, Reading, Massachusetts (1956).chap. 9, section 9.7.

[34] H.C. CORBEN and P. STEHLE, "Classical Mechanics", 2nd edition, Wiley, New York (1960).chap. 11, section 64.

[35] L. PESQUERA and P. CLAVERIE, "Derivation of Fokker-Planck equations through response theory", to be published.

[36] P. JULG, (a) Folia Chimica Theoretica Latina, 6, 99 (1978); (b) Results to be published.

[37] (a) L. PESQUERA, "The anharmonic oscillator in Stochastic Electrodynamics (SED) : the problem of "radiation balance" at each frequency", communication in these Proceedings.
(b) L. PESQUERA and P. CLAVERIE, "The quartic anharmonic oscillator in Stochastic Electrodynamics, to be published.

[38] G.N. WATSON, "A Treatise on the Theory of Bessel Functions", Cambridge University Press (1966).

[39] T.W. MARSHALL, "On the sum of a family of Kapteyn series" (submitted to Z.A.M.P. (J. Appl. Math. Phys.)).

[40] L. PESQUERA, P. CLAVERIE and A. DENIS, "Stochastic Electrodynamics of non-linear systems. III. Accurate stationary solution for the hydrogen atom" (to be published).

THE KEPLER PROBLEM IN
STOCHASTIC ELECTRODYNAMICS

T.W. MARSHALL
DEPARTMENT OF MATHEMATICS
MANCHESTER UNIVERSITY
MANCHESTER, ENGLAND

Stochastic electrodynamics is the Brownian motion of a charged particle in a random electromagnetic field with spectrum proportional to $\hbar\omega^3 \coth(\hbar\omega/2kT)$. If the deterministic force field is simple harmonic, the properties of the system resemble closely those of the quantum-mechanical oscillator, but the extension to non-linear systems has proved to be a complex problem. The Kepler problem is an important test for stochastic electrodynamics, and there is some evidence that the same SO(4) symmetry which simplifies the treatments of both the classical and quantum-mechanical systems also reduces the complexity of the problem in stochastic electrodynamics.

The Fokker-Planck equation for the Kepler system is obtained, and the requirements for a physically meaningful solution are discussed -

in particular the question, recently raised by Boyer, of radiation balance. It will be argued that the requirement of radiation balance is simply another form of the Equipartition law, which is as obsolete in stochastic electrodynamics as in quantum mechanics.

1. Stochastic electrodynamics as a form of Brownian motion

We must note certain features of stochastic electrodynamics which are peculiar to that theory. The orders of the damping and stochastic forces are such[1] that we are in the extreme underdamped, rather than the extreme overdamped limit. This means that the description of the process, used so fruitfully by Einstein and Smoluchowski, as Markov in configuration space, is not possible in stochastic electrodynamics. Moreover, the stochastic force in stochastic electrodynamics bears no resemblance to white noise, and therefore the more general model of Uhlenbeck and Ornstein, in which the process is Markov in phase space, is also not applicable.

Nevertheless, if the dynamical variables are suitably chosen, the description of stochastic electrodynamics as approximately Markov, or quasi-Markov, is possible. Consider the unperturbed motion of a conservative dynamical system consisting of a particle in a potential field $V(\underset{\sim}{x})$:

$$m\ddot{\underset{\sim}{x}} = -\text{grad } V(\underset{\sim}{x}). \tag{1.1}$$

This has three constants of the motion which we denote (ξ_1, ξ_2, ξ_3). One of these is the total energy of the system, and the other two are obtained, in general, by solving the Hamilton-Jacobi equation. As an example, consider the case where V is a function of $r = |\underset{\sim}{x}|$ only. Then ξ_2 and ξ_3 may be taken as the magnitude of the angular momentum and a component of the angular momentum respectively. The process $\xi(t)$ may be shown[2] to be quasi-Markov, in the sense that successive increments $\Delta\xi$ are independent, provided the corresponding time intervals Δt are large compared with both the natural periods of the unperturbed system and the memory of the stochastic force.

We shall confine attention to systems for which the Hamilton-Jacobi equation is separable. For such systems the variables (ξ_1, ξ_2, ξ_3) are conveniently taken as the <u>action variables</u>. The advantage of this choice is that, in the quasi-Markov approximation, the conjugate, or angle, variables are uniformly distributed over the interval $(0, 2\pi)$. This leads to rather simple expressions for the drift and diffusion coefficients in the Fokker-Planck equation.

2. The Fokker-Planck equation

We now take account of the frictional and stochastic forces, so that (1.1) is replaced by a generalized Langevin equation

$$m\ddot{\underset{\sim}{x}} - \text{grad } V(\underset{\sim}{x}) + \underset{\sim}{G} + \underset{\sim}{F}(t), \tag{2.1}$$

where $\underset{\sim}{G}$ is the Lorentz radiative reaction

$$\underset{\sim}{G} = \frac{2e^2}{3c^3}\dddot{\underset{\sim}{x}}, \tag{2.2}$$

and $\underset{\sim}{F}(t)$ is the stochastic force resulting from the particle's interaction, in the dipole approximation, with the vacuum field

$$<\underset{\sim}{F}(t)> = 0, \tag{2.3}$$

$$<F_i(t)F_j(t+\tau)> = e^2\delta_{ij}\ B_E(\tau). \tag{2.4}$$

Then, subject to the approximation referred to in the previous section, $\xi(t)$ is a diffusion process with the Fokker-Planck equation

$$\frac{\partial\rho}{\partial t} = \frac{\partial}{\partial\xi_i}\left(-A_i\rho + B_{ij}\frac{\partial\rho}{\partial\xi_j}\right). \tag{2.9}$$

The coefficients in this equation are given, through second order perturbation theory, by

$$A_i = \overline{G_j(t)\ \frac{\partial\xi_i}{\partial p_j}(t)}, \tag{2.6}$$

$$B_{ij} = \tfrac{1}{2}e^2\int_{-\infty}^{\infty} B_E(\tau)\overline{\frac{\partial\xi_i}{\partial p_k}(t)\ \frac{\partial\xi_j}{\partial p_k}(t+\tau)}d\tau. \tag{2.7}$$

In (2.6) and (2.7) the bar denotes time-averaging over t, and the time variation is that of the unperturbed motion. These time averages may be carried out by using certain properties of the contact transformations in classical mechanics[3], namely

$$\frac{\partial\xi_i}{\partial p_k} = \frac{\partial x_k}{\partial\eta_j} \tag{2.8}$$

where η_i is the angle variable associated with ξ_i, and the multi-periodic expansion

$$x_k = \sum_{\underset{\sim}{n}} x_{k\underset{\sim}{n}}(\xi)e^{i\underset{\sim}{n}\cdot\underset{\sim}{\eta}}. \tag{2.9}$$

The use of these relations leads to expressions for the drift and diffusion coefficients in terms of the "oscillator strengths" of Old

Quantum Theory. The similarity of stochastic electrodynamics to that theory has already been noted by Boyer[4].

3. Some special cases

We shall suppose for the remainder of this article that the correlation of the stochastic field is that of the zero-temperature vacuum

$$B_E(\tau) = \frac{\hbar}{3\pi c^3}\int_{-\omega_c}^{\omega_c} |\omega|^3 e^{i\omega\tau} d\omega \tag{3.1}$$

where ω_c is some unknown cutoff frequency which is large compared with the natural frequencies of our system.

We next investigate the effect on the drift and diffusion coefficients of various kinds of symmetry in the potential $V(\underset{\sim}{x})$.

In the case of the central field, as we have already noted, ξ_2 and ξ_3 may be expressed in terms of the angular momentum. In that case we find[3] that

$$A_3 = \frac{\xi_3}{\xi_2}A_2, \quad B_{13} = \frac{\xi_3}{\xi_2}B_{12}, \quad B_{23} = \frac{\xi_3}{\xi_2}B_{22}, \tag{3.2}$$

which means that, provided we limit attention to those states having rotational symmetry, the Fokker-Planck equation reduces to

$$\frac{\partial\rho}{\partial t} = \frac{1}{\xi_2}\frac{\partial}{\partial\xi_i}\,\xi_2(-A_i\,\rho + B_{ij}\,\frac{\partial\rho}{\partial\xi_j}), \tag{3.3}$$

the summation now being limited to $(i,j) = (1,2)$.

Further simplification follows for certain special central fields. These are potentials which give rise to Hamilton-Jacobi equations which are separable in more than one system of coordinates, and which lead to the 'accidental' degeneracies of quantum mechanics. We consider three such potentials, for each of which the total energy H has a simple expression in terms of (ξ_1,ξ_2) (for more details see ref.3):

(i) the rigid rotator $V = \delta(r-r_0)$, $H = \frac{\xi_2^2}{2I}$, $I = mr_0^2$,

(ii) the linear oscillator $V = \frac{1}{2}m\omega_0^2 r^2$, $H = \omega_0(2\xi_1+\xi_2)$,

(iii) the Kepler system $V = -\frac{2e^2}{r}$, $H = -\frac{mZ^2e^4}{2(\xi_1+\xi_2)^2}$.

The 'accidental' degeneracy occurs because the natural frequencies $\omega_i = \partial H/\partial\xi_i$ are simply related, that is (i) $\omega_1 = 0$ (ii) $\omega_1 = 2\omega_2$ (iii) $\omega_1 = \omega_2$. In each case there is a corresponding simplification in the drift and diffusion coefficients.

(i) The only non-vanishing coefficients in (3.3) are A_2 and B_{22}. The system is effectively one-dimensional, and the stationary solution of the Fokker-Planck equation is

$$\rho_0 = \frac{1}{2\pi^2\hbar^2}\exp[-2\xi_2/\hbar]. \tag{3.4}$$

(ii) The stationary solution in this case is rather well known

$$\rho_0 = \frac{1}{\pi^3\hbar^3}\exp[-2H/\hbar\omega_0]. \tag{3.5}$$

In both the above cases the condition of <u>detailed balance</u> is satisfied, that is

$$A_i\rho_0 - B_{ij}\frac{\partial\rho_0}{\partial\xi_j} = 0 \qquad (i = 1,2), \tag{3.6}$$

which means that the process is invariant under time reversal.

(iii) The Fokker-Planck coefficients are much more complicated in this case[3]. However, the 'accidental' degeneracy does result in some simplification. In the above cases the system emits and absorbs radiation at a single frequency (ξ_2/I and ω_0 respectively). A general separable system absorbs and emits any frequency from the set $\underset{\sim}{n}.\underset{\sim}{\omega}$, and for a central-field system this set is reduced to $n\omega_1+\omega_2$. Since, as we have seen, the Kepler system has $\omega_1 = \omega_2$, the set is further reduced in this case to $n\omega_1$. The Kepler system is therefore the simplest to study after the rigid rotator and the linear oscillator.

No solution has yet been obtained even for the stationary Fokker-Planck equation in this latter case. It has, however, been established from the Fokker-Planck coefficients that such a solution could not possibly satisfy the detailed balance condition, and therefore in this case the process is not time-reversal invariant.

4. Discussion

The crucial property possessed by the rigid rotator and the linear oscillator is that, in a given state of the system, they interact with only a single frequency of the radiation field, whereas a general dynamical system is continually transferring energy in the radiation field among the various harmonics of its natural frequencies.

Boyer[4] has shown that, if the zero-point spectrum is replaced by the Rayleigh-Jeans spectrum

$$S_E(\omega) = \int_{-\infty}^{\infty} B_E(\tau)e^{-i\omega\tau}d\tau = \frac{4kT\omega^2}{3c^3}, \tag{4.1}$$

then the detailed balance condition is satisfied for any separable system, and the stationary solution of the Fokker-Planck equation is the Maxwell-Boltzmann distribution. Moreover, if we impose the condition, known as radiation balance, that both the dynamical system and the radiation field are in equilibrium, then (4.1) gives the only possible form of the radiation spectrum.

If radiation balance were a required condition in stochastic electrodynamics, this would lead to a picture of the atom not greatly differing from the Rutherford atom before Bohr. The Maxwell-Boltzmann distribution for the Kepler system gives a distribution concentrated, with probability 1, at $r = 0$. This means that the Rayleigh-Jeans spectrum is too weak, at high frequencies, to prevent the radiative collapse of such an atom. On the other hand, qualitative arguments[1] suggest that the zero-point field can prevent such a collapse.

The radiation field in which every dynamical system finds itself is the unavoidable residue of interaction with the "rest of the world" after all immediately neighbouring systems have been removed. In principle it represents the most long-range interaction in the universe, apart possibly from gravitational waves. That nearby material objects can modify the spectrum of this field has already been demonstrated in studies of the Casimir effect and dispersion forces[5,6].

It is therefore, I suggest, quite reasonable to suppose that a dynamical system is in equilibrium in this field, while the field itself is not in equilibrium. After all, the dynamical system and the "rest of the world" have vastly different relaxation times!

Such a world surely corresponds much more closely to that in which we live than the world of radiative equilibrium pictured by Boyer. If this argument is accepted, then it is easy to see why time-reversal invariance is not, in general, satisfied in stochastic electrodynamics.

5. References

1.Claverie P.and Diner S.,Stochastic electrodynamics and quantum theory,*Int.J.Quant.Chem.*,12,Supp.1,41,1977.

2.Marshall T.W.,Brownian motion and quasi-Markov processes,article submitted to *Proc.Roy.Soc.*

3.Marshall T.W. and Claverie P.,Stochastic electrodynamics of non-linear processes,article submitted to *J.Math.Phys.*

4.Boyer T.H.,Statistical equilibrium of nonrelativistic multiply periodic classical systems and random classical electromagnetic radiation,*Phys.Rev.A*,18,1228,1978.

5.Marshall T.W.,A classical treatment of blackbody radiation, *Nuovo Cim.*,38,206,1965.

6.Marshall T.W.and Henry L.L.,A classical treatment of dispersion forces,*Nuovo Cim.*,41,188,1966.

THE ANHARMONIC OSCILLATOR IN STOCHASTIC ELECTRODYNAMICS (S.E.D.) : THE PROBLEM OF "RADIATION BALANCE" AT EACH FREQUENCY.

Luis PESQUERA

Laboratoire de Chimie Quantique - Institut
de Biologie Physico-Chimique, 13, rue
Pierre et Marie Curie - 75005 Paris - France

Abstract

The balance between absorbed and emitted power at each frequency ("radiation balance") is investigated in Stochastic Electrodynamics (S.E.D.). The case of a slightly anharmonic oscillator (with a $\beta.X^4$ *perturbing potential) is examined in full detail. The results show that the "radiation balance" is not exactly satisfied as soon as* $\beta \neq 0$, *a property which disagrees with the quantum results (moreover the maximum absorption frequencies are not the same as in quantum theory). It seems that this lack of "radiation balance" is related with the occurence of more than one single harmonic in the Fourier expansion (with respect to time) of the deterministic motion.*

Stochastic Electrodynamics (S.E.D.) is a theory which has been proposed as a possible alternative to quantum theory. Basically S.E.D. is classical electrodynamics with radiation damping and a stochastic electromagnetic field [1].

The theory is a well-defined problem of mathematical physics, but his solution is rather difficult, due to the non-white character of the random background field. It is just recently that the non-standard techniques required to treat non-linear problems in S.E.D. have been worked out [2].

We apply these techniques to the case of a slightly anharmonic oscillator (with a $\beta.X^4$ perturbing potential) to obtain the stationary probability density. Then, we use it to calculate the "emission coefficient", and, using the Kubo linear response theory, the absorption coefficient at each frequency. Knowing both coefficients it is possible to study the balance between absorbed and emitted power at each frequency. This work is a continuation and extension of a preliminary study by P. Julg [3].

We first recall the equation of motion in S.E.D. for a non-relativistic charged particle (with mass m and charge e). In one dimension, this may be written as the pair of coupled stochastic differential equations [4]

$$(1) \qquad m\dot{x} = p + \tau f(x) + e\mathscr{A}^{(\tau)}(t) \quad , \dot{p} = f(x)$$

where f(x) is the external force, τ is the (small) parameter $\frac{2e^2}{3mc^3}$ and $\mathscr{A}^{(\tau)}(t) = \int^t dt' \int_0^\infty E(t'+\tau s)\exp[-s]\, ds$, E being the random electric background field.

A Fokker-Planck equation may be obtained for the probability density of the so-called Markovian approximation to the process (x,p), solution of the system (1) [See 2, Sec. 4]. For the stationary probability density W, it is possible to reduce this Fokker-Planck equation to the following one [5]

$$(2) \qquad -G^{EE}\frac{dW}{dE} + G^{E}\, W = 0$$

where E is the energy and G^E (G^{EE}) is the reduced drift (diffusion) coefficient.

For a multiperiodic system these coefficients may be obtained under the form of a series other all harmonics [6]. In the one-dimensional case the system is simply periodic, and these coefficients are given by

$$(3) \qquad G^E = \frac{-4\pi e^2}{3c^3}\,\omega^3 \sum_{n=-\infty}^{+\infty} n^4 |x_n|^2 \;, \qquad G^{EE} = \pi e^2\,\omega \sum_{n=-\infty}^{+\infty} S_E(n\omega)\, n^2 |x_n|^2$$

where ω is the fundamental frequency of the system, S_E is the spectral density of the random electric background field (in S.E.D.

$S_E(\omega) = \frac{2\hbar}{3c^3}|\omega|^3$ [2, Sec 3A]), and X_n are the Fourier components of the deterministic motion.

We consider now an anharmonic oscillator with a potential $U = (1/4)m\beta x^4 + (1/2)m\omega_o^2 x^2$, where β is a small parameter. In order to obtain W by integrating (2), we calculate $\frac{G^E}{G^{EE}}$. For the energies such that $y = \frac{E\beta}{m\omega_o^4} \ll 1$, we have [7] :

(4) $$\frac{G^E}{G^{EE}} = - \frac{2}{\hbar\omega_o} [1 - \frac{3}{4} y + \frac{129}{128} y^2]$$

where we have dropped terms of order β^3.
From (2) and (4) we obtain through order β^2 :

(5) $$W(E) = N \exp\left[-2E/\hbar\omega_o\right] [1 + \frac{3}{4} \beta\frac{E^2}{\hbar m\omega_o^5} + \beta^2 \frac{E^3}{\hbar m^2\omega_o^9} (- \frac{43}{64} + \frac{E}{\hbar\omega_o} \frac{9}{32})]$$

where $N = \frac{1}{\hbar\pi} [1 - \frac{81}{256} (\frac{\hbar\beta}{m\omega_o^3})^2]$ is the normalization constant in the phase space. In actual fact this expression of W(E) is only valid for energies such that $E^3\beta \ll 1$, but it may be used for obtaining the mean energy $\bar{E}$, because the contribution of the range energies $E^3 \beta \gtrsim 1$ to $\bar{E}$ is negligible. The reason of this is that W(E) behaves as $\exp[-E^{3/4}]$ for $y = \frac{E\beta}{m\omega_o^4} \gg 1$.

Through order β^2, $\bar{E}$ is given by :

(6) $$\bar{E} = \frac{\hbar\omega_o}{2} [1 + \frac{3}{8} \frac{\hbar\beta}{m\omega_o^3} + \frac{141}{256} (\frac{\hbar\beta}{m\omega_o^3})^2]$$

what differs, even in the sign, from the quantum result at the second order in β [3].

Knowing W(E), we may apply the Kubo linear response theory for obtaining the absorption coefficient (ratio of the absorbed power to the incident power) $a(\omega)$ [8]. It may be shown that for one dimentional periodic system [6]

(7) $$a(\omega) = \frac{-8\pi^3}{3} e^2 \omega \sum_{n=o}^{\infty} [|Xn|^2 \frac{dW}{d\omega'}]_{\omega'=\omega/n} = \frac{-8\pi^3}{3} e^2\omega \sum_{n=0}^{\infty} [|X_n|^2 \frac{G^E}{G^{EE}} \frac{dE}{d\omega'} W]_{\omega'=\omega/n}$$

where we have used (2).

In our case, namely an even potential for which the fundamental frequency has a minimum value ω_o, only a finite number of odd terms contribute to $a(\omega)$. Furthermore we can restrict ourselves to just one term, because $a(\omega)$ is not negligible only for frequencies very close to $(2n+1)\omega_o$. The essential point is that for frequencies such that $\omega-\omega_o \gg \beta$ (e.g. $\omega-\omega_o \gtrsim \beta^{n-1/n}$, $n > 1$), $E(\omega) \gtrsim \beta^{-1/n} \gg 1$, and then $W(E) \lesssim \exp[-\beta^{-1/n}]$, whereas $|\frac{G^{E}}{G^{EE}}|$, $\frac{dE}{d\omega}$, $|X_n|^2$ are bounded by powers of β.

Therefore the first line of the spectrum (frequencies very close to ω_o) is given by $a_1(\omega) = \frac{-8\pi^3}{3} e^2\omega\,[\,|X_1|^2 \frac{G^E}{G^{EE}} \frac{dE}{d\omega'} W]_{\omega'=\omega}$. Now we need to know E as a function of ω. When ω is close to ω_o this is given by

$$(8) \qquad E(\omega) = \frac{m\,\omega_o^4}{\beta} \frac{4}{3} \left(\frac{\omega-\omega_o}{\omega_o}\right) \left[1 + \frac{23}{12}\left(\frac{\omega-\omega_o}{\omega_o}\right) + \frac{107}{72}\left(\frac{\omega-\omega_o}{\omega_o}\right)^2\right]$$

where we have dropped terms of order $\left(\frac{\omega-\omega_o}{\omega_o}\right)^4$.

Using (4),(5), (8) and the expression of $|X_1|^2$ [6] we obtain for frequencies such that $\omega-\omega_o \ll \beta^{2/3}$:

$$(9) \qquad a_1(\omega) = \frac{128}{27} \frac{\pi^2 m\omega_o^5 e^2}{\beta^2 \hbar^2} N \exp\left[-\frac{8}{3}\frac{\gamma\Delta}{\beta}\right]\Delta\left[1 + \frac{17}{4}\Delta + \frac{719}{144}\Delta^2 - \frac{34}{9}\frac{\gamma\Delta^2}{\beta} - \frac{33}{2}\frac{\gamma\Delta^3}{\beta} + \frac{578}{81}\frac{\gamma^2\Delta^4}{\beta^2}\right]$$

where $\gamma = \frac{m\omega_o^3}{\hbar}$ and $\Delta = \frac{\omega-\omega_o}{\omega_o}$.

The maximum of this first line, namely $\omega_1 = \omega_o + \frac{3\,\hbar\beta}{8m\omega_o^2}$ is not the same as the one in quantum mechanics :

$$\omega_1^{Q.M.} = \omega_o + \frac{3}{4}\frac{\beta}{m\,\omega_o^2} \quad [3].$$

In a similar way, we obtain the absorption coefficient for frequencies close to $3\omega_o$ ("second line") :

$$(10) \qquad a_3(\omega) = \frac{-8\pi^3}{3} e^2\omega\,\left[\,|X_3|^2 \frac{G^E}{G^{EE}} \frac{dE}{d\omega'} W\right]_{\omega'=\omega/3}$$
$$= \frac{8\pi^2}{81} \frac{m\omega_o^5 e^2}{\beta^2\hbar^2} N \left(\frac{\omega-3\omega_o}{3\omega_o}\right)^3 \exp\left[-\frac{8}{3}\frac{\gamma}{\beta}\left(\frac{\omega-3\omega_o}{3\omega_o}\right)\right]$$

The absorbed power for the first line I_{a_1} is given by :

(11) $$I_{a_1} = \int_{\omega_o}^{\omega_o+A} a_1(\omega)\rho(\omega)d\omega = \frac{\hbar\omega_o^3 e^2}{3mc^3}\left[1 + \frac{9}{4}\frac{\beta\hbar}{m\omega_o^3} - \frac{567}{512}\left(\frac{\beta\hbar}{m\omega_o^3}\right)^2\right]$$

where $\rho(\omega) = \frac{\hbar}{2\pi^2 c^3}\omega^3$ is the energy spectral density of the random field [2, sec 2] and $A \gg \beta$, but $A \ll \beta^{2/3}$ (domain of validity for the expression (9) of $a_1(\omega)$), because the contribution of the frequencies $\omega > \omega_o + \beta^{2/3}$ to Ia_1, it may be shown to be negligible, due to the behavior of W(E) for large energies.

Similarly, for the second line :

(12) $$I_{a_3} = \int_{3\omega_o}^{3\omega_o+A} a_3(\omega)\ \rho(\omega)d\omega = \frac{\hbar\omega_o^3 e^2}{3mc^3}\left[\frac{729}{512}\left(\frac{\beta\hbar}{m\omega_o^3}\right)^2\right]$$

The total absorbed power I_a is given, through order β^2, by the two first lines, because the contribution of the other lines to I_a is, at least of order β^4, due to the fact that $|X_{2n+1}|^2 \simeq y^2|X_{2n-1}|^2$, $(y = \frac{E\beta}{m\omega_o^4})$.

Finally, it may be pointed out that, in contrast with quantum theory, the linewidth is of order β.

Thus far we have studied the absorption of energy. Now we turn to the other aspect of the "radiation balance" : the emission of energy.

For the one-dimensional periodic systems the "emission coefficient" (emitted power at a frequency ω) $I_e(\omega)$, is given by [6]

(13) $$I_e(\omega) = \frac{8\pi}{3c^3}e^2\omega^3\sum_{n=o}^{\infty}\left[|X_n|^2\frac{dE}{d\omega'}W\right]_{\omega'=\omega/n}$$

In a similar way that for $a(\omega)$, it may be shown that only for frequencies very close to $(2n+1)\omega_o$ the emission is not negligible. Therefore for frequencies close to ω_o ("first line") :

(14) $$I_{e_1}(\omega) = \frac{8\pi}{3c^3}e^2\omega^3\left[|X_1|^2\frac{dE}{d\omega'}W\right]_{\omega'=\omega} = \frac{64}{27}\frac{m\omega_o^5 e^2}{\beta^2\hbar c^3}N\omega^3\Delta\exp\left[\frac{-8}{3}\frac{\gamma\Delta}{\beta}\right]\cdot$$
$$\left[1 + \frac{17}{4}\Delta + \frac{881}{144}\Delta^2 - \frac{34}{9}\frac{\gamma\Delta^2}{\beta} - \frac{33}{2}\frac{\gamma\Delta^3}{\beta^2} + \frac{578}{81}\frac{\gamma^2\Delta^4}{\beta^2}\right]$$

We note that for these frequencies there is not radiation balance at each frequency. The emitted power $I_{e_1}(\omega)$ and the absorbed power $I_{a_1}(\omega) = a_1(\omega)\ \rho(\omega)$ are not the same at each frequency. In fact for

$\omega-\omega_o \lesssim \beta$, $I_{e_1}(\omega) - I_{a_1}(\omega) \lesssim \beta$.

For the "second line" (frequencies close to $3\omega_o$)

$$(15) \qquad I_{e_3}(\omega) = \frac{4}{243} \frac{m\omega_o^5 e^2}{\hbar\beta^2 c^3} N \omega^3 \left(\frac{\omega-3\omega_o}{3\omega_o}\right)^3 \exp\left[\frac{-8}{3}\frac{\gamma}{\beta}\left(\frac{\omega-3\omega_o}{3\omega_o}\right)\right]$$

and therefore the radiation balance is again not satisfied. For $\omega-3\omega_o \lesssim \beta$, $I_{a_3}(\omega) - I_{e_3}(\omega) = 2\, I_{e_3}(\omega) \lesssim \beta$.

These results show that the anharmonic oscillator obsorbs energy at the frequencies close to $3\omega_o$, and radiates net energy at the frequencies close to ω_o.

From the total emitted power for the two first lines

$$(16) \qquad I_{e_1} = \frac{\hbar\omega_o^3 e^2}{3mc^3}\left[1 + \frac{9}{4}\frac{\beta\hbar}{m\omega_o^3} - \frac{81}{512}\left(\frac{\beta\hbar}{m\omega_o^3}\right)^2\right], \quad I_{e_3} = \frac{\hbar\omega_o^3 e^2}{3\,mc^3}\left[\frac{243}{512}\left(\frac{\beta\hbar}{m\omega_o^3}\right)\right]^2$$

We recover the conservation of energy $I_{a_1} + I_{a_3} = I_{e_1} + I_{e_3}$. The amount of energy flowing towards the oscillator at the frequencies close to ω_o exactly balances that flowing away at frequencies close to $3\omega_o$.

This result is in agreement with the one obtained by Boyer [9] for an anharmonic oscillator with potential in X^3, which is found to be in equilibrium only with the Rayleigh-Jeans law of radiation.

If we consider finite temperature (instead of zero temperature) this lack of radiation balance, for any spectrum different from the Rayleigh-Jeans one, means that it is not at all evident to fulfill Kirchhoff's law in theories such as SED (by contrast, this law is actually satisfied in quantum theory [10]).

Finally, it may be pointed out that this lack of radiation balance in S.E.D. seems to be related with the occurence of more than one single harmonic in the Fourier expansion (with respect to time) of the deterministic motion.

BIBLIOGRAPHY

[1] For the general principles of S.E.D., see e.g. T.H. Boyer, Phys. Rev. D11, 790 (1975), S. Diner (this volume) and references therein.
[2] P. Claverie and S. Diner, Int. J. Quant. Chem. XII, Suppl.1, 41 (1977).
[3] P. Julg, Folia Chimica Theoretica Latina, 6, 99 (1978), and results to be published.
[4] E. Santos, Nuovo Cimento, 19B, 57 (1974).
[5] H. Haken, Rev. Mod. Phys. 47, 67 (1975).
[6] P. Claverie (this volume).
[7] For the calculation of X_n, see L.D. Landau and E.M. Lifshitz, Mechanics, (Pergamon Press, Oxford, 1960).
[8] R. Kubo, J. Phys. Soc. Jpn. 12, 570 (1957).
[9] T.H. Boyer, Phys. Rev. A18, 1228 (1978).
[10] G. Nienhuis, Physica, 66, 245 (1973).

[illegible]

[1] For the general properties of [illegible], see e.g. [illegible] 111, 790 (1970), [illegible]
[2] [illegible] (1979).
[3] [illegible]
[4] [illegible] (1976).
[5] [illegible] (1979).
[6] [illegible]
[7] [illegible] Press, Oxford [illegible]
[8] [illegible]
[9] [illegible] (1978).
[10] [illegible]

PART II

EVOLUTION AND IRREVERSIBILITY

REVERSIBILITY VERSUS IRREVERSIBILITY IN THE PHYSICAL UNIVERSE

R. THOM
Institut des Hautes Etudes Sciéntifiques
Bures sur Yvette, France

1. Time-Reversibility

What do we mean when we say that the laws of physics are time-reversible? There is an immediate answer: if in the mathematical expression of physical laws, we change t into $-t$, then we get the same system or an equivalent one. This purely formal criterion needs some clarification. Let us suppose that the considered law is expressed as a differential system. In general, this system is not of first order (more frequently of second order). By introducing new variables (in general put equal to first order derivatives), the p-variables,one gets a new differential system of first order: a flow defined in a manifold M (phase space) by a vector field X. We may define the same flow in the product $M \times \mathbb{R}$ of M by the time axis $\mathbb{R}$. Then time reversibility can be expressed as follows.

There exists an automorphism A of order two (A^2 = identity) of the manifold $M \times \mathbb{R}$, such that

$$A(m, t) = (m', -t)$$

with the property that $AX = -X$.

Instead of $-t$, we could as well write $k - t$, or eventually $k - \lambda t$.

For physical laws put into the form of Partial Derivatives Systems, the construction could be the same, except that in such cases the reduction gives rise to a standard form of Pfaffian system $\Sigma u_i dx_i = 0$ in a space of P.D.E. ("jets" in modern terminology), the automorphism A then leaves invariant this Pfaffian system, hence all its local - global - or - partial - solutions.

Is there any specific consequence of reversibility for the local properties of a system? The answer is no. For let us consider the following examples:

1) $\ddot{q} = -\omega^2 q$ (linear oscillator). In phase space, we have a central singular point ($q = p = 0$) hence any orbit (except the origin) is periodic.

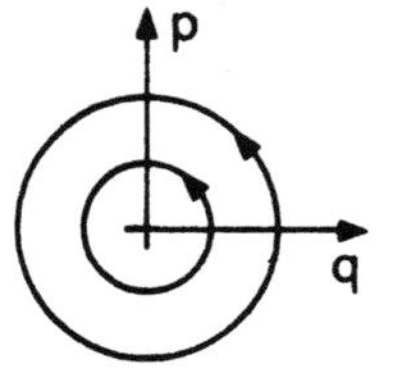

2) $\ddot{q} = \omega^2 q$ (repeller). Here in phase space (q, p) we have a saddle point. All trajectories (but the origin and its separatrices) go to infinity, such a system is inherently unstable.

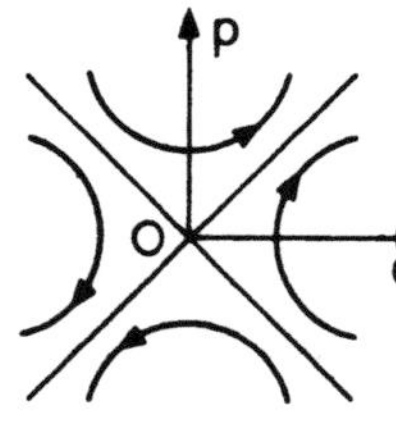

3) Consider, on the 2-sphere $x^2 + y^2 + z^2 = 1$ in $\mathbb{R}^3$ (Oxyz) the flow defined by $dm/dt = \text{grad}(z)$.

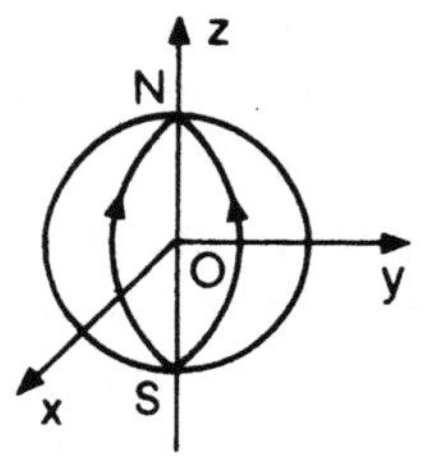

It is a typically irreversible flow, as any trajectory which is not a singular point (North Pole $x = y = 0$, $z = 1$, South Pole $x = y = 0$, $z = 1$), goes from the South Pole to the North Pole. But such a flow is time-reversible as the symmetry $A : (x, y, z) \to (x, y, -z)$ transforms the flow in its sign opposite.

More generally, any flow X in a manifold M can be embedded in a time reversible flow. It suffices to take another example of the M manifold, let M', and on the disjoint union $P = M \cup M'$ define a flow $\tilde{X}$ by $\tilde{X}|M = X$, $\tilde{X}|M' = -X$, then the automorphism $A(P) \to P$ defined by $A|M \to M'$, $A|M' \to M$ obviously reverses the sign of $\tilde{X}$ on **P**.

One should not confuse, for a dynamics, the property of being time-reversible, which locally says nothing, with another very important property, namely ***recurrence***. Let us recall, in that respect, the classical definition of G.D.Birkhoff:

In a flow (M, X) a point m is *wandering*, if a small ball B_m with center m has the property that its temporal image $h_\tau(B_m)$ do not interest B_m, if τ is large enough. By definition, wandering points form an open set in M; the complementary invariant closed set Ω, the *non-wandering* set, carries all the asymptotic properties of the dynamics. In particular, any point ω in Ω has the "return" property that there are always points ω_i nearby ω, and time transforms h_{τ_i} such that $h_{\tau_i}(\omega_i)$ can be made inferior to any distance ε_i tending to zero. In fact the relation between τ_i and ε_i, what we could describe as the "return spectrum" around ω, contains a lot of information about the flow around ω. In some cases, this can be defined as a global property of some time transformation h_τ. For instance, let us consider the flow defined on the two dimensional terms $T^2 = R/z \times R/z$, $x \times y$, by the linear relation $dy/dx = \alpha$. Such a flow can be characterized by a discrete time flow acting on the circle y by a rotation $r_\alpha y_0 \to y_0 + \alpha$.

If α is rational, then r_α is of finite order k and r_α has an inverse in positive time $r_\alpha^{k-1} \circ r_\alpha = 1$.

If α is irrational, then r_α has only a quasi-inverse, a "return spectrum": to any $\varepsilon_i \to 0$, there corresponds a N_i such that $(r^{N_i} \circ r - \text{identity}) < \varepsilon_i$.. As well known, many arithmetic properties of the irrational α can be extracted from the return-spectrum: an algebraic quadratic irrational, as the golden number $e(e^2 = 1 + e)$ — badly approximated by rationals — has a very different return spectrum as the rotation r_α where α is a Liouville transcendental number (strongly approximated by rationals) [1].

An interesting problem is to characterize those recurrent dynamics (for an analytical flow around the origin $0 \in \mathbb{R}^{2n}$) such that the return spectrum is the same at all of its points. I conjecture that only the complex rotation $\mathbb{C}^n \underset{f}{\to} \mathbb{C}^n$ defined by $f(z_j) = e^{i\theta} z_j$ satisfies this condition. This could be the ultimate motivation of the Quantum-Mechanical formalism and the introduction of unitary-Hilbertian theory.

2. Equivalence Between Systems. Structural Stability

Two flows (M, X), (M′, X′) are said to be equivalent if there exists a diffeomorphism $\varphi : M \to M'$ carrying the vector X at (m_0) to the vector X′ at point $m_0' = \varphi(m_0)$. This definition is in general too strong and has to be relaxed in the following way: the two flows are h-equivalent, if there exists a homeomorphism $h : M \to M'$ carrying any orbit of the X-flow onto an orbit of the X′-flow. A flow (M, X) is said to be structurally stable, if any perturbation X′ of X, such that $\|X' - X\|_m$ in the $\mathscr{C}^m$ topology is sufficiently small, makes (M, X′) h-equivalent to (M, X).

On any compact manifold M structurally stable systems are known to exist: "generic" gradient systems, and generic "Morse-Smale systems" in particular. But it is known also that such systems are not dense in the full function space $X^m(M)$ of flows on M.

Given any flow (M, X) if we integrate the flow from time t_0 to time $t_0 + \tau$ we get a diffeomorphism h_τ (time-transform) which is an equivalence for (M, X) onto itself. This means that from the viewpoint of the topologist, nothing happens in the time evolution of a flow, as the end state is equivalent to the original one. But if we consider the recurrence properties of h_τ (and of the flow itself) then obviously something may happen.

In a gradient dynamic like is Example 3 (grad z in the unit-sphere of R^3), every point tends to the North Pole as its final asymptotic state. Hence we have a clear irreversibility. In an hamiltonian dynamic, the time transform h_τ is a canonical transformation (symplectomorphism compatible with X). But if we have a flow defined, as usual, on the covector-bundle $T^*(V)$ of some configuration space V (as base space), then the time-transform h_τ is no longer compatible with the projection $\pi : T^*(V) \to V$. But this projection is very important from the viewpoint of macroscopic phenomenology as it is always easier to observe the position of an object than its speed (the reverse may be more usual in Microphysics ...).

The fact that h_τ "mixes q and p" is an obvious factor of macroscopic irreversibility, it is tied to a progressive loss of the possibility of description of the states of our system. In Kaplecian dynamics the situation is not as bad, because all elliptical trajectories are closed, hence there exists some kind of stability for configuration space description; any trajectory leaving a point q in phase space comes back to this point later indefinitely many times. This accounts quite a lot for the apparent macroscopic stability of the solar system. In the general n-body system, such property does not hold, and apparent "chaos" ensues very rapidly.

3. Attractors

In a flow (M, X), if we follow a trajectory γ starting from a point m_0, γ has a ω-limit set, consisting of all the points x such that $\| m_0 (t)x \|$ can be made arbitrarily small for t tending to infinity. In general this invariant limit set A (attractor) has the property that it admits an open neighbourhood U (basin of the attractor) with the following properties:

a) any trajectory starting from a point u of U admits A as ω- limit set;

b) any trajectory having as α-limit set (for $t \to -\infty$) a point of A is in A;

c) almost any trajectory in A is dense in A.

There exists always in U a Lyapounov function $L : U \to R^+$ vanishing in A, positive on U–A, and decreasing along the trajectories in U. Such a function plays the role of a local negentropy for the attractor A.

As a result, if we perturb slightly the field X, the perturbed flow $X' = X + \delta X$ still enters a level hypersurface of $\mathcal{L}$, $\mathcal{L}^{-1}(\varepsilon)$ in U. Hence X admits also some attractor (A) inside $\mathcal{L}^{-1}(\varepsilon)$. An attractor cannot explode under perturbation. But it may *implode* giving rise to lower dimensional attractors. It is believed that this implosive process has to stop, hence leading to "stable" configurations of attractors (although their internal topology may vary). This problem of characterizing the structure of "stable" attractors is one among the most difficult of qualitative dynamics. It is believed that such attractors have to carry a smooth invariant measure allowing to define some kind of local thermodynamics. Examples of such are already known, in particular the Anosov systems, which seem to be the only structurally stable systems for discrete symplectomorphisms (Newhouse's theorem) [2].

But the important remark to be made here is that, if we associate to any trajectory (γ) of a flow (M, X) its ω-limit attractor, *we so to speak annihilate the effect of time of this dynamic,* as the dynamic X/A is obviously of recurrent type (almost any trajectory in A is dense in A). Hence on A we have a dynamics in which "nothing really happens", as in the case of hamiltonian dynamics on its energy hypersurface.

This requirement of considering recurrent and conservative dynamics, to stabilize the object may also have as a consequence the isolation, or freezing of "external" coordinates which won't any longer be allowed to vary, or to vary only very slowly. If for instance we have a flow on R^n, having at the origin O an isolated singularity, what happens if we require that every trajectory, from $t=-\infty$ to $t=+\infty$, remains in a compact domain of $R^n - 0$? If we postulate moreover that the flow is analytic around O, then it is to be expected that such a singularity could not exist unless there is a "first integral" for X, as the energy for an hamiltonian flow. This conjecture can very easily be proved for the linear part of the flow (note also that n has to be even if we want to avoid the existence of fixed points of a flow on any even dimensional sphere). A further condition of "indecomposibility" of the system could lead us to the complex rotation case already considered at the end of § 3.

4. Separation of Systems. The Origins of Hamiltonian Formalism

The problem of describing how it is that in nature we have to do with systems which can be considered as "isolated" despite the fact that they never cease to interact is one of the most difficult question that a general description of the Universe has to face. Perhaps one of the motivations for the appearance of hamiltonian formalism in Mechanics is due to the fact that it allows the splitting of systems to be relatively stable. Suppose we have a compound system $M_1 \times M_2$, with hamiltonian dynamic of type $H = H_1(m_1) + H_2(m_2) + W(m_1, m_2)$, coupling hamiltonian.

If we suppose that the dynamic H_2/M_2 is very fast in comparison with the one in M_1, and that the fast dynamic H_2 is ergodic in M_2, then we may consider that the true dynamic in $M_1 \times M_2$ can be mapped onto the dynamic of M_1, provided with a modified hamiltonian

$$\tilde{H}_1(m_1) = H_1(m_1) + \int_{\bar{M}'(m_1)} W(m_1, m_2)dm_2$$

due to addition of coupling terms. The slow variation due to change in the initial position m_2^0 will be interpreted by the inhabitants of M_1 as "noise", and they undoubtedly will devise stochastic procedures to describe it.

Of course the consideration of Liouville measure of the energy hypersurface in (M_2) will lead to considering some kind of local entropy (the log of this measure).

Another motivation for Hamiltonian Dynamics is the need for intersubjective consensus in the description of a system. Any kind of conceptual apparatus used to describe the state of the system has to be accepted by all observers, in particular by the same observer when time passes on, and the observer has to compare his vision at time (t_0) with his vision at a later time $(t_0 + \Delta t)$. Hence these two visions have to be matched by an isomorphism, in principle depending only on Δt, and not of the position (nor of the state) of the observer himself. The hamiltonian formalism appears then as a way of expressing on an extended manifold (V) "parametrizing the position of observers" ... the construction referred earlier for the singularity O of R^n. One permits V to glide on itself, but nearby trajectories not on V have to stay all the time in a compact set of a tubular neighbourhood $T'(V)$ of V, compact in any fibre $M : T'(V) \to V^*$.

5. Transformation Between Systems. Algebraic Models

This leads to the following general consideration: inasmuch we decide that a dynamics is in a fixed asymptotic regime, we have to consider the *internal phase space* of this dynamic, let M_u denote such a phase space. Then the dynamics also depends on external parameters forming a space U of slowly varying parameters. The system (M_u) may disappear when its phase space undergoes a topological change, for instance crossing a singularity (as an example, crossing the origin for the hamiltonian $H = \Sigma(q_i^2 + p_i^2)$ of the linear oscillator).

*) Moreover, the hamiltonian formalism also appears when we have a medium subject to constraints which may tear this medium following an hypersurface (H) as locus of discontinuities. If at any point of V we have a cone $T_V(x)$ to which (H) must be tangent, then the hamiltonian formalism describes for any such (H), the field of directions on which it is tangent to the cones; this describes a flow for the discontinuity of a tangent vector field in the medium.
Two observers, if they have to compare their visions of the world should have first the same vision of space, let us say an Euclidean space E. Then if an observer O' sees the space E moving along a flow X_1, the other observer sees E moved by the flow X_2, then X_1 and X_2 are such that $X_1 - X_2$ is tangent to the hypersurface (H) which separates the two visions.

Let us suppose that all these M_u belong to the same algebraic family; in the terminology of Algebraic Geometry (u) is a family of ***moduli*** parametrizing these varieties. Then when the topological type of M_u changes by z crossing a singularity (ζ), this singularity is located as a point ζ (more or less constructible or fictitious) of the boundary of the modular space U. Then the system is destroyed (by ***bifurcation***) and new systems arise from this "catastrophe." These new systems (P_i) will have themselves modular spaces U(P_i) which will meet at ζ forming at least a part of the "unfolding" of the singularity (ζ).

Such conjectural considerations may be given some mathematical consistency only in case of dynamics, defined by gradients, then called elementary catastrophe theory [3]. But there is some hope to see this framework valid outside this field, each time we ***have a global analytic structure acting as a model for the whole bifurcation process.*** Systems as (P_i), which belong so to speak to the same morphological level of organization as M, may be said "satellites" of (M). This division of observable systems into domains of a kind of combinatorial syntax corresponds also to the intuitive nature of hierarchical level of organization. For we do have to explain in a fundamentally irreversible universe, not only the stability of systems, but also the stability of the syntax ruling the organization of their transformations.

The nature of the bifurcation associated with a singularity such as (ζ) may of course vary. In our macroscopic world, there is a strong tendency to see these catastrophes governed by the "elementary catastrophe theory". In Microphysics however, where the ontological status of space time is somewhat subject to doubt, it looks as if a purely unitary formalism of S-Matrix type may be a better model (the control space being no longer space time, but space of momenta of the incoming or outgoing particles).

6. Irreversibility in Bifurcations. The Problematics of Initial Data. Types of Irreversibility.

When a system (S) appears, it owes its birth, in general, to some anterior catastrophic process, arising out from some satellites systems (S_1) . . (S_u). But to obtain (S) we must in general follow a well defined procedure (by a linguistic text known as preparation protocol). As a result, when (S) is prepared, it appears in general in a fixed well-defined state, and any other initial state in its internal phase space has to be obtained by a later procedure. Hence despite the fact that all points in M are theoretically reachable, it is by no means certain that all are indifferently reachable. For instance, it is much easier to prepare an object at rest, than the same object moving at very high speed. (We have to prepare the bullet at rest before putting it inside the gun).

As in the case of a perfect delay catastrophe [3] following the usual wiggle diagram in dimension one the new regime is created at the foot of the "fast trajectory" jumping from the metastable regime (ρ_1) to the lower lying stable one (ρ_0). Such points, like α , which are obtained by "canalized procedures" are the most easily amenable to linguistic description. Prepared states are by definition described by a linguistic test, hence they form an at most countable set in the phase space M of our system.

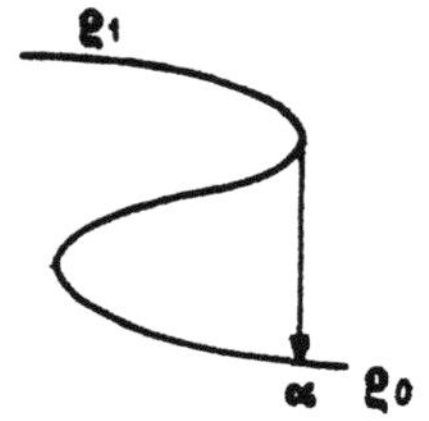

Now if the dynamic in M is of ergodic mixing type (which is the general case if this dynamic is structurally stable), the linguistic description loses very rapidly its efficiency. For a small ball B_{m_0} around m_0 is very rapidly distorted: this is particularly the case for K-systems (Anosov flows), the boundary sphere ∂B becomes increasingly extended in one direction, contracted into the complementary plane, in phase space, and increasingly distorted in the configuration space V (if $M = T^*(V)$). Hence in the space of maps $C^\infty(S^{m-1})$ there is little doubt that the images $\pi h_\tau(S^0)$ have steadily increasing topological complexity allowing to define some kind of entropy increasing along these very particular objects. A "distal" flow like the linear flow on a torus won't have in general such distorting property.

Hence, when we speak of *irreversibility* in dynamical systems, we should carefully distinguish between two types of irreversibility:

a) a *true* irreversibility which is either going to an attractor in a flow, or a preferred way of crossing a bifurcation.

b) a *fictitious* irreversibility, which is the one described by the second principle: it expresses the fact that the linguistic description looses progressively its accuracy. A state description has to be replaced after "coarse graining" by a thermodynamic description, defined on an equivalence class of micro states generated by thermodynamic functions (which cannot be defined everywhere on M).

The second type of irreversibility is fundamentally *internal* to *the description* of a system closed in itself, with all interaction with the outside being neglected. In contrary the first irreversibility when it deals with the preferred ways of crossing a bifurcation involves several systems. In some sense, the type a irreversibility re-creates the efficiency of linguistic description which was lost due to the effect of type b irreversibility. Dissipation in a dynamical system is also irreversibility of first type.

There has been, in thermodynamical thinking, the permanent tendency to confuse these two types of irreversibility, because of the fact that with respect to thermodynamical description of an isolated system, entropy is a Lyapunov increasing function for time evolution. But these attempts had for only effect to confuse the issue. Up to now, there is no way to measure quantitatively the singular character of a given initial data. Those theoreticians who believe that the present obvious irreversibility of time is due to the slow smoothing out of an initially extrememy singular initial data (the "big bang") cannot be refuted. But their viewpoint is no more than pushing the dust under the carpet.

Let us summarize our ideas on the following diagram

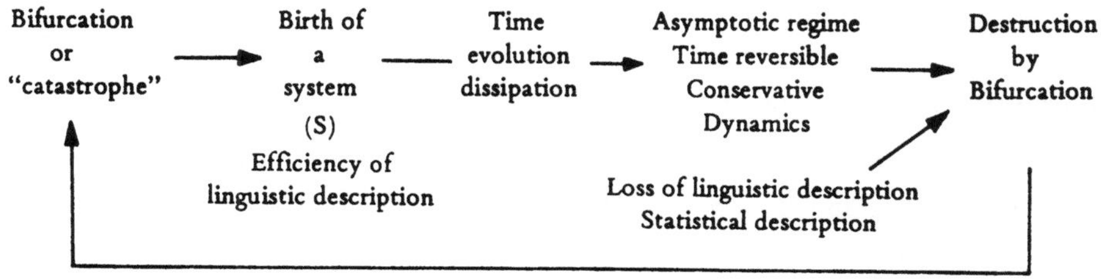

What is then the permanent factor of irreversibility which moves around the diagram? Of course, in nature, the flux of energy has constant direction, and, following the slow or abrupt gravitational collapse of stars, it has a fundamentally irreversible character (from gravitational energy to electromagnetic, and then to thermal energy). But the main reasons for irreversibility may not be only thermodynamical. For instance, I have long observed the fact that in the "cusp-bifurcation," involving the splitting of the single minimum of $V = x^4 + x^2$ into the double minimum of $V = x^4 - x^2$, the evolution 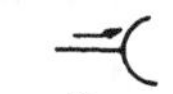 which splits a regime is much more difficult to obtain (in nature) than the reverse  which fuses two minima. This (according to my view) has something to do with the symbolic of the arrow ..., which, I believe, is by no means a cultural accident. Hence, I believe, it is quite possible that some purely morphological factors are at work to favorize some channel in a bifurcation at the cost of some others.

In microphysics, as hinted earlier, it is not the scheme of catstrophe theory which is valid, but an S-matrix unitary formalism. But here again the notion of "channel" suggests that the crossing of the collision singularity is not entirely arbitrary, and has a tendency to follow well defined pathways. The main problem here is to decide whether such canalization effects are only due to singularity of initial data, or may be caused by reasons of more profound nature. This favorizing effect may in fact be quite compatible with classical time reversal (the J-symmetry of QCT). Time reversibility would be "de jure" valid, but effective irreversibility may well be a de facto truth.

It may be difficult for physicists to accept that second type irreversibility might be a purely linguistic affair. Is not the impossibility of the "perpetuum mobile" a well known established fact? Here we should not forget that the definition of temperature automatically excludes some microstates; at the cost of this exclusion and of a huge equivalence of relation on the complementary space, we are able to get "apparent" irreversibility. The situation is no more paradoxical than, by excluding a point on the circle

S' we may transform the constant kinetics on S' which is recurrent into an irreversible one just by excluding a point from S'.

7. Linguistic vs Physical Description

The description of systems given above suggests that these systems may be considered as forming a huge graph Γ, a vertex of this graph corresponds to a collision bifurcation between such systems, where some disappear and others appear. Such a presentation is in fact common to quantum field theory (where vertices describe collisions), statistical mechanics, and usual language according to my own visualisation of linguistic activity Cf. [4] (and also to chemical Kinetics, where a vertex corresponds to a chemical reaction); the topological invariance of the phase space (M) of a system corresponds to the fixity of the meaning of a concept, described grammatically by a Noun. Vertices denoting collisions correspond to grammatical verbs. And quantum states of systems, or stable regimes of systems, correspond to adjectives. Bosons not easily localized, correspond more to adjectives than fermions which by Pauli's principle are more noun-like.

In that respect, one of the most obvious discrepancies between Quantum Physics and everyday language deals with the fundamentally irreversible character of instructions described by verbs: to create, to annihilate, to absorb, to exist ... and the corresponding quantum transformations which due to unitary formalism have a more reversible character.

Conclusion

As already stated, we believe that our universe exhibits a fundamental irreversibility of time, the reversibility of some dynamics (Hamiltonian for instance) and of fundamental physical laws reflect the global constraint of insuring a permanence of the conceptual apparatus needed to describe a given system, or the global permanence of space-time. If matter and radiation are considered as pathologies of space-time, – which they are, – because they break the global homogeneity of the vacuum, then the laws of fundamental physics express the fact that the impact of such pathologies on the geometry of space time is strictly controlled (think for instance on Lavoisiez's law about the invariance of mass in a chemical reaction).

Such a view nevertheless does not solve the basic question: where is the origin of irreversibility? If systems, particles, etc. may be visualized as watermills moved (in a periodic way) by the flow of the Heraclitean river of irreversible time, the question remains:

where is the origin of every phenomenology, which is fundamentally irreversible in character? Perhaps a realist theory of space time has to associate to any point of space-time an event, that is, a local punctual explosion of this point, unfolding along the light cone from the apex of the cone. Already, the simple observation of macroscopic caustics proven in optics, proves the existence of wavefronts of coherent light which cannot originate but in such a punctual explosion. In that respect, space time won't exist as a global ether, but more physically as the topological quotient, by a fluctuating identification, of a lot of distinct universes associated to such punctual explosions. Observation could also be defined as the time opposite process of the implosion of an universe in the "eye" of the observer. We would that way recover the still unsettled problem of the irreversibility of Huyghen's principle, about advanced or retarded waves ...) and, as I have suggested elsewhere, the Quantum Mechanical phase could be interpreted as a breaking of intersubjective consensus about one observation by several observers of the same punctual event [5]. In such a vision of the world, our local time irreversibility could be considered as due to a local excess of "explosion" processes over "implosions" ...

References

[1] Avez, A., *Méthodes ergodiques en Mécanique classique,* (part on Return spectrum of a rotation), Gauthiers-Villars 1966.

[2] Théorème de S. Newhouse, in Abraham R., Narsden J. Foundations of Mechanics Benjamin 1978 (second edition).

[3] Thom R., *Stabilité structurelle et Morphogenese* chap. V. Benjamin 1972.

[4] see [3] and Thom R. Analysis and catastrophes, Proceedings International Congress of Semantics, Vienna June 1979, to appear.

[5] Thom R., L'espace et la réalité physique, Orbetello Congress of the International Academy of Sciences, April 1979, Agazzi ed.

QUANTIZATION AS A STABILITY PROBLEM

By Georges LOCHAK
C.N.R.S. and
Fondation Louis de Broglie

I, rue Montgolfier F-75003 Paris

> "The view that a physical process consists of continual jump-like transfers of energy parcels between microsystems cannot, when given serious thought, pass for anything but a sometimes convenient metaphor."
>
> Erwin Schrödinger

I. Introduction.

Modern conceptions about structure of matter are dominated by quantum theory i.e. by the idea that the energy of microsystems only changes by jumps (quantum transitions), where microsystems suddenly go from a stationary state to another, exchanging between them energy parcels.

Present microphysics is based on detailed description of stationary states (quantum states), while transitions are only submitted to statistical calculations.

Thus, theoretical expectations often brilliantly verified by experiment are got from the description of possible stationary states and from the computation of the probabilities between them. But a transition as an individual process is never described.

So, the astonishing stability of stationary states cannot be explained because such an explanation requires the answer to the question: what happens when a perturbation deviates a system from a stationary state and what is the process which leads it to go back to its previous state or, in the contrary, to move to another one?

Such a question is meaningless in quantum theory. Physical universe is presently described like a sort of dice game in which dice would never roll on the table and would seem to us standing on one side , or suddenly on another one, with probabilities we would be able to calculate precisely.

Can physics stay in such a position? Is Universe really understable only by stationary states whithout explaining how they appear and keep on? Can we seriously maintain the view that it is possible to describe by stationary states the new particules presently discovered, although some of them are so fugacious and short-lived that their creation is almost indistinguishable from their desintegration and their energy hardly defined?

Isn't it better to consider them not as true stationary processes but as steps in transitory processes? Aren't the future progresses in microphysics to be found in the investigation of those quantum transitions presently ignored by the theory?

Isn't microphysics to be based on irreversibility rather than on stationarity? Such is the question asked by the de Broglie group twenty years ago. And since then, all our efforts have been directed, in various ways, towards the achievement of the following program:

I) Connect the preeminence of the quantum states with general stability properties of the corresponding movements, with respect to the perturbations caused by the environment.

2) Explain by structural stability properties of dynamical systems the fact that the quantum pattern remains only a little altered even in vast assemblies of interacting microsystems.

3) Give a space-time description available for quantum transitions as irreversible transient processes.

4) Reconcile, in this way, the quantum theories and the general theory of dynamical systems and stability of motion, whose great advances in our century still remain almost ignored by to-day's microphysics which is essentialy based on hamiltonian scheme.

In trying to explain the stationary states, instead of postulating them, we hope to contribute to instore the reign in microphysics, of calculations that would be physically more significant and mathematically better founded than is now a part of the mathematical machinery of quantum mechanics which often consists rather in heuristic recipes than in a true theory.

We can take as ours this statement of René Thom:

" Pour bien décrire, il faut comprendre. Le temps des miracles de la mécanique quantique des années 1925-1930 semble hélas révolu."(1).

No doubt that with such a program there's many a slip 'twixt the cup and the lip. But we are strongly convinced that the future advances in the field of theoretical microphysics will, to a large extent, rely on a reconciliation with the theory of dynamical systems.

In the following few pages, I shall try to give a brief account of some problems, de Broglie school has attacked in the past years. Some points will be exposed with more details in this same book, by F. Fer, D. Fargue and J. Vassalo Pereira.

It must still be added that there are, in fact, in the Foundation Louis de Broglie, much more diverse and even contradictory tendencies. By the way, some of them can be found in this volume.

The diversity of theoretical approaches encountered in the Summer Seminar organized by the C.I.S.M. is similar to the diversity of our own Seminar and "Annales de la Fondation Louis de Broglie": that is why we are so glad to take part in it and that is why Louis de Broglie readily accepted the honorary chairmanship.

2. Some remarks on non-linear mechanics.

In most experiments, atomic systems seem to us possess only a discrete set of physical states. After a perturbation, they seem to be able only to jump, suddenly, from one of these states to another.

Are these facts the sole privilege of microscopic world? Are they really unknown in the macroscopic world? Numerous examples show that it is not so.

Consider for example the pendulum of a wall clock, ignoring the mechanism that keeps it in motion. We never see but two states of oscillation: either it beats with the frequency and the well determined amplitude corresponding to the normal motion, or it is motionless.

Suppose that the pendulum being found in one of these two states, we perturb it by small shocks and then leave it unobserved for a few minutes before resuming the experiment.

1) " A good description needs a good understanding. The era of the miracles of quantum mechanics in the years 1925-1930 seems unfortunately over ..."

We know we shall find the pendulum again in one of these same two states: either the perturbation was too small and it will be found in its initial state, or the perturbation was strong enough to produce a change of state (i.e. the clock will be going if it was before motionless or we shall have stopped it if it was previously in motion).

Had we carry on our observation just after the perturbation, the slowness of the pendulum's movement would have allowed us to see that the transition is in fact far more elaborate than the one we have just described.

The pendulum does not reach instantly its stable steady or motion states, but first passes through a more or less complicated transient state.

This transient state is always the more difficult to observe as the inertia of the pendulum is smaller: when obvious on a wall-clock, we shall need accurate instruments to detect it on a wrist-watch.

The smaller the inertia will be, the more clearly the movement will appear as quantized, i. e. simple and discrete.

The observation of this transient process will become still more delicate if instead of a mechanical clock we use a quartz clock, or a fortiori an electronic vibrator which seems also to have only a discrete set of states with well determined frequencies and amplitudes and the ability of jumping from one of these states to another.

In fact, what does happen in this kind of phenomena?

Such a system as a clock or an electronic vibrator is a non-conservative system. It constantly dissipates energy into heat, through the friction forces, the Joule effect, and so on ...

At the same time, the system will regain this energy from an energy source consisting in a relaxing spring, a weight slipping down or an unloading battery. But it doesn't regain this energy in an arbitrary manner. in fact the losses themselves govern the discharge of the source through a mechanism whose extreme generality was made clear by Norbert Wiener: the feed-back.

It is the feed-back that tends to establish an equilibrium in the energetical exchanges between the source and the heat losses.

When this equilibrium has been established, the system seems to be in a very simple stationary state, as is shown by the permanence of the movement's frequency and amplitude. But such a stationary state (called in the simpler cases Poincaré limit cycle) is far more complicated than it appears at the first glance.

The stationary states like Poincaré limit cycles are caracterised by the fact that they are gone through by a permanent flow of energy and a permanent flow of negentropy (more precisely, the mean value of these flows on a period of the cycle is a constant of the motion).

The stability of such a state is secured by the modifications of these flows under the influence of perturbations.

If the system is given an excess of energy by a perturbation, the feed-back will slow down the discharge of the source so that the energy can be dissipated into heat.

On the contrary, if a certain amount of energy is taken out from the system, the feed-back will call for a compensation from the source.

Finally, if it happens that the perturbation is too important, it might be so, that the system will not be able to return to its initial state by any compensation, but will start oscillating on a different mode (or stay on an equilibrium) whose stability will be ensured again by the feed-back process.

This process may appear in various ways. In a clock as the escapement, in an oscillating circuit as a triode or a transistor; but many other feed-back mechanisms can appear in phenomena which all refer, though very different, to the same scheme, as for instance an electric ring, a violin string excited by a violinbow, periodic hormonal cycles, the ball governor of a steam engine, the periodic radiation of certain varying stars (the so called cepheids), the vibration modes of a laser, the heart beatings, some chemical reactions, the α rythm of the brain, most of the wind instruments etc...

All these systems are of the <u>self-oscillating</u> type and possess the following characteristic features:

I) They possess only a discrete set of stable periodic motions (limit cycles) whose periods and amplitudes are well defined.

2) They "forget" the initial conditions and so, enjoy a property of ergodicity. Apart a discrete set of unstable states, not observed in the experiments, any movement of the system which is not a stable limit cycle, will asymptotically tend to one of these cycles which, thus, enjoy the Liapunov's asymptotic stability property.

This stability can be so strong - and transient states so short- that the system practically never happens to appear in another state than the ones defined by the limit cycles.

During the perturbation it seems as it passed instantly from a cycle to another. Otherly said, the system looks "quantized".

3)The systems are dissipative and maintain or reestablish their periodic movements by taking energy from a non periodic source, through a feed-back process.

4) The working of this feed-back demands that, either the supply of the source of energy, or the dissipation of heat, or both, depend on the state of the system; this makes appear a non-linearity property in the equations describing the evolution. That is, the superposition of two stable states is not a stable state.

5) The self-oscillating system is seen to be connected with two energy sources.

From one of them, it takes energy but also negentropy, and just in the same way as for a heat engine, it keeps up its motion only by a loss of energy and negentropy that is furnished to the second source as a compensation, in the form of heat.

The dissipation of energy given under the form of heat to the "cold source" is no less indispensable to set up and to maintain the limit cycle, than the supply of energy by the "hot source".

In contradiction with a "true" heat engine, this compensated heat given by the system to the "cold source" is not necessary to transform in work the heat given to the system by the "hot source", because here, what we call "hot source" already directly furnishes work!

In the present case, the dissipation of heat and the "cold source" are necessary to preserve the stability of the system.

The dissipated heat is the price that must be payded for the feed-back: it is the price of the stability.

We can sum up this point in saying: "No stability without dissipation of energy into heat". Or: "No stability without generation of entropy".

By this proposition we have identified the asymptotic stability in the Liapunov's sense (which is the stability of limit cycles) with the thermodynamic stability.

This principle is still strongly confirmed by the striking analogy between the monotonous variation of the Liapunov V-function which caracterises the asymptotic stability of a limit cycle and the monotonous increasing of generated entropy in thermodynamical systems.

We can so understand why the self oscillating systems are irreversible and why the equations that describe them are not invariant by time inversion.

Thus, the non-linear mechanics provides us with examples of physical systems that hide, under a quantized-looking form, the existence of irreversible transient processes capable of describing a phenomenon analogous to a quantum transition.

Relying on the analysis of such examples, Andrade e Silva, Fer, Leruste and Lochak have proposed to try to construct a non-linear theory of quantum transitions.

These reasonings have been put afterwords under a thermodynamic form by Andrade e Silva and Lochak, following works of Louis de Broglie about hidden thermodynamics (or "thermodynamics of isolated particle").

3. Two simple mathematical models.

We now give two mathematical examples with certainly no physical meaning but illustrating nevertheless, what has just been said.

a) Consider the non-linear oscillator:

$$\ddot{x} + F(w)\,\dot{x} + x = 0$$

where w is the energy of the harmonic oscillator:

$$w = \frac{1}{2}\left(\dot{x}^2 + x^2\right),$$

and F a function looking like on the following diagram:

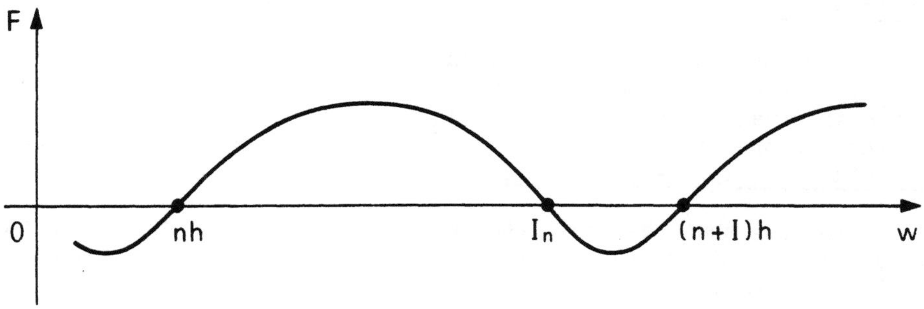

Then $F(w) = 0$ will stand for $w = nh$ $(n \in \mathbf{N})$ and F increases in the neighbourhood of these points. It is seen that, when $w = nh$, the equation becomes:

$$\ddot{x} + x = 0$$

and the motion will be that of the harmonic oscillator with quantized energy $w = nh$.

When $w = I_n$, the equation still reduces to that of the harmonic oscillator with now a motion with energy $w = I_n$.

Let us look, however, at what would happen for arbitrary initial conditions.

Suppose first that:

$$nh < w < I_n .$$

Within such an interval, $F(w) > 0$ and the term $F(w)\dot{x}$ will act as a braking term; the magnitude of the oscillation will decrease and we obtain a transient regime during which the function w will decrease and tend to nh.

Suppose now that we have:

$$I_n < w < (n + I)h .$$

Then $F(w) < 0$, $F(w)\dot{x}$ acts as a "negative brake" which increases the amplitude of the oscillation and in the course of this new transient regime, w will tend to $(n + I)h$.

In particular, let us suppose that we slightly perturb the oscillator when it is found in a state $w = nh$: if in the course of the perturbation, w doesn't cross the value I_{n-I} or I_n, the oscillator will go back to the state nh. Thus, the states $w = nh$ are asymptotically stable.

If, on the contrary, the oscillator would be found in a state $w = I_n$, then whatever small the perturbation, w would be in one of the two intervals (nh , I_n) or $(I_n , (n + I)h)$ and the oscillator will tend to one of the two states $w = nh$ or $w = (n + I)h$. Then the states $w = I_n$ are unstable.

It follows that the oscillator will be always seen in a neighbourhood of the quantized states $w = nh$ in case it is submitted, all the time, to small perturbations.

It's no use, of course, to look for a possible feed-back mechanism in the above described process since it is only a mathematical model. It exhibits, however, a system with a discrete set of stable limit cycles, identical with the quantized states of the harmonic oscillator.

b) The second example generalized the preceding one. Consider a multiply periodic hamiltonian system:

$$\dot{p}_i = -\frac{\partial H}{\partial q_i} \quad ; \quad \dot{q}_i = \frac{\partial H}{\partial p_i} \quad ; \quad H = H(p_k , q_k),$$

written in the (w_k , J_k)-system (angle and action variables):

$$\dot{J}_k = -\frac{\partial H}{\partial w_k} \quad ; \quad \dot{w}_k = \frac{\partial H}{\partial J_k} = \nu_k(J_I, \ldots, J_N); \; H = H(J_I, \ldots, J_N).$$

Then, the Bohr-Sommerfeld quantization conditions consist in setting

$$J_k = n_k h \qquad (n_k \quad I, 2, 3, \ldots)$$

Let us change now the above system of equations to:

$$\dot{J}_k = - F(J_k) \quad ; \quad \dot{w}_k = \frac{\partial H}{\partial J_k} ,$$

where F is the preceding function.

It is quite evident that the Bohr quantum states are stable solutions of theses equations. In fact, the sign chosen for F will make J decrease to nh if, after a perturbation, J takes a value slightly above nh ; and will make J increase when J takes a value below.

There, we must notice that the attractors are no longer limit cycles, but a set of trajectories, filling the manifolds of the phase space defined by the equations:

$$J_k(p_i , q_i) = n_k h \qquad (n_k \quad I, 2, 3, \ldots).$$

Effectively, in case of degeneracy, these manifolds will define the usual quantum states, only if we replace the classical equations $\dot{J}_k = 0$ by the new equations $\dot{J}_k = - F_k(J_k)$ for the sole restricted set of J_k which correspond to the classical <u>adiabatic invariants</u>.

So, if a point in the phase space describes a classical trajectory on such a stable manifold and if it is slightly scattered from it by a perturbation, the point will return asymptotically to the same manifold, but generally not to the same trajectory.

Thus, it is seen that even in such a simple model, without any allusion to the wave properties of matter, we have already lost the notion of individually observable trajectories in the atom.

In addition, it is to be noticed that the function F which appears here as governing the stability is nothing but a Liapunov's V function and we sée that F behaves like an entropy function. Reciprocally, one can prove that if it is possible to associate an entropy to a dynamical system, it is necessarily a Liapunov's V function.

We can so perceive a possible interpretation of quantum states as kinds of thermodynamically stable states and we understand better the meaning of adiabatic invariance in quantum mechanics, which is connected with the second principle of thermodynamics. These questions were carefully analysed by the de Broglie's group and we shall, further, briefly return to them.

4. The field equations problem.

We are clearly led to this problem by the wave properties of matter and we have tried to clarify it, from the beginning, with the help of mathematical models similar to the preceding ones. Nevertheless, they are of course more complicated.

Since then, the main angles for attacking the problem were the following:

a) The first idea was, as we said, the one of <u>hidden thermodynamics</u>, due to Louis de Broglie.

It consists in trying to rebuilt the whole wave mechanics from the hypothesis of a subquantum medium, a kind of thermostat, with which the particules would interact. The quantum transitions would then be ruled according to the Carnot principle and the stationnary states must be caracterized by constant values of entropy. The formula of de Broglie for this limit case are:

$$h\nu_c = kT \quad ; \quad \frac{A}{h} = \frac{S}{k} ,$$

where h and k are respectively the constants of Planck and Boltzmann, ν_c and A the cyclic frequency and the hamiltonian action of the <u>inner</u> vibration of the particle , when T and S denote the temperature and the entropy of the hidden thermostat.

These formula are relativistically invariant and based on a reasoning analogous to the one which led Louis de Broglie to the basic formula of wave mechanics:

$$h\nu = mc^2$$

where m denotes the mass of the particle and ν the frequency of the associated wave.

It is known that the wave mechanics was born from the hypothesis that the phase accordance between the inner oscillation of the particle with frequency ν_c and the oscillation of the wave with frequency ν must be realized for all the gallilean observers, though the relativistic variances:

$$\nu_c = \nu_o \sqrt{I - \beta^2} \quad \text{and} \quad \nu = \nu_o / \sqrt{I - \beta^2}$$

are opposite to each other §.

That is the reason for imposing the relation $Vv = c^2$ between the speeds of the wave and the particle, whence the formula $\lambda = h/mv$ can be deduced. In short, wave mechanics started from a reasoning on the frequency ν in the frame of geometrical optics. Next step was made by Schrödinger (to the wave optics) who so discovered the basic equation of wave mechanics.

On the contrary hidden thermodynamics starts not from the wave frequency ν , but from the inner frequency ν_c ; the invariant relation so postulated is not between ν and m but between ν_c and T; the new theory begins not in the frame of geometrical optics but in the frame of equilibrium thermodynamics; the fundamental next step should be to lead theory out of equilibrium, to reach the description of quantum transitions.

b) The adiabatic invariance.

As well known, the first results in wave mechanics were obtained, already in the frame of old quantum mechanics, that is to say at the geometrical optics approximation, before the discovery of the Schrödin-

§ Let us recall that in the proper system of the particle the wave is postulated to be stationary and synchronous with the inner oscillation.

In this system the two frequencies ν and ν_c are equal and they are defined by the equality:

$$h\nu_o = m_o c^2 .$$

ger equation.

In the same way, we can try to apply the ideas of hidden thermodynamics in the frame of the present quantum mechanics, before seeing the future development of the theory.

According to the above exposed ideas, we must consider quantum states, not as the only possible states but mereby as the most probable states: it may be expected that quantum mechanics is, to the future theory, what is thermodynamics of equilibrium states to the thermodynamics of irreversible processes.

So, instead of defining quantum states as did Schrödinger (i.e. as normal modes of vibration of the wave: represented as a product $f(t)g(x)$ of a function of time by a function of space), we must look for a thermodynamical criterion to define these states.

This criterion will be the adiabatic invariance.

Really, starting from von Neumann's entropy:

$$S = -k \operatorname{Tr}(\varrho \operatorname{Log} \varrho) \quad ; \quad \varrho = \text{density operator},$$

it may be shown that a mixture of states with a given energy, is an equilibrium mixture (therefore a mixture with maximum entropy) only when these states are adiabatically invariant.

Now, in a conservative system, the only adiabatically invariant§ states are the stationary states , i.e. the usual quantum states.

Thus, we have a thermodynamical explanation for the preeminence of quantum states in the nature, that is an answer to the question: "Why are the stationary states by far more numerous in most of known experiments, than are the coherent superposition states?"

Of course, it must be borne in mind that even with this principle, we cannot get a general process able to represent the breaking out of a superposition state and so, able to describe the creation of a mixture from a coherent state.

Nevertheless, and although the adiabatical principle is valid only

§ Let us remind of the definition of adiabatical invariance of stationary states of a system depending on a parameter α : when ψ is written as

$$\psi = \Sigma\, a_k(\alpha) \exp\left(-\frac{i}{\hbar} E_k t\right) | k > ,$$

i.e; as an expansion on stationary states, then we have, for an infini-

at thermodynamical equilibrium, explaining so why the quantum states so frequently appear in the nature, but not how they appear, we may expect more generality from this principle than from the simple condition of stationarity (i.e. $\psi = f(t)g(x)$).

In order to prove that, we must find a case in which the new principle is still valid, while stationary states no longer exist.

Such an example is furnished by microsystems submitted to an external field periodically varying with the time.

For such a system, energy is no longer conservative and there are obviously not any stationary states (not any normal mode $f(t)g(x)$). However, the adiabatical principle still operates:

It is possible, for an assembly of such systems, to define a von Neumann's entropy of the same form as precedently and to find the conditions of thermodynamical equilibrium.

These conditions are formally the same as the well -known conditions for an equilibrium in the case of conservation of energy §§. But we have shown that these conditions cannot be formulated in the most general case, by using an arbitrary basis of orthogonal solutions of the Schrödinger' equation.

We have shown (see Bibliography) that the existence of a thermodynamical equilibrium requires the existence of adiabatically invariant states. The equilibrium is necessarily obtained in the form of a mixture of such states. We have called them: permanent states.

It was also shown that had we tried to define an equilibrium by a mixture of non invariant states, it would be possible to violate the second principle of thermodynamics in modifying the entropy of an assembly of microsystems by an infinitely slow variation of a parameter.

So, to apply our thermodynamical criterion, one needs to find adiabatic solutions (i.e. permanent states) of the Schrödinger's equation. It is just what was done for systems depending periodically on the time. In this case, a permanent state appears as a product of a stationary-like vector, by a unitary operator $T(t)$ which depends periodically on t:

$$p_k = T(t) \exp\left(- \frac{i}{h} \mu_k t \right) | k > .$$

tely slow variation of α :

$$| a_k(\alpha(t)) - a_k(\alpha_0) | = O(\alpha).$$

§§These conditions can be found in the famous book of von Neumann and in the book of Louis de Broglie on the Theory of Measure .

This particular form is a consequence of the Floquet's theorem, which is well known in the theory of linear differential equations with coefficients periodically depending on time.

We have demonstrated, for the permanent states, an adiabatical theorem generalisating the analogous theorem known in the usual case of stationary states.

It is shown that the so defined permanent states are univoquely defined and they constitute an orthogonal basis, beeing proper states of a hermitian operator $\mathscr{R}(t)$ which represents a constant of the system: the so-called reduced energy. The μ_k, that appear in the formula given above for the permanent state, are the proper values of this new constant of the motion.

The permanent states were observed in experiences realized in the domain of herzian spectroscopy and laser optics: their adiabatic properties were verified as well as the probabilities of transitions between them and the frequencies emitted by these transitions.

The analogue of the Gibbs distribution was calculated for the thermodynamical equilibrium of a statistical ensemble of microsystems in a periodically varying field. The density operator for the equilibrium state, that is for the most probable distribution of permanent states, takes the form:

$$\varrho_o = \zeta^{-I} \exp(-\mathscr{R}(t)/kT); \quad \zeta = \mathrm{Tr}\exp(-\mathscr{R}(t)/kT).$$

Just as the reduced energy, ϱ_o is diagonal in the system of permanent states and the preceding formulae become:

$$\varrho_{o\,nn} = \zeta^{-1} \exp(-\mu_n/kT); \quad \zeta = \sum_n \exp(-\mu_n/kT).$$

It's not difficult to recognize here the generalization of the canonical ensemble of Gibbs: the conservation of energy is merely replaced by the conservation of reduced energy.

These formulae have been verified in the domain of the Nuclear Magnetic Resonance. Indeed, we have been able to give so, a general physical foundation to the idea of spin temperature in the rotating frame.

It was shown that all the results of the theory of spin temperature (theory which is experimentally verified), are obtained with the only one hypothesis that in a rotating, or more generally, in a periodically varying field, the nuclear spins relax on permanent states.

This hypothesis, expressed in the above given formulae, defines a temperature which is nothing but the so called "temperature in the rotating frame". In our theory, this result is obtained not by the particular and artificial transformation of the rotating frame (as done in usual calculations): the equilibrium temperature is defined on the basis of a general principle (the adiabatical principle) and the results expressed in a quantum mechanical invariant manner wich needs not any special reference system to describe the phenomena.

The fact that theory of spin temperature is experimentally verified seems to us to be an argument in favour of our assertion: quantized states of matter (i.e. usual quantum states, or the more general permanent states) may be considered as, simply, the most probable states, or those which appear in the thermodynamical equilibrium.

The first successes obtained by the adiabatical principle leads us to hope that adiabatical invariance will play a fundamental role in the future theory of quantum transitions.

The importance of this idea was perceived by Norbert Wiener who wrote to us twenty years ago, when he looked into our first works (where nothing was said about this principle): "You must introduce the adiabatical invariance in your theory".

As was suggested by Giacomo Della Riccia, one can look at a quantum transition as at a breaking of adiabatical invariance.

It seems also, that this invariance must play for the theory of quantum transitions a role analogous to that of the Bohr's correspondance principle in the old quantum theory.

In Bohr's theory, this was de facto the correspondance between the geometrical and the (future) wave mechanics; in our case, it might be the correspondance between equilibrium and non equilibrium thermodynamics.

Further ideas about adiabatical invariance are developed in this volume by J. Vassalo Pereira.

c) In the last two points, we shall give some short information about the problem of finding a physical model, in order to describe properly speaking the process of quantum transition.

The first idea in this direction is due to F. Fer. He started from the Schrödinger's equation, assuming the usual boundary conditions for the wave function. He then introduced in the equation a feed-back term which is nothing else but the reaction of the radiation emitted by the atom.

In other words, the dissipative element needed for the stability is considered to be due to the radiation itself. A very important feature of this theory is the hereditary character of the phenomena, related to the finite velocity of the electromagnetic interactions. We shall say no more about this subject because it's developed in this volume by D. Fargue.

d) The escape of de Broglie waves through a potential barrier as a possible element of stability of quantum states.

Contrary to the problems of adiabatical invariance and radiation self-action about which much was published by our group, the idea we shall give now a short survey, is still just an idea, not yet a theory, and nothing was still published by us about it. Nevertheless, this question seems to us interesting enough to be discussed here.

Ordinary, since the first famous Schrödinger's papers, the theory of quantization is based on a boundary problem. This problem is formulated with regard to an isolated system, with a perfect attractive potential which is supposed acting infinitely far, and most frequently the boundary conditions are imposed at the infinite.

But why should we be allowed to impose boundary conditions at the infinite, supposing so, that the attractive potential (of an oscillator, or an atom, for instance) keeps there the same expression?

In fact, an atom is never alone; in its vicinity, there always will be (besides the surrounding radiation field) other atoms which shall perturb stochastically the attractive potential.

So, the quantization theory may be set up on strong basis, only if this theory remains valid when <u>certain</u> values postulated for the potential, are perturbed by a <u>stochastic</u> potential representing the surrounding atoms.

The strict boundary conditions seem so to be given up. On the contrary, we have to prove that whatever the wave function will be at a given instant, the most probable state after a certain time will be represented by a superposition of quasi-stationary states, the frequencies of which will be close to the usualy calculated quantum frequencies, while their wave functions shall be close to the usual wave functions, i.e. with correct boundary properties (although they were not postulated at the origin).

In other words, our question may be formulated as follows:

> The usual quantization theory is a theory of stationary states of an idealy <u>closed system</u> (the isolated atom). In this view, the presence of other atoms is regarded as a perturbation of this ideal situation (evidently, we must except the cases when this per-

turbation is so strong that the same nature of the problem is modified: condensed matter, intense electric field, inelastic collisions and so on).

BUT, all the atoms we observe are or in the presence of other atoms, or they have been so (as far as we know!).

So, must we not, on the contrary of what is usually done, look at the atom in the way of quantization as at an open system?

If an atom comes near the first one, we have indeed a new system with two attractive centers separated by a potential barrier. By this fact, arises the possibility of an escape of de Broglie waves from one center to another by tunnel effect.

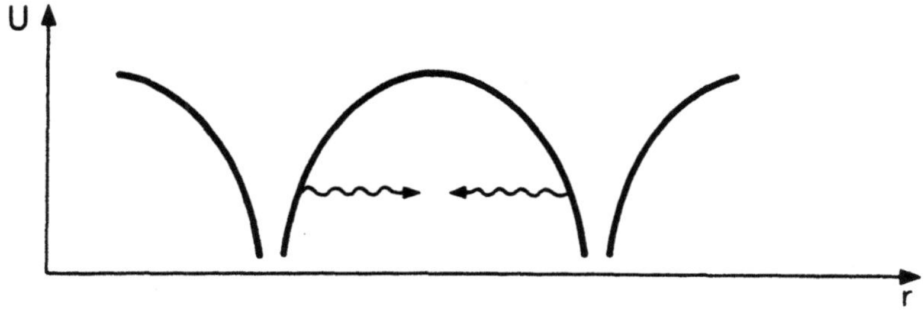

Isn't possible this escape to be a dissipative element which should be added to the electromagnetic emission? As we shall se, it seems even possible that this element plays an essential role in the arising and in the stability of quantum states.

If would be so, then the usual theory of stationary states of bounded, isolated (and so closed) systems, would be a mathematical model, just valid when the statistical equilibrium between all different attractive centers is reached.

Let see now some arguments in support of this hypothesis. The fundamental argument is the fact that the passage of a wave through an imperfect potential well is selective for the frequencies.

The privileged frequencies are those which are in the neighbourhood of the proper frequencies of the corresponding perfect well, i.e. the well with perfectly impassable edges.

For instance, it's precisely this "resonant" light which goes more easily through the air included between two glass slides. The same phenomenon is responsable of the irisation of thin layers, observed for example on roman glass-wares or on a drop of petrol at the surface of water.

But over all, this phenomenon is observed on Newton's rings and in the Pérot and Fabry interferometer. As well known, the latter is used in the laser and it was shown (by Fox and Li) how the spectral lines are narrowed by successive partial reflections on imperfect mirrors.

In other words the interferometer acts as a frequency filter. These phenomena are also known for electrons. The finest example is the Ram - sauer effect which is observed in the collision between a rare gas (ar- gon) and a very low energy electron beam.

As may be expected, the diffusion of electrons increases if we re- duce their velocity: the electrons, of course, become less and less pene- trating. But suddenly, when the velocity is reduced enough, until a quite definite value, the gas becomes transparent for the electron beam and the diffusion almost vanishes! Why?

It may be shown that this effect is due to the simple fact that the observed critical velocity is that for which appears inside the atom a normal mode of the electron's de Broglie wave §.

The theoretical calculation of the Ramsauer's effect is rather com- plex, but the effect is already comprehensible on the following suggesti- ve example. Consider a limited harmonic oscillator with the potential:

$$U(x) = \frac{Kx^2}{2} \quad , \quad x \in [-l,+l] \quad ; \quad U = 0 \quad , \quad |x| > l.$$

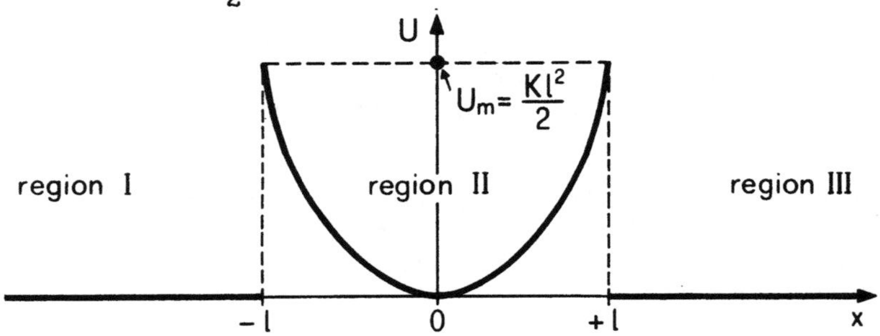

Let seek for a wave propagating from $-\infty$ to $-\infty$. We shall have in regions I, II, III:

$$\psi_I = e^{-ikx} + A\, e^{+ikx} ,$$

where the Ist term is the incident wave and the 2nd one is the par- tially reflected wave;

$$\psi_{II} = B_I\, \psi_I + B_2\, \psi_2 ,$$

§ It is curious to note that this effect was observed by Ramsauer in Göttingen in I922 (before the discovery of wave mechanics!). Unhappily the theorists thought then, it was an artefact and neglected it. But in every case, theory springs not from the experience but from human genius!

where ψ_I and ψ_2 are solutions of the usual equation of the oscillator;

$$\psi_{III} = C\ e^{-ikx}.$$

$|C|^2$ is the intensity of the wave transmitted by the oscillator. As the intensity of the incident wave was taken for unity, $|C|^2$ will be also the transparency coefficient T of the imperfect oscillator.

The result of the calculation is quite different according as the value of energy is near to the quantum values $(n + I/2)\hbar$ or is far:

(I) Resonance case: $E = (n + I/2)\hbar\omega$; we have:

$$|C|^2 = T_I = \frac{4\mu^2}{(I+\mu^2)^2} \quad ; \quad \left(\mu = \frac{E}{U_m} \quad , \quad U_m = \frac{Kl^2}{2} \right)$$

and the intensity is like this graph:

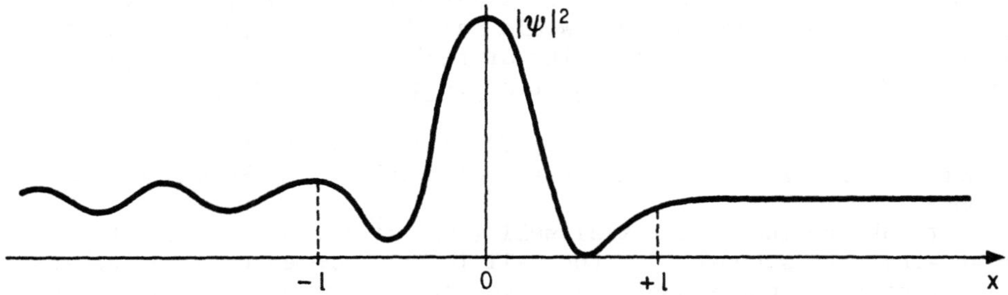

(2) Non resonance case: $E \neq (n + I/2)\hbar\omega$; we have now:

$$|C|^2 = T_2 = F(E/\hbar\omega)\ l^{4E/\hbar\omega}\ e^{-2(m\omega/\hbar)l^2}\ \frac{4\mu^2}{(I+\mu^2)^2}\ ,$$

where the coefficient $F(E/\hbar\omega)$ doesn't depend on l. Because of the exponential factor, T_2 strongly decreases (for a given energy) when increases l. The new graph of the intensity is:

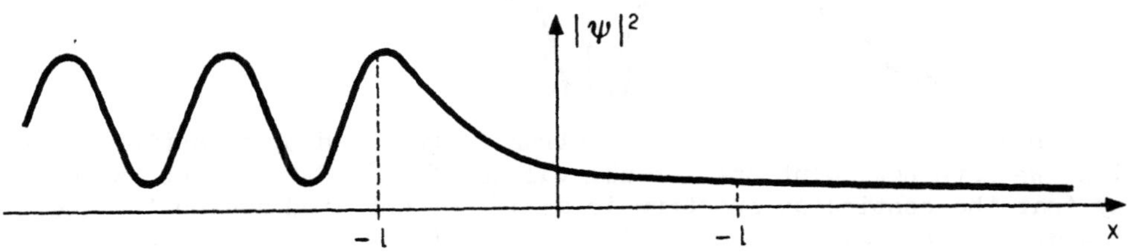

The non quantum like waves (i.e. non resonant) are much more flattened than the quantum like waves (i.e. resonant). In return, non quantum like waves are more reflected on the barrier: they are more diffused.

These facts make us understand qualitatively the Ramsauer's effect but their interest goes farther. The calculation can show, indeed, that if $l \to \infty$, then the amplitude of quantum like waves (i.e. $E = (n + 1/2)\hbar\omega$) tends to the amplitude of the usual stationary waves for all $x > -l$ (but don't forget that $l \to \infty$!).

Thus, we can say that the quantum like waves tend to the stationary waves. In particular, the boundary values go close to the usually imposed conditions. And here, they are not the result of a postulate but the result of the action of a physical system.

Inversely , it is shown that for the non quantum like waves, $\psi \longrightarrow 0$ for $x > -l$. So when $l \to \infty$, non quantum like waves disappear and just remain quantum like waves.

If, instead of sending on the oscillator a monochromatic plane wave, we should have sent a wave packet (i.e. a superposition of quantum like and non quantum like waves), the oscillator would have play the role of a filter: the non quantum waves would have been principally reflected and the higher intensity around the attractive center (i.e. inside the atom) would be that of a superposition of quantum like waves.

This result was established by Titchmarsh for a quite different and far more general form of "half open" potential wells. We may resume as follows:

Let take an ideal potential well (as x^2 for an oscillator or r^{-1} for an atom) and suppose that the well is half opened by a perturbating field: a mathematical example was given above, but a physical one is furnished by the Stark effect because a uniform field added to the Coulomb field gives the following graph we shall use a little farther:

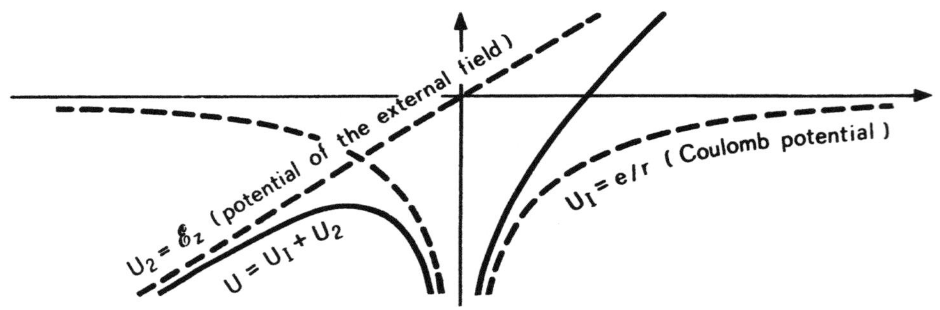

In all cases, the discrete spectrum is replaced by a continuous one. (i.e. we have now solutions for all values of E). But in fact, we can say that the continuous spectrum is as if concentrated around the ideal discrete proper values, because it is shown (see Titchmarsh) that, in a Fourier expansion of an arbitrary wave packet, the weight of these quantum like waves is dominating. And "higher" are the edges of the well, more the quantum like waves dominate.

All these properties are well known, but in our opinion little utilized. They seem to us so remarkable, that we suggest as a new aim to the theory, to develope the attempt of describing the above process of appearing of discrete values and boundary properties and to try to introduce this process in the same basis of the theory of quantization, instead of the usual postulates relied to the eigenvalue problem.

Of course, it may be objected that in such a perspective the atom will lose its stability as at last any wave will leave it! To this objection, we can give several answers:

α) For that we are not responsible, the nature is! This instability results from the unavoidable interactions between microsystems and from the wave properties of matter: of that we are not masters.

β) The only stable atom would be an idealy isolated one. This is nearly realized (during a short time) in some experiments, for instance in atomic beams. But we see then already constituted atoms and they were born in a collective system.

γ) Inside the matter, the escape of waves from an atom is compensated by the incidence of waves escaped from other atoms. So they appear stable on an average. The electrons are only more or less delocalized.

δ) It is interesting to briefly recall the case of condensed matter. In a cristal, indeed, the atomic levels are deformed and widened until to become the Brillouin zones.

The travelling time of an electron between two neighbouring atoms may be estimated to be about 10^{-15}s. So, the electrons are strongly delocalized.

It helps to understand the very illuminating process of beam-foil spectroscopy. Let take the precise following case: a proton beam with energy 0,1 Mev falls on a carbon pellicle 100 Å thick:

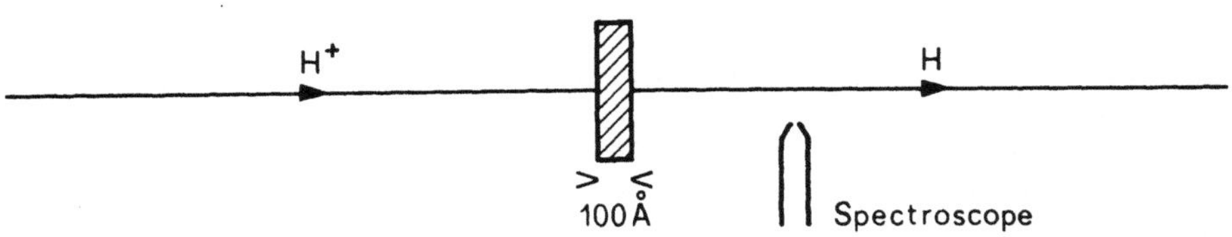

It is found that, going through the carbon pellicle, an appreciable proportion of protons have captured an electron and, on leaving, they appear as hydrogen atoms (most frequently, highly excited), the spectral lines of which may be observed in a spectroscope.

Now, what is the time a 0,1 Mev proton needs for crossing 100 Å ? : about 10^{-15} s. , i.e. a time of the same order an electron needs to cross the potential barrier between two neighbouring atoms.

We understand why the potential well surrounding a proton had a good chance to capture and keep a bounded electron. But after this process is over, the hydrogen atoms in the beam emerging from the carbon, are nearly isolated. They can live a long time: the mean life time of an excited state (about 10^{-8}s.) . This is just what is found by measuring the distance from the carbon pellicle at which the different spectral lines are seen going out.

ε)In conclusion let us try to appreciate the solidity of an atom in an exterior field. We shall see it in the case of Stark effect on hydrogen atoms.

How do we see experimentally that electrons of a certain energy level (more exactly here, a <u>quasi</u>-level) are leaving the atom? We see it when are disappearing the spectral lines corresponding to the falling from this level to the lower levels, because it means the migration time through the potential wall becomes smaller than the mean life time of the state ($\sim 10^{-8}$ s.) : it is the so called <u>self-ionisation</u>.

It is found for the first Balmer lines, that the depopulation of the excited states begins with a field of about <u>10^6 V/cm</u> (with an excellent agreement between theory and experience).

This might seem to be a weak field, if comparing to the Coulomb field of the proton on the first Bohr's orbit ($\sim 10^9$ V/cm), but it must be compared not to this field, but to the field taking part in a low energy atomic collision: then the nucleus field is <u>screened</u> by electrons and at a distance of the order of Bohr's radius the resulting field is even already slightly negative, i.e. repulsive. This is experimentally confirmed by the fact that very few ionisations occur in a gas at ordinary temperatures.

BIBLIOGRAPHY

Alaoui, A. and Lochak, G., La théorie des états permanents, <u>Ann. Fond. L. de Broglie</u>, 2, 87, 1977.
Alaoui, A. and Lochak, G., Sur le concept de température de spin dans le référentiel tournant et la théorie des états permanents, <u>C. R. A. S.</u>, 280, 589, 1975.
Andrade e Silva, J., Fer, F., Leruste, Ph. and Lochak, G., on the general subject: Stability and quantization, <u>C. R. A. S.</u>, 251, 2305, 2482, 2662, 1960; <u>Cahiers de Physique</u>, 15, 210, 1961; 16, 1, 1962.

Andronov, A., Vitt, A. and Khaïkin, C., <u>Theory of oscillators</u>, Pergamon, London, 1966. (translated from Russian)

Blokhintsev, D., <u>Eléments de mécanique quantique</u>, Dunod, Paris,

Broglie, L. de, Passage des corpuscules électrisés à travers les barrières de potentiel, Ann. Inst. H. Poincaré, 3, 349-446, I933.

Broglie, L. de, La thermodynamique de la particule isolée, Gauthier-Villars, Paris, I964.

Broglie, L. de, Jalons pour une nouvelle microphysique, id. , I978.

Fer,F., L'irréversibilité, fondement se la stabilité du monde physique, Gauthier-Villars, Paris, I977.

Haken, H., Synergetics, Springer, Berlin-Heidelberg, I977.

Lochak, G., L'irréversibilité en physique, Seminar "Philosophie et Mathématique" (E.N.S. Paris), printed in Bull. Inf. IEN-ENSM , N°3I, I979 (reprints available), to appear in Dialogos.

Lochak, G., Sur l'invariance adiabatique et la notion d'entropie en physique statistique, Ann. Fond. L. de Broglie, I, 56, I976.

Lochak, G., and Vassalo Pereira, J., On adiabatical invariance in classical and quantum mechanics, C. R. A. S., 282, 32I, 657, II2I, I976.

Titchmarsh, E.C., Eigenfunction expansions associated with second order differential equations, Clarendon, Oxford, I946 (t. I), I958 (t. II).

ADIABATICAL INVARIANCE IN MICROPHYSICS

José VASSALO PEREIRA
Faculty of Sciences, University of Lisbon

Adiabatical Invariance made its first appearance in Physics in the Solvay Meeting of 1911 (Réunions Solvay[1]), when the President – who was Henrik Anton Lorentz himself – asked to the attendants what would become the oscillations of a classical pendulum if one undertakes to shorten its length very slowly. The answer to that question was immediately given by Einstein: under such conditions the quotient between the total energy E and the frequency ν should remain the same.

Now, nobody ignores what was happening in the Solvay Meeting of 1911: there was brought together for the first time the cream of theoretical physicists trying to set right what was left undamaged in Physics after the earthquake of quanta and relativity – which was not much, as we know. So at first glance one might ask what such an academic discussion around classical pendulum had to do in such a place. The reason is that in doing so, Einstein and Lorentz were touching at the very foundations of quanta : in such a system, the recent-born theory of quanta postulated that the only possible states were those with *discrete* values of energy given by $E = E_n = n h \nu$ $(n = 1,2,...)$; now, by varying a parameter of the system (such as the length), the E and ν then become functions of time, and even if the variation is performed slowly enough (in order to avoid transitions between the states) it is by no means evident that the above fundamental relation of quanta remains true.

So it would be very annoying if any of the classical integrals, such as E/ν that were "ad hoc" quantified (that is, put equal to a multiple of h) did not remain unvaried in a slow changing of any parameter! In order that the theory of quanta might be aware of inner contradictions, Einstein's answer must be true.

As a matter of fact, at that time nobody knew if it was true or not. That was a question in Classical Mechanics which had never arisen before, and this is why physicists and mathematicians had never provided the corresponding answer. But once their attention has been thus directed to this point and after some theorems and calculations they could ascertain that Einstein's answer *was* fortunately true. Actually, mathematicians went much farther, and proved the invariance of other classical constants of motion (Sommerfeld[2], Burgers[3]), particularly the so-called Schwartzchild integrals, $\oint p\, dq$ (of which the expression above is but a particular case) and which were essential in the theory for it was by means of them that quantification was introduced in Physics (by putting them equal to a multiple of h).

In fact, even today mathematicians are still providing more and more sophisticated demonstrations and this is by no means a useless task for, without exaggeration, one may say that even the most elaborated hypothesis which appear in such demonstrations are physically meaningful. We will briefly give here an example:

For instance, right from the beginning physicists suspect that in order that these constants of motion might remain unvaried under a modification of a parameter it was not enough to require a "very slow" changing. And indeed it wasn't, far from it. For if, for instance, in an oscillating pendulum we modify its length as slowly as we wish but only in certain precise values of the phase, then in a sufficiently long interval of time it is possible to make the amplitude take any previously chosen value. This is the well known phenomenon of "parametric resonance" : we still have a "very slow" evolution but the result is somewhat unlike the expected one!

So a new condition was introduced in the theory which seemed (and indeed was) necessary in order to avoid such difficulties : one requires the "slow" variation to be performed regardless of the phase or, perhaps better, the variation is performed in the course of time with equal probability of occurring continuously in any value of the phase (Juvet[4], Pars[5]). In this way probabilities got into a theory which apparently had no need of statistical reasonings, and since we know (and physics repeatedly shows) that "statistics" means "thermodynamics" we may rightly expect this *mechanical* theory to lead straightforwardly to thermodynamics! Let us say a few words about it.

When Ehrenfest first began to study these subtle "very slow" transformations, Einstein suggested to call them Adiabatic Transformations (Ehrenfest[6]). Of course, he had second motives in doing so : for Adiabatic Transformations were a well known concept within usual phenomenological thermodynamics, where they denote the change of thermodynamical state of a macroscopic system performed without interference of heath – no absorbing, no transfer. Now this kind of transformations we have been talking about concerns purely *mechanical* systems! So, in attaching the same denomination to both, Einstein obviously intended to establish an ambitious connection between them. In fact, it was a challenging program of research that was implicitly suggested to the forthcoming physicists : to elucidate the conditions under which mechanical trasnformations of the kind just considered could be identified to a (usual) thermodynamical adiabatic one, that is, to clarify the mechanical interpretation of heath!

Let us say at once that this program is not yet fulfilled – as far as such a program can be achieved (perhaps it cannot...) – but that the subtlety and craft of mathematicians have arrived very far in that direction (Fer[7]). Obviously, it is no longer enough to have a "very slow" variation in order to assert that the heath is absent from the transformation, but rather a certain number of elaborated conditions. Thus, by means of more and more precise mechanical hypothesis we lay a siege more and more close to the fundamental thermodynamical concept, the heath!

Of course, this was not the first time that physicists tried to reduce – or at least circuscribe – thermodynamical concepts by mechanical ones (Poincaré[8]). The theory of Adiabatical Invariance is but a link (among others) connecting with more or less risks these two distant domains, and we may say that by means of it some preterit essays acquired a revival of interest.

This was the case, for instance, of a very mysterious, powerful and somewhat forgotten formula, $\delta Q = 2\nu\delta(\overline{T}/\nu)$, which were deduced by Boltzmann (by no less enigmatic considerations) relating the heath δQ supplied to a vibrating elementary *mechanical* system of frequency ν and mean kinetic energy $\overline{T}$ (Boltzmann[9], Brillouin[10]). In fact, by means of this controversible formula, the adiabatical invariance of E/ν was automatically proved (for if $\delta Q \doteq 0$ then $\overline{T}/\nu$ = constant; now in the case of the pendulum, $\overline{T} = (1/2)E$ so that we actually obtain $E/\nu =$ ct.). In this manner, by a sort of dialectical effect, this result thus reinforced the credibility of that profound but questionable formula.

More or less the same thing happened with Boltzmann-Planck's law $S = k \log \Omega$ which at that time had not yet turned into the unwavering dogma it became since then. The question was : since in a *thermodynamical* adiabatical transformation we have $\delta Q/\overline{T} = dS = 0$, then the *mechanical* quantity Ω (the volume of the phase space, that is, the number of Planck's "complexions") *should be* an adiabatical invariant. And again the pre-established harmony worked, for the "a posteriori" demonstrations showed that it was actually true (Kahan[11], Perrin[12]). That is, the element of volume of phase space, the probability measure over which we build everything we know in the world of statistical physics, is an adiabatical invariant.

Now let us move up from the early of quanta to quantum mechanics as it stands now. The fact that is at the basis of this transition is the appearing of Schrodinger's equation. With it, the way in which the possible physical states were provided changed drastically and at the same time the reach and importance of Adiabatical Invariance.

Before Schrodinger, in the theory of quanta, the quantifiable physical quantities were adiabatical invariants, so that Adiabatical Invariance was at the very basis of the theory; but now quantum states were provided by a much more elegant method : they were the solutions of a differential equation and by imposing some "natural" conditions to these solutions (continuity, regularity, null boundary conditions, etc) we get a discrete, "quantified" set of eigenfunctions ψ_k which are the allowed states.

Of course, one could still ask for the Adiabatical Invariants, and that was what Born and Fock did (Born, Fock[13]), and they found that for a quantum system with Schrodinger equation $H\psi = i\hbar\partial_t\Psi$, the Adiabatical Invariants were the Fourier components of the states vector Ψ ever the eigenfunctions ψ_t of H, that is, the constants c_k in $\Psi = \Sigma\, c_k\, \psi_k$ (This is improperly and almost always shortened by saying that the ψ_k "are" adiabatical invariants). But such a result now sounds somewhat academical and by no means lies at the foundations of schrodingerian quantum physics.

Yet *even here,* in spite of all formal beauties (and true progress) of Schrodinger equation we are not in a very different situation : after all, the general solution of $H\psi = E\psi$ is a superposition $\Sigma\,\alpha_k\,\psi_k$, and after all in order to agree with experimental evidence we must more or less implicitly postulate that a quantum system is always found in one of the eigenfunctions ψ_k and *never* in a superposition of them.

So we must still *select* among several possibilities as it was the case in the ancient theory of quanta. The stationary states, those which we do actually meet in nature, are not determined without some ambiguity or, which turns to be the same, without privileging the eigenstates (that is, the "normal modes" ψ_k) against their superpositions $\Sigma\,\alpha_k\,\psi_k$. Certainly one may argue that this is a "natural" choice, that they are "the most important ones", but such periphrasis by no means can contradict the fact that we make use here of an *extra* postulate, completely allogeneous to the rest of the theory.

So it seems that there again it would be useful to possess a criterion separating the allowed states from the others, and this is the way Adiabatical Invariance comes up again : for, in accordance with the Born-Fock theorem, the eigenstates ψ_k are adiabatical invariants, while their superpositions $\Sigma\,\alpha_k\,\psi_k$ obviously are not.

We have spoken of a criterion of Adiabatical Invariance for the stationary states. But so that this criterion may be a useful one, we must not restrict to certify that it does provide what is already known, but rather to test it in situations different from those already known.

The question could be the following:

In the case where no normal modes, that is, no stationary states exist, are still the most probable states adiabatical invariants? If the answer is *yes* (and let me tell at once that indeed it is) then we could hardly resist the strong temptation of postulating that in every situation of equilibrium the allowed quantum states (that is, the most probable ones) must always be adiabatical invariants. Therefore the property of Adiabatical Invariance retrieves the fundamental importance it had apparently lost since the good old days of the theory of quanta, and thus becomes a true, essential criterion for further research!

The result that lies at the basis of this optimism is Lochak's theory of permanent states (Lochak, Alaoui[14]) in which is considered the non conservative case of an hamiltonian explicitly depending on time and periodic. In such conditions it can be proved that a certain integral of motion exists (the "reduced energy" R which plays in this theory a role similar to that of the energy H in the usual conservative case) whose eigenfunctions χ_k — obviously non stationary! — "are" adiabatical invariants in the sense it has been explained above. Lochak next proceeded to statistical applications in Nuclear Magnetic Resonance and Hertzian Spectroscopy, taking as most probable states the "permanent" states χ_k ("possessing" the property of Adiabatical Invariance) and thus reach the agreement with experimental results.

Now, starting from the well-established, confirmed theory we have been outlining, let us look into the future and answer the crude question: *what is Adiabatical Invariance good for?*

The complete answer must be split into two : the first, which we may call the objective one, has been just given : The Adiabatical Invariance provides a criterion which we postulate as valid in all types of equilibrium : the most probable states (those which actually are to be found in experiments) are adiabatical invariants. Such criterion has been tested successfully in the theory of permanent states, and we have good reasons to believe that this theory can be extended to much more general situations.

But from a purely qualitative point of view the preceding results are good for something else : in a certain way, they are forerunning sign of a complete change in the methods of research in microphysics, which will tend more and more to become *essentially thermodynamical.*

In fact, if the part of thermodynamics in ordinary quantum physics is so unimportant, this is due to the fact that "orthodox" quantum physics is mainly the theory of stationary states — where, as Schrodinger would say, nothing happens ... — and that in such conditions the second principle of thermodynamics loses the essential part of its strength : it is no longer a principle of evolution, but rather of conservation.

No doubt in a more profound description of microphysical reality — beyond stationarity and equilbrium, describing quantum transitions, etc — the role of the principle of evolution of thermodynamics will be essential.

Now one may ask whether this can already be seen, or at least foreseen : the answer is yes and the evidence is just provided by Adiabatical Invariance. For we may reasonably expect the intermediate evolution between two situations of equilibrium to be ruled by that principle of evolution (entropy increasing? free energy diminishing?) since at the beginning and end of such evolution (the "attractors" alpha and omega) we find adiabatical invariants, as we have seen.

This qualitative, almost subjective statement is perhaps the most interesting feature of the theory of Adiabatical Invariance as it stands. The *essentially thermodynamical* substrate of microphysical world, from which we know but little more than the emersed part of stationarity, is already manifest even without going further : by means of Adiabatical Invariance of the states we meet at equilibrium.

To state it briefly :

Up to now, very few attempts have been done to explore that "subquantum" world (de Broglie [15], Andrade e Silva, Fer, Leruste, Lochak [16]), and quantum transitions remain refractory to a mathematically correct, physically meaningful description; but on the other hand, no demonstration has been given yet of the impossibility of such a description.

In this "Götterdämmerung" atmosphere, near the end of an exhausted and long since accomplished theory, and at the threshold of a different, long since expected one, we may reasonably prophesy that this new one shall be essentially thermodynamical – or never shall.

BIBLIOGRAPHY

[1] La Théorie du Rayonnement et les Quanta, Réunions Solvay, 1911, Gauthier – Villars, Paris, 1912.

[2] Sommerfeld, A., Atombau und Spektrallinien, F. Viewcg and Sons, Braunschweig, 1951.

[3] Burgers, J.M., Phil. Mag., 33, 1917, p. 514.

[4] Juvet, G., Mécanique Analytique et Théorie des Quanta, Blanchard, Paris, 1926.

[5] Pars, L.A., A Trieatise on Analytical Dynamics, Heinemann, London, 1965.

[6] Ehrenfest, P., Phil. Mag. 33, 1917, p. 500.

[7] Fer, F., Journal de Physique et le Radium, 23, 1962, p. 973.

[8] Poincaré, H., Thermodynamique, Gauthier-Villars, Paris, 1908.

[9] Boltzmann, L., Vorlesungen über die Prinzipe des Mechanik, 2, Barth, Leipzig, 1897.

[10] Brillouin, L., Les Tenseurs en Mécanique et Elasticité, Masson, Paris, 1960.

[11] Kahan, T., Physique Théorique, Tome I, vol. 2, Presses de France, Paris, 1960.

[12] Perrin, F., Mécanique Statistique Quantique, Gauthier-Villars, Paris, 1939.

[13] Born, M., and Fock, V., Zeitsch.f.Phys., 51, 1928, p. 165.

[14] Lochak, G., and Alaoui, A., Annales de la Fond.L.de Broglie, 2, 2, 1977, p. 87.

[15] de Broglie, L., La Thermodynamique de la Particule Isolée, Gauthier-Villars, Paris, 1964.

[16] Andrade e Silva, J., Fer, F., Lochak, G., Leruste, P., Cahiers de Physique, 15, 1961, p. 210; 16, 1962, p.1.

EVOLUTION OF NON-EQUILIBRIUM STABLE THERMODYNAMIC PROCESSES

P. Glansdorff

Université libre, Bruxelles

1. INTRODUCTION - The laws of thermodynamics are concerned with the macroscopic properties of systems involving a large number of interacting particles, when considered at the microscopic scale. Any evolution of the mean behaviour of such collective processes is characterized by two very important effects : the existence of fluctuations and the appearance of irreversibility. Indeed, the latter never occurs in the dynamic laws governing the trajectories of only a very small number of constitutive particles, whereas it plays a prominent part as soon as the system exhibits a higher degree of complexity. On the other hand, the general stability problem of a non-equilibrium macroscopic state is strongly dependent on the existence of fluctuations. Moreover, this influence is particularly important around states located far from equilibrium.

In what follows, we derive this property as well as some others connected to the general problem of evolution and self-organization, from the most recent developments that have been obtained in the non-linear region of the thermodynamic theory of irreversible processes. However, to avoid misunderstandings and ambiguities, we first recall the main basic definitions and the meaning of the additional restrictive hypothesis, used in this theory.

2. EQUILIBRIUM - The state of equilibrium is characterized by zero-values for all the currents everywhere within the system. For instance, mechanical (electrical) equilibrium, corresponds to the cancellation of the mean mass (electrical charge) velocity v at each point of the system (maxwellian microscopic distribution in the classical case of a dilute gas) ; thermal equilibrium and diffusion equilibrium, mean respectively, zero values for the heat diffusion flows namely : $W = 0$, $\Delta_\gamma = 0$, $(\gamma = 1, \ldots, c)$; chemical equilibrium denotes the absence of any rate of reaction, i.e. $w_\varrho = 0$ $(\varrho = 1, \ldots, r)$.

It must be kept in mind, that equilibrium by no means implies uniformity. Indeed, a head of water, or an isothermal column of gas subject to a gravitational field, are examples of equilibrium but non uniform systems on account of the change of pressure with height. Reciprocally, a uniform and permanent state kept up by one or several homogeneous chemical reactions in some open system, cannot be interpreted as an equilibrium state, but rather as a stationary state, according to the definition recalled hereafter [1].

3. STATIONARY STATE - This denomination, or alternatively those of steady state, defines any behaviour independent of time. For instance, the final temperature distribution within a wall separating two media, each of which is maintained at constant temperature resp. T_1 and T_2, represents a non uniform stationary solution of this heat conduction problem.

In this respect, we would like to warn the reader against frequent misuses in the common language, regarding the above words. As an example, expressions such as ecological or social equilibrium should not be used as they are inaccurate in describing a situation more or less close to a steady state, but by no means in agreement with the definition of an equilibrium state.

4 - NON EQUILIBRIUM PROCESSES - A stationary state belongs to the category of non equilibrium states and therefore corresponds to a time independent irreversible process. However this category also contains all the various time dependent processes as e.g. oscillations, periodic or cyclic phenomena, wave and quasi-wave propagations and so on. The study of all these evolutions leads to space like as well as time like and mixed problems.
The general method for solving such problems implies the use of respectively [3]:

- the conservation equations of mass for each component.
- the conservation equation of momentum if convective effects occur.
- the conservation equation of energy i.e. the local formulation of the first law of thermodynamics.

Besides, this system of equations has to be closed with the help of the additional constitutive equations, which introduce the explicit values of the fluxes involved in terms of the selected variables and their space derivatives. As an example, the well known Fourier or Fick law, related respectively to the heat flow and the diffusion flow, as well as the chemical kinetic equations related to the reaction rates, often occur as the suitable constitutive relations.

Finally, it remains yet to specify the external constraints

prescribed for the system.
As a rule, and according to the case, this requirement leads either to initial value problems, or to boundary value problems ; mixed problems also occur. Let us also recall that the constraints are generally given in terms of the variables themselves, and (or) their derivatives. Sometimes also the exchange of fluxes with the outside world are directly prescribed, as e.g. in the case of an isolated system (zero value of all external exchanges) or also of a closed system (zero value of external mass exchange).

5 - THERMODYNAMIC METHOD - As may well be imagined, such an analytic procedure for describing the evolution and the stability characteristics is extremely difficult except in very elementary situations. This is already a first reason for having recourse to the possibilities afforded by the thermodynamics of irreversible processes. Indeed in many cases this theory permits the association of an evolution criterion to the above equations, and hence the derivation of at least approximate solutions by optimization of some suitable function or functional. Besides, even in the non equilibrium region for which a similar potential does not exist, the thermodynamic method provides us with stability criteria especially suitable for investigating the physical meaning of the stability conditions. Indeed this insight seldom appears from the analytic procedure above mentioned (kinetic stability).
On the other hand, the thermodynamic stability criteria subsequently discussed, give rise directly to the use of some basic mathematical procedures regarding the stability problem such as Liapounoff functions. This allows the introduction of the important concept of stable or unstable branch as well as those of bifurcation which may occur at some critical state. In such a case an abrupt change of behaviour may occur when the system passes from one branch - as e.g. the thermodynamic branch arising from the equilibrium state -

to another more stable one with the appearance of an entirely different organization. If this new order so obtained is more elaborate (structured !) than the former, one says that it corresponds to that of dissipative structure. [3]

6 - THE LOCAL EQUILIBRIUM ASSUMPTION - Unfortunately, the above properties have not been reached without making some restrictions. Indeed, their field of validity is appreciably less than that of the whole domain of irreversible processes, due to the necessity of introducing the so called local equilibrium hypothesis. However, in spite of this limitation, the remaining part is still of major interest as it concerns much of chemical non equilibrium and even non linear processes, together with transport phenomena such as heat flow, diffusion and convection effects.
In addition the important question relating to the possibility of a physical interpretation of biological, in agreement with the second law of thermodynamics, may also be investigated on the same basis [4].

According to the local equilibrium hypothesis, the variables required to describe a non equilibrium state are locally the same as in the case of equilibrium. Therefore, the validity of the well known Gibbs formula subsists, as an equation of state defining the local entropy, even in the case of irreversible chemical reactions. Its local form may be written as :

$$T ds = de + p dv - \sum_{\gamma} \mu_{\gamma} dN_{\gamma} \tag{1}$$

where s denotes the specific (per unit mass) entropy, e the specific energy, v the specific volume μ_{γ} the specific chemical potential of the component γ $(\gamma = 1, \dots c)$, and N_{γ} the corresponding mass fraction with the separate condition $\sum_{\gamma} N_{\gamma} = 1$.

Clearly, it would be misleading to identify this local equilibrium hypothesis with those of a system locally in equilibrium. In fact, as pointed out in the kinetic theory, the use of the Gibbs law merely supposes that the local behaviour of the system at the microscopic scale, corresponds to distribution functions close to the well known Maxwell-Boltzmann equilibrium form (Prigogine, 1949). However, the parameters included then occur as functions of space and time. In short, such situations correspond to prevailing values of the elastic collisions which tend continuously to restore the thermodynamic equilibrium perturbed by the gradients of the transport processes, as well as by the chemical affinities. The non elastic collisions being much less numerous than the former, the corresponding rates of reaction remain moderate.
When these conditions are fulfilled, the macroscopic scales of space and time are generally of an order of magnitude larger than the corresponding microscopic scales relating respectively to e.g. the mean free path, and the mean free period of the molecular thermal motion.

As a result, processes such as rarefied molecular flows, or interacting fermions at very low temperature, are to be excluded from this local description, due to the lack of collisional effects in such cases. Likewise, non equilibrium processes depending on, either additional variables than those of the Gibbs law (hidden variables), or even derivatives of these variables, are also beyond the scope of this local equilibrium hypothesis. For instance many problems in the field of rheology as e.g. viscoelasticity and plasticity, belong to this category.

Henceforth, we shall focus exclusively our attention on the region of irreversible processes covered by the local equilibrium hypothesis. This enables us to use Gibbs law as a suitable definition of the local entropy. Then the corresponding extensive value S supposing that s obeys the additive principle, may be written as

$$S = \int_V \varrho\, s\, dV \tag{2}$$

where V denotes the volume of the whole system, and ϱ the local density $(\varrho \equiv v^{-1})$.

7 - THE SECOND LAW OF THERMODYNAMICS AND THE ENTROPY PRODUCTION - The change of entropy per unit of time, i.e. dS/dt, may be interpreted as depending on two separate contributions. One of them results from the irreversible processes generated inside the system, and called therefore the entropy production $P[S] = d_i S / dt$. Its sign, prescribed by the second law of thermodynamics, has to be non negative, the zero value concerning the equilibrium state. The second corresponds to the entropy flow $\Phi[S] = d_e S / dt$, arising from the external world ; whence the symbolic relation:

$$\frac{dS}{dt} = P[S] + \Phi[S] \; ; \qquad P[S] \geqslant 0 \tag{3}$$

for the entropy balance equation.

In the particular case of a closed system, the entropy flow results solely from the external heat flow dQ/dt which gives rise to the value $\Phi[S] = T^{-1}\, dQ / dt$ Then the entropy balance equation reduces to the well known Carnot-Clausius differential inequality :

$$dQ_n = TdS - dQ \geqslant 0 \; ; \quad dQ_n = TP[S]\, dt \tag{4}$$

In this formulation, the non compensated heat dQ_n introduced by Clausius is used instead of the entropy production. Anyway, one deduces directly from ineq. (4), that in the case of an isolated system $(dQ = 0)$, the equilibrium state corresponds either to the maximum of its entropy, or to the absolute minimum of its entropy production.

However, to go further, we have need of a more explicit formulation of the above entropy balance equation, expressed in terms of the variables adopted to describe the state of the system. To be specific, we first consider the simple case of a uniform medium, involving one single chemical reaction such as $0 = \Sigma\, \nu_\gamma \mathcal{M}_\gamma$ where $\mathcal{M}_\gamma$, denotes the molecular or atomic mass of the species γ, and ν_γ, the corresponding stoechiometric coefficients (> 0 , for the created components ; < 0 for those destroyed by the reaction). Then, let us adopt as independent variables, the temperature T , the pressure p and the chemical variable ξ introduced by Dalton's law, written as

$$\xi = \frac{n_\gamma - n_\gamma^0}{\nu_\gamma} \qquad (m_\gamma = \mathcal{M}_\gamma n_\gamma)$$

where n_γ^0 denotes the number of moles at the initial state of the chemical reaction $(\xi = 0)$.

Taking into account, the first law of thermodynamics written as :

$$dQ = dH - V dp \qquad (H = E + pV)$$

in terms of the enthalpy H , and then expanding dH and dS in terms of the increments dT , dp and $d\xi$, we obtain after regrouping terms, the following expression for the Carnot-Clausius inequality :

$$A_{p\xi}\, dT + A_{T\xi}\, dp + A_{Tp}\, d\xi \geqslant 0$$

Let us now observe that, according to the theory of Th. De Donder, in a uniform medium, all irreversible processes arise necessarily from the chemical effects, which implies the identities :

$$A_{p\xi} = A_{T\xi} = 0$$

Therefore, any change of the physical variables such as $T, p, \xi \rightarrow T, V, \xi \rightarrow \ldots \rightarrow x, y, \xi$, give rise to the series of equalities :

$$A_{Tp} = A_{TV} = \ldots = A_{xy} = A$$

after some elementary manipulations. This last result allows us to interpret the quantity A as a new state function, and moreover to write the Carnot-Clausius inequality under the more specific form

$$A\,d\xi \geqslant 0 \qquad (5)$$

Indeed one sees that the sign of the state function A prescribes that of the rate of reaction $w = d\xi/dt$, and may be interpreted as giving the direction of the chemical process for the state T, p, ξ.
These two properties justify the denomination of chemical affinity A introduced by Th. De Donder (1922) together with the inequality (5) of the second law.
Clearly, the extension to a system of r independent chemical reactions, is straightforward. It gives rise to the more general relation

$$P[S] = T^{-1} \sum_{\varrho} A_{\varrho} w_{\varrho} \geqslant 0 \qquad (\varrho = 1, \ldots, r) \qquad (6)$$

This may involve some coupling effects $(A_i w_i < 0)$, provided that the global sign subsists [1].

According to the above presentation in the space T, p, ξ, one may also deduce by simple manipulations the following expression of the chemical affinity

$$A = -\left(\frac{\partial H}{\partial \xi}\right)_{Tp} + T\left(\frac{\partial S}{\partial \xi}\right)_{Tp} = -\left(\frac{\partial G}{\partial \xi}\right)_{Tp} = -\sum_{\gamma} \nu_{\gamma} \mathscr{M}_{\gamma} \mu_{\gamma} \quad (7)$$

often useful. The first equality displays explicitly the distinction between the concept of affinity and that of the reaction heat emitted, i.e. $-(\partial H / \partial \xi)_{Tp}$, contrary to the former Berthelot assumption valid only at very low temperatures. The second equality presents the affinity as a derivative of the free enthalpy $G = H - TS$, with respect to the chemical variable ξ. The last equality establishes the connection between the affinity and the chemical potentials μ_{γ} of Gibbs, which in turn are derivatives of the free enthalpy with respect to the masses involved m_{γ}.

To obtain now the explicit formulation of the entropy balance equation in the general case of a non uniform medium moving with a local mean barycentric velocity $\mathbf{v}$, involving diffusion velocities Δ_{γ} as well as an heat flow density $\mathbf{W}$, let us introduce in the local expression (1) of Gibbs law, the mass and energy balance equations, written respectively in the local form [2], [3]

$$\frac{\partial \varrho_{\gamma}}{\partial t} = \sum_{\varrho} \nu_{\gamma\varrho} \mathscr{M}_{\gamma} w_{\varrho} - \frac{\partial}{\partial x_j} \left[\varrho_{\gamma} \Delta_{\gamma j} + v_j\right] \quad (8)$$

and

$$\frac{\partial(\varrho e)}{\partial t} = -P_{ij} \frac{\partial v_i}{\partial x_j} + \frac{\partial}{\partial x_j} \left[W_j + \varrho e v_j\right] \quad (9)$$

Summations over the repeated subscripts i and j, are implied. To obtain eq. (9), we have supposed for the sake of simplicity, the equality of all the fields of forces, i.e. $F_\gamma = F$. Also, we observe that in the case of a fluid, the pressure tensor $\mathbf{P}$, contains separately the reversible hydrostatic pressure p, and an irreversible part p_{ij}, which yields :

$$P_{ij} = p\,\delta_{ij} + p_{ij} \quad ; \quad \delta_{ij} = \begin{cases} 0 & i \neq j \\ 1 & i = j \end{cases} \qquad (10)$$

As a result, the entropy balance equation written for a non uniform open system, bounded by a fixed surface Ω, in the frame of reference selected, takes the form :

$$\frac{\partial S}{\partial t} = P[S] + \Phi[S] \qquad (11)$$

with the following explicit values of the entropy production :

$$P[S] = \int_V \left[W_j \frac{\partial T^{-1}}{\partial x_j} - \sum_\gamma \varrho_\gamma \Delta_{\gamma j} \frac{\partial(\mu_\gamma T^{-1})}{\partial x_j} - p_{ij} \frac{\partial v_i}{\partial x_j} + \sum_\varrho w_\varrho T^{-1} A_\varrho \right] dV \geq 0 \qquad (12)$$

and of the entropy flux (n_{in} = internal normal) :

$$\Omega[S] = \int_\Omega \left[\mathbf{W} T^{-1} - \sum_\gamma \varrho_\gamma \mathbf{\Delta}_\gamma (\mu_\gamma T^{-1}) + \varrho\, s\, \mathbf{v} \right] \cdot \mathbf{n}_{in}\, d\Omega \qquad (13)$$

without defined sign [2], [3].

As pointed out by the inequality (12), the entropy production of a non uniform system contains in addition to the chemical contribution, already present in the uniform case (6), the effects of all the transport phenomena, i.e. the heat conduction, the diffusion, the viscosity and the convection associated to the barycentric velocity.

Besides, to emphasize the bilinear form (12) of the entropy production depending on the fluxes J_α (say) and the generalized forces X_α (say), one often uses the more compact formulation

$$P[S] = \int_V (\Sigma J_\alpha X_\alpha) dV \geq 0 \qquad (14)$$

or even symbolically : $J_\alpha \quad X_\alpha$, for the sake of simplicity.

8. THE EVOLUTION CRITERION - To begin with, let us focus our attention on the region of non equilibrium state satisfying the usual stability conditions of equilibrium, well known in classical thermodynamics as :

$$c_V > 0 \qquad \text{(thermal stability)}$$

$$\chi = -(\partial v/\partial p)_T / v > 0 \qquad \text{(mechanical stability)}$$

$$\sum_{\gamma\gamma'} \mu_{\gamma\gamma'} \delta N_\gamma \delta N_{\gamma'} > 0 \qquad \text{(diffusion and chemical stability)}$$

involving the specific heat at constant volume c_V, the isothermal compressibility χ , and the partial derivatives $(\partial \mu_\gamma / \partial N_{\gamma'})_{Tp}$ enclosed in the definite quadratic form $\sum_{\gamma\gamma'}$.

Alternatively, these conditions may be rewritten in the form of the unique criterion :

$$\delta^2 s = -\frac{1}{T}\left[\frac{c_v}{T}(\delta T)^2 + \frac{\varrho}{\chi}(\delta v)^2_{N_\gamma} + \sum_{\gamma\gamma'} \mu_{\gamma\gamma'}\,\delta N_\gamma\,\delta N_{\gamma'}\right] \leq 0 \qquad (15)$$

prescribing that the second differential of the entropy, deduced in the space e, v, N_γ, from the Gibbs formula (1) has to be a negative definite quadratic form.

As pointed out hereafter, this property appears to be of particular interest in order to build an evolution criterion valid in a large region, linear or not, of the irreversible processes. Going back to the masses and energy balances (8) and (9), we multiply the two sides, respectively by the derivatives $-d_t(\mu_\gamma T^{-1})$ and $d_t T^{-1}$. Then we add term by term and integrate over the whole system. We consider at first the case of a strictly dissipative system (absence of convective effects : $\mathbf{v} = 0$; $d/dt = \partial/\partial t$). After integration by parts, we get for fixed boundary conditions: $(d_t T)_\Omega = (d_t \mu_\gamma)_\Omega = 0$, the following result

$$-\int_V \frac{\varrho}{T}\left[\frac{c_v}{T}\left(\frac{dT^{-1}}{dT}\right)^2 + \frac{\varrho}{\chi}\left(\frac{dv}{dt}\right)^2_{N_\gamma} + \sum_{\gamma\gamma'} \mu_{\gamma\gamma'}\frac{dN_\gamma}{dt}\frac{dN_{\gamma'}}{dt}\right] dV$$

$$= \int_V \left\{ W_j \frac{d}{dt}\left(\frac{\partial T^{-1}}{\partial x_j}\right) - \sum_\gamma \varrho_\gamma \Delta_{\gamma j}\frac{d}{dt}\left(\frac{\partial(\mu_\gamma T^{-1})}{\partial x_j}\right) + \sum_\varrho w_\varrho \frac{d(A_\varrho T^{-1})}{dt}\right\} dV$$

The sign of the l.h.s. is prescribed by that of $\delta^2 s$ given by the inequality (15), since in the two cases we deal with the same type of negative definite form (same values of the coefficients). Besides, by a straight comparison between the r.h.s. and the expression (12) of the entropy production, we find for the evolution criterion sought, the inequality (Glansdorff and Prigogine, 1954) [3].

$$\frac{d_X P}{dt} = \int_V \sum_\alpha J_\alpha \frac{dX_\alpha}{dt}\, dV \leq 0 \qquad (16)$$

Moreover, it is easy to verify on the conservation equations used above, that after introducing the values $(J_\alpha)_{st}$ related to the steady state, one finds, if it can be attained under the prescribed constraints :

$$\left(\frac{d_X P}{dt}\right)_{st} = \int_V \left(J_\alpha\right)_{st} \frac{dX_\alpha}{dt}\, dV = 0 \tag{17}$$

In the present case, the equality sign in (16) corresponds to the steady state. Then, this criterion may be extended directly to processes involving convection effects as well. To this end we have to take into account the contribution of the momentum balance written as [2], [3].

$$\frac{\partial(\varrho v_i)}{\partial t} = \varrho F_i - \frac{\partial}{\partial x_j}\left[P_{ij} + \varrho v_i v_j\right] \tag{18}$$

Multiplying the two sides by $T^{-1}\, \partial v_i / \partial t$, performing then the same manipulations as above on the suitable forms of the masses and energy balances, and taking into account the supplementary boundary condition, i.e. $(\partial_t v_i)_\Omega = 0$, one derives the same type of evolution criterion. Obviously, the new values J'_α (say) of the fluxes and X'_α (say) of the forces, contain additional convective terms.

Going back to the criterion (16), let us now emphasize its connection with the properties of the entropy production. Indeed, by splitting into two parts the time change of $P[S]$ as derived from eq. (14), it becomes :

$$dP = d_X P + d_J P \tag{19}$$

where $d_X P = J_\alpha\, d X_\alpha$, and $d_J P = X_\alpha\, d J_\alpha$, in the symbolic notation. Accordingly, the criterion (16) is directly associated to the properties of the entropy production (12), since the

change of the forces X_α is always such as to decrease its own. value. Nevertheless, the sign of $d_J P$ and consequently of dP itself are by no means prescribed.
On the other hand, it must be kept in mind that the bearing of the evolution criterion under its form (16), is quite independent of the type of kinetic relations between the fluxes and the forces. It is therefore appropriate to the study of non linear processes. However, above all, the main point to underline is that as a rule, the inequality (16) does not occur as an exact differential expression. Therefore, two possibilities arise, namely:

1°) $d_X P$ is reducible to an exact differential, with or without using and integrating factor ; whence

$$d_X P = d\Phi \leq 0$$

In such a case the evolution near a steady state is governed by the new kinetic potential Φ which consequently plays a part similar to that of the thermodynamic potentials for an equilibrium state, except that generally the potential Φ appears as a functional (non uniform systems) rather than a simple function (uniform systems).

2°) $d_X P$, is no longer reducible to an exact differential. This is the general case in the non linear region. It forbids the existence of any uniform scalar potential associated to the evolutions considered. However rotations become then possible in the space of states around a steady state as e.g. the well known nonlinear predator-prey cycle around a steady state, discussed by Lotka and Volterra.

Anyway, it must be kept in mind that all hypothesis relating to the very existence of some general potential has to be ruled out.

9 - THE STABILITY CRITERIONS - Even when there exists no potential, the evolution criterion (16) allows to derive the stability conditions for a non equilibrium state, thanks to a method already used in the classical equilibrium case. We ensure stability through the requirement that any process, real or virtual, starting from the state of reference cannot be in agreement with the second law for the constraints prescribed. According to (16) such a condition for small perturbations δ, and unperturbed boundary conditions $(\delta_\Omega = 0)$, takes the form :

$$\delta_X P \geq 0 \tag{20}$$

For example, in case of a steady state, the stability criterion of a dissipative system, may be written as :

$$\int_V \left(\sum_\alpha \delta J_\alpha \, \delta X_\alpha \right) dV \geq 0 \tag{21}$$

taking into account the property (17) of a steady state.
This second order expression, often useful in the non linear stability theory has been called the excess entropy production. The associate excess entropy balance equation may be derived by the same line of reasoning used in the previous section, but applied now to the perturbation equations instead of the conservation equations (8) and (9) themselves. Multiplying respectively by $-\delta(\mu_\gamma T^{-1})$ and δT^{-1}, adding the results and integrating by parts, one obtains after the cancellation of the boundary term, the excess entropy balance equation for small disturbances :

$$\frac{\partial}{\partial t}(\delta^2 S) = \int_V \left(\sum_\alpha \delta J_\alpha \, \delta X_\alpha \right) dV \tag{22}$$

Prescribing now the positive sign to the r.h.s. and keeping in mind that the local value $\delta^2 s$ is a negative definite quadratic form in the space e, v, N_γ, we may consider the quantity $\delta^2 S$ as a Liapounoff functional, and derive from it the global stability criterion as [3], [4]

$$\delta^2 S < 0 \quad ; \quad \frac{\partial}{\partial t}\left(\delta^2 S\right) \geq 0 \tag{23}$$

for all values of the time $\left(t > t_0\right)$, according to Liapounoff's second method for the linear stability. If the equality is ruled out from (23), the criterion of stability reduces to the asymptotic stability, which means that the global disturbance vanishes for $t \to \infty$.

On the other hand eqs (22) and (23) enable us to recover the stability condition (21) formerly obtained through the procedure already used in classical thermodynamics. In each case the excess entropy production has to be positive. However, with the Liapounoff's second method, the stability condition then becomes valid not only around steady states, but also around non steady situations such as e.g. cycles or oscillating motions.

In the case of systems involving convection effects, the local entropy s considered above has to be replaced by the function:

$$z = s - \frac{1}{2} T_0^{-1} v^2$$

where only the quantities s and v are to be varied, T_0 being the initial value of the reference state, i.e. a non varied constant: $\delta^2 z = \delta^2 s - \frac{1}{2} T_0^{-1} \delta^2 v^2$. This gives rise to the enlarged global stability criterion [3], [4].

$$\delta^2 Z < 0 \quad ; \quad \frac{\partial}{\partial t}\left(\delta^2 Z\right) \geq 0 \tag{24}$$

for all values of time $\left(t > t_0\right)$.

10 - <u>THE THERMODYNAMIC BRANCH</u> - Thermodynamic equilibrium is characterized by the absence of irreversible fluxes and forces, namely

$$J_\alpha^{eq} = 0 \quad ; \quad X_\alpha^{eq} = 0$$

Therefore, each equilibrium state is surrounded by a region where the dependence of the J_α in terms of the X_β's corresponds to the linear part of its expansion and thus may be written as [2] :

$$J_\alpha = \sum_{\alpha\beta} L_{\alpha\beta} X_\beta \quad ; \quad \left[L_{\alpha\beta} = \left(\frac{\partial J_\alpha}{\partial X_\beta} \right)_{eq} \right] \tag{25}$$

Such relations are called the linear phenomenological laws, and the constants $L_{\alpha\beta}$, the corresponding phenomenological coefficients. In addition, the matrix $\| L_{\alpha\beta} \|$, is symmetrical as pointed out by Onsager (1934), that is :

$$L_{\alpha\beta} = L_{\beta\alpha} \quad (\alpha \neq \beta) \tag{26}$$

The so called linear thermodynamics presents some typical properties with respect to the evolution and stability problems. First of all, the entropy production (14) takes here the positive definite quadratic form [2] :

$$P[S] = \int_V \left(\sum_{\alpha\beta} L_{\alpha\beta} X_\alpha X_\beta \right) dV \geq 0 \tag{27}$$

which prescribes a positive value to the diagonal coefficients $L_{\alpha\alpha}$, while the sign of the mutual values is not prescribed by the condition : $(L_{\alpha\beta})^2 < L_{\alpha\alpha} L_{\beta\beta}$, as is the case for the thermodiffusion coefficient. Then, the stability criterion (21) for small disturbances becomes :

$$\int_V \left(\sum_{\alpha\beta} L_{\alpha\beta}\, \delta X_\alpha\, \delta X_\beta \right) dV \geq 0 \tag{28}$$

which is identically satisfied. Indeed, the comparison between the

expression (27) of the entropy production and the condition (28), shows that the two definite positive quadratic forms are of the same type.
On the other hand, taking eq. (26) into account, it results from eqs (16) and (19), that in the region of linear thermodynamics, one has :

$$\frac{d_X P}{dt} = \frac{d_J P}{dt} = \frac{1}{2}\frac{dP}{dt} \leq 0 \qquad (29)$$

which means that in the present case, the sign of $d_J P$, is no longer indeterminate[3].In addition the last term introduces the well known theorem on the minimum entropy production at the steady state for given constraints (Prigogine, 1945). This is one of the main characteristics of linear thermodynamics. Indeed, it establishes that if it is true that equilibrium corresponds to an absolute minimum of the entropy production as recalled in section 7 (zero value of $P(S)$), the same property subsists around a steady state (positive value of $P(S)$). Moreover, from the beginning, we have taken into consideration the case of a stable equilibrium state $(\delta^2 S < 0)$ and we derive now the stability of the steady states in its vicinity, thanks to eq (28).

We may thus conclude that, for given constraints, the behaviour of a system near a steady state in the linear region is quite similar to that which occurs near equilibrium[1].In other words, all these stable evolutions belonging to the so called thermodynamic branch, obey to the same type of behaviour order $\rightarrow$ disorder, consistent with the constraints prescribed, as pointed out formerly by Boltzmann in the case of the vicinity of equilibrium with his law of probability (Fig. 1).

$$S = k \log W$$

The region of validity of linear thermodynamics is large in the case of transport phenomena, practically as large as our basic local equilibrium hypothesis discussed in section 6. On the contrary, as soon as chemical reactions, occur with or without transport phenomena, the situation changes entirely and this reduces greatly the interest of linear thermodynamics. Obviously, this situation enhances extensively the interest of the non linear domain of the thermodynamic of irreversible processes for chemical problems [4].

11 - NON LINEAR THERMODYNAMICS - As pointed out here above, this region is always separated from the equilibrium state, by a finite distance which corresponds to the extent of the linear domain.
This is also the reason why the main features of this region are often called the far from equilibrium properties. Anyway the evolution and stability problems become here very much more complicated. Nevertheless, the thermodynamic analysis gives useful information on the part played by the distance from thermodynamic equilibrium, and accordingly interpreted as a universal parameter in a great number of physical problems. Besides, it opens the way to the relationship with fluctuation theory. The first important investigation in the field of non linear thermodynamics is relatively recent and corresponds practically to the appearance in 1954 of the evolution criterion (16) [3].

Application to specific situations shows that in the non linear domain, that is far from equilibrium,the stability criterion is no longer identically satisfied. This may arise e.g. when convective effects occur, such as the onset of free convection in a horizontal layer of a fluid at rest, heated from below (Bénard problem, critical Rayleigh number), or from the laminar-turbulent flow transition (critical Reynolds number). Also autocatalytic and more generally feedback effects in a sequence of coupled chemical reactions may generate a similar instability.

This last observation suggests a comparison between the manifesta-

tions of order by fluctuations in macroscopic physics, and those of biological processes in which autocatalytic reactions are often present due to enzymatic effects. This has been clearly demonstrated today for many applications in this field as e.g. the glycolytic cycle for which the stability of the thermodynamic branch is no longer ensured. The demonstration that such evolutions actually occur on other stable branches beyond a critical state, is a considerable success, in favour of the latter comparison [3], [4] (Figs 2 ; 3).
Briefly speaking, the stability criterion based on the sign of the excess entropy production and recalled in section 9, is no longer fulfilled at the critical marginal state where it vanishes. This occurs because in the non linear region, the above quantity can often be written as a difference between two groups of positive terms, pointing out the underlying competition between the stabilizing action of dissipation and the destabilizing action arising from non linear contributions such as convection or catalytic processes.

The stability condition becomes then symbolically :

$$\{\text{Dissipation}\} - K\,\{\text{Convection or Catalysis}\} \geq 0 \qquad (30)$$

The constant K represents the parameter of the constraint associated to the non-equilibrium state ; it is usually a non-dimensional ratio. The critical marginal state, corresponds to a change of branch. It is attained for the value K_{cr} for which the inequality (30) reduces to an equality.

12 - DISSIPATIVE STRUCTURES - Some other non linear models chosen for their relative simplicity have been now investigated to explain the mechanism through which the catalytic effects in the constitutive equations lead to a critical state beyond which an entirely new behaviour

may appear. Then, i.e. for stronger values of the constraints, the system still evolves according to the second law, but it generally does not comply any more to the minimum entropy production theorem of the linear region. On the contrary, as it happens frequently, an internal organization generated by symmetry breaking appears inside the system, which thus becomes more orderly than the states belonging to the thermodynamic branch. Such organized dynamic situations have been named dissipative structures. (Figs 2 ; 3 ; 4).

They correspond either to temporal periodicities as in case of chemical clocks or limit cycles in homogeneous medium, or to spatial periodicities as for the Bénard convection cells. Spatio temporal structures as chemical waves and quasi-waves also occur in non homogeneous systems (low value of the diffusion coefficient). So far, several examples of each case, have been investigated.

One observes, e.g. on the Bénard cells, an apparent similarity between these new symmetry breaking and the classical phase-transitions at equilibrium such as crystals. However, the distinction is essential as the former subsists uniquely thanks to the permanent constraints and the suitable irreversible exchanges with the outside world, while the crystal may be preserved without any exchange with its surrounding.

13 - BIFURCATIONS - At the first marginal state for which the condition (30) reduces to an equality, the thermodynamic branch becomes unstable while the new behaviour belongs to a new stable branch. Such situations are investigated in the bifurcation theory, a chapter of the mathematical analysis devoted to the non linear differential equations [4] Different types of bifurcations are actually well known, but the question become quickly cumbersome in case of several constraint parameters, and several dimensions in the space of the state variables. A classical example, related to one single parameter and one dimension is exhibited on the figure 5 . Successive critical states, corresponding for

instance to a sequence of bifurcations may also occur, due to a progressive increasing of the external constraints, characterized by more than one parameter. This situation allows the system to reach after several bifurcations a highly organized state. This possibility brings still nearer the concept of dissipative structure to that of the biological order and confers to it at the same time, a historical background. Then the deviation far from the equilibrium state appears as a measure of order generated from the non-linearity of the evolution laws and from the existence of the associated bifurcation points. In other words, ordered behaviour exhibited by the desorder $\longrightarrow$ order processes, is quite consistent with the second law of thermodynamics, provided that the distance from the equilibrium state is sufficiently large. Indeed, in its neighbourhood the classical degradation prescribed by the relation (29) about the minimum entropy production is of a general validity in the linear region. Then the Darwin-Clausius conflict so much discussed during the 19th centry, is now resolved.

14 - <u>FLUCTUATIONS</u> - The mechanism responsible for instabilities with creation of dissipative structures, giving rise to the physical interpretation of the bifurcation, is that of fluctuations. Every physical macroscopic medium undergoes necessarily local fluctuations. Around equilibrium, as well as in the whole linear region, fluctuations are and remain always only of a microscopic scale, since each fluctuation is immediately followed by a regression. However, near a state of irreversibility namely around a bifurcation, the system exhibits a markedly coherent behaviour, often associated with long range fluctuations and the breaking of spatial symmetries (R. Brout, 1965). On the other hand, in such regions neighbouring a critical state, the basic local equilibrium assumption becomes itself questionable and another approach to the evolution problem has to be considered. So far, the more suitable method to investigate the behaviour of fluctuations around a non equilibrium state seems to be the stochastic analysis.

This procedure has been already used successfully in various non equilibrium situations such as some reaction-diffusion problems. The main feature proceeds from the possibility for fluctuations to be amplified up to the macroscopic scale where they occur as giant fluctuations, stabilized by the energy and mass flows, prescribed by the external constraints. As a result, the deterministic laws of Thermodynamic governing the behaviour of the dissipative structures are more or less tempered by the stochastic aspect of the fluctuations, which introduces an element of randomness.

We may thus conclude that these recent developments in the field of thermodynamics of irreversible processes, give new impetus to mathematical theory of bifurcations and the stochastic theory of fluctuation as well.

But these topics go beyond the scope of the present lectures, and only some examples are given.

GENERAL REFERENCES

[1] I. PRIGOGINE, Introduction to Thermodynamics of Irreversible processes (1961)

[2] S.R. DE GROOT and P. MAZUR, Non-Equilibrium Thermodynamics, North-Holland, Amsterdam (1962)

[3] P. GLANSDORFF and I. PRIGOGINE, Thermodynamics of Structure, Stability and Fluctuations, J. Wiley, London (1971).

[4] G. NICOLIS and I. PRIGOGINE, Self-Organization in Non Equilibrium Systems, J. Wiley-Interscience, New York, London (1977).

[5] H. HAKEN, Synergetics, Springer, Berlin (1977).

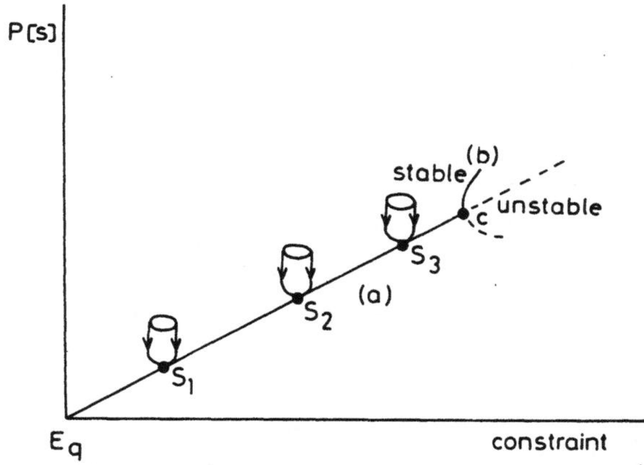

Fig. 1 – The thermodynamic branch starting from equilibrium and the stable steady states – S_1, S_2, S_3, ...

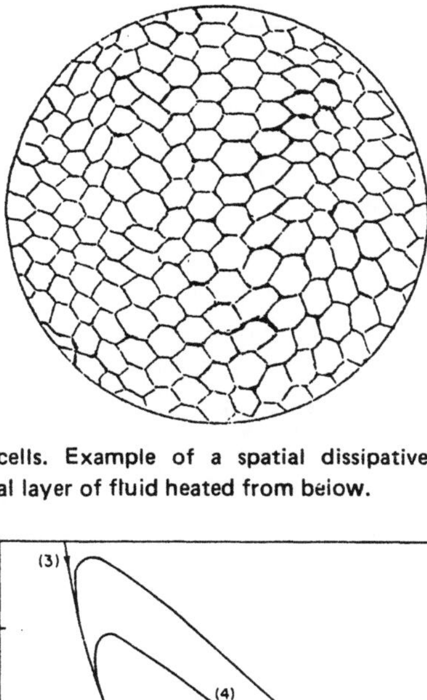

Fig. 2 – Bénard cells. Example of a spatial dissipative structure in a horizontal layer of fluid heated from below.

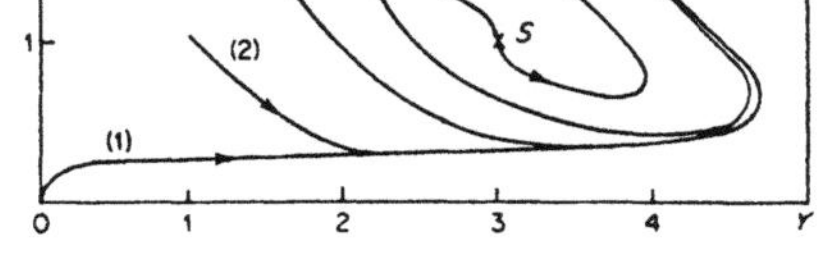

Fig. 3 – Limit cycle. Example of a temporal dissipative structure in a homogeneous biological system plotted in the X–Y plane. Phosphofructokinase reaction (X and Y are concentrations).

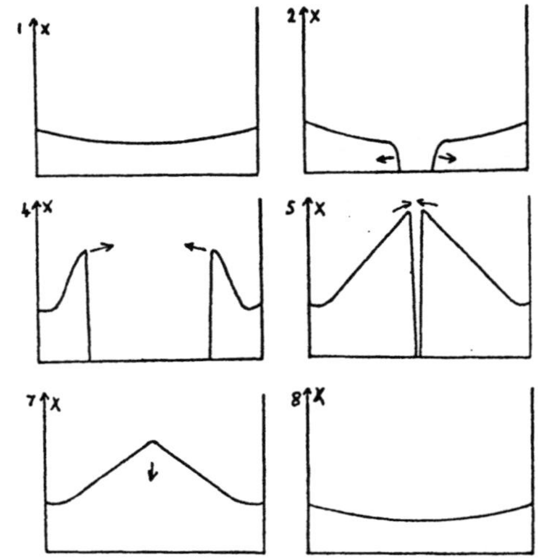

Fig. 4 — Chemical waves and quasi-waves propagation. Example of a spatio-temporal dissipative structure in a non-homogeneous chemical system for different values of t.

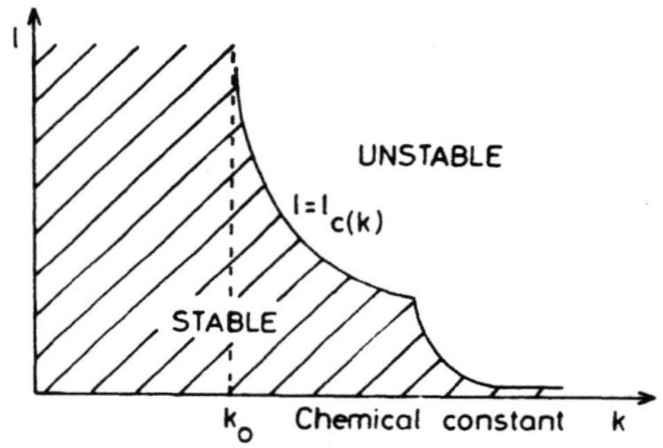

Fig. 5 — Example of bifurcation for a constraint $\lambda = \lambda_{cr}$ large enough to be far from equilibrium.

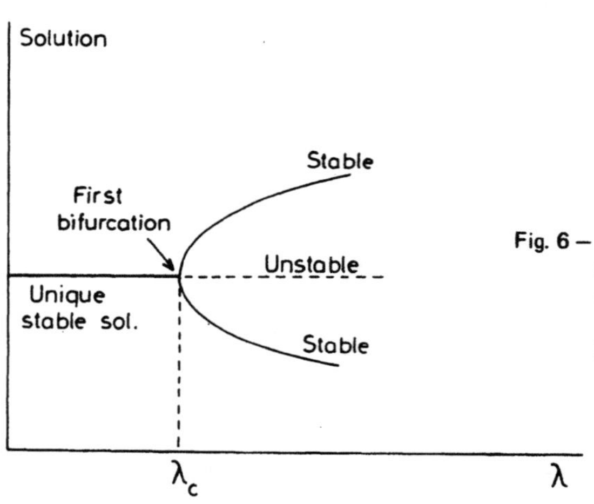

Fig. 6 — Frequency of the fluctuations versus the constraint. In the neighborhood of equilibrium all the fluctuations regress (stability). Beyond the critical state, the high wave-lengths become destabilizing.

INTRINSIC STOCHASTICITY AND IRREVERSIBILITY OF CLASSICAL QUANTUM SYSTEMS

M. Courbage
Bld du Triomphe, C.P. 231
Bruxelles (Belgium)

This contribution is devoted to the question of the relation between the deterministic laws of dynamics and probabilistic descriptions of physical processes. This problem has been the object of many attempts. It is generally accepted that probabilistic processes can arise from deterministic dynamics through a process of "coarse-graining", "contraction of the description" or by introducing extra dynamic approximations like the "molecular chaos". In this short communication I will summarize recent results obtained by B. Misra, I. Prigogine and myself [1], [2], [3] on an alternative approach to this problem which consists in obtaining stochastic processes from deterministic dynamics by a similarity invertible linear transformations acting on the space of the distribution functions, when the dynamics is inherently random or unstable. In this case the irreversibility is obtained by a change of representation without any supplementary hypothesis. Here, irreversibility is described by its main feature, the second law of thermodynamics, which we express as the existence of a functional of the statistical state of the system $\varrho(\underline{p},\underline{q})$, $\underline{p}=(p_1,\dots,p_N)$, $\underline{q}=(q_1,\dots,q_N)$. It is clear that such a functional cannot be expressed by $\int \varrho_t(\underline{p},\underline{q}) \lg \varrho(\underline{p},\underline{q})\, d\underline{p}\, d\underline{q}$ or $\int |\varrho_t(\underline{p},\underline{q})|^2\, d\underline{p}\, d\underline{q}$ which remain constant under the Liouvillian evolution. However such a functional can be expressed as a quadratic form of $\varrho_t(\underline{p},\underline{q})$, in other words by considering an operator $M \geq 0$ such that

$$\Omega(\varrho_t) = \int \varrho_t(\underline{p},\underline{q})\,(M\varrho_t)(\underline{p},\underline{q})\, d\underline{p}\, d\underline{q}$$

$$= < \varrho_t \,, M\varrho_t >$$

decreases monotonically and tends to $< \varrho_{m.c.} \,, M\varrho_{m.c.} >$ for all initial state $\varrho_0(\underline{p},\underline{q})$ where $\varrho_{m.c.}$ is the microcanonical distribution and the above scalar product is defined on the Hilbert space of the square integrable functions on the phase space . As shown by Misra [4], a necessary condition for the existence of such M is the Mixing. In other, certain form of instability is necessary for the dynamics in order to display the irreversibility. A stronger property is the inherent randomness of the dynamics. This concept is defined by the existence of a transformation Λ (on the statistical states) which associates to an initial ϱ_0 a new state $\tilde{\varrho}_0 = \Lambda \varrho_0$ evolving under a Markov process i.e. $\tilde{\varrho}_t = (\Lambda U_t \Lambda^{-1})\, \tilde{\varrho}_0$, $\Lambda U_t \Lambda^{-1}$ being a semi group of strong Markov process where U_t is the unitary evolution of the states. In this case the functional Ω is expressed by

$$\Omega(\varrho_t) = < \Lambda\varrho_t \,, \Lambda\varrho_t > \; = \; < \tilde{\varrho}_t \,, \tilde{\varrho}_t >$$

and M is merely $\Lambda^*\Lambda$. This problem has an exact solution at least for a class of ergodic systems.

To explain the notion of inherent randomness by avoiding as much as mathematical details (see [1], [3]) we consider in the phase space of classical dynamical systems, the constant energy surface Ω the dynamical group T_t :

$$\omega \in \Omega \longrightarrow T_t \omega = \omega_t \in \Omega$$

and finally the invariant normalized measure μ (e.g. the Liouville measure). The evolution of any square integrable function $\varrho \in L^2_\mu$ is given by

$$\varrho_t(\omega) = \varrho(T_{-t}\,\omega) = (U_t\,\varrho)(\omega)$$

and U_t is a group of unitary operators on L^2_μ. In the case of the hamiltonian dynamics, $U_t = e^{-itL}$ where L is the Liouville operator $L\varrho = i\left[H, \varrho\right]_{B,P}$.

Now, a Markov process in Ω is associated to a semi group $W(t)$, $t \geq 0$ of contraction operators on L^2_μ (i.e. $\|W_t\|_{oP} \leq 1$ such that

i) W_t preverves the positivity (i.e. if $f(\omega)$ is positive then $W_t\,f$ is)

ii) W_t preserves the equilibrium distribution 1 i.e. $W_t\,1 = 1$

The transition probability of ω to $\Delta \subset \Omega$ is then given by :

$$P(t, \omega, \Delta) = (W_t\,\varphi_\Delta)(\omega)$$

where φ_Δ is the characteristic function of Δ. As well known a probability measure $\tilde{\varrho}_{(\Delta)}$ given by the density $\tilde{\varrho}(\omega)$ (i.e. $\tilde{\varrho}_{(\Delta)} = \int_\Delta \varrho(\omega)\,d\mu$ evolves by the Markov process following :

$$\tilde{\varrho}_t(\Delta) = \int \tilde{\varrho}(\omega)\,P(t, \omega, \Delta)\,d\mu$$

or equivalently :

$$\varrho_t(\omega) = (W_t^* \tilde{\varrho})(\omega)$$

We require moreover that the microcanonical ensemble 1 must be invariant under the Markov process and that for all initial state ϱ , $\| W_t^* \varrho \|^2$ decreases monotonically as t goes to $+\infty$ and tends to $\| W_t^* 1 \|^2$.

A dynamics U_t is intrinsically random when it possesses a Λ transformation such that it has essentially the following properties :

1) Λ preserves the set of states
2) $\Lambda U_t \Lambda^{-1}$ extends to a Markov processes W^*_t
3) Λ preserves the equilibrium : $\Lambda 1 = 1$

An important consequence of the incorporation of the second law in the frame of the dynamics is the change of the usual description. Now observables are no longer functions on Ω , but operators on L^2_μ not necessarily reducible to functions ; in particular neither Λ nor M can be given by fonctions on Ω . The space L^2_μ plays a fundamental role as a support of the observables. This expresses the fact that, on account of the instability one cannot approximate a trajectory with the motion of small volume elements having finite extension in Ω , and therefore trajectories are no longer observable.

The main result of this approach is that the class of intrinsic stochastic dynamical systems is not empty and it contains all classical systems that are Bernouilli systems (with discrete time). The evolution is then given by the transformation B and $T_n = B^n$ In this case Λ can be constructed explicitly as follows :

There exists a complete family of orthogonal projection operators $\{E_n\}$ in the orthocomplement of $1, n$ extending from $-\infty$ to $+\infty$, such that :

$$U E_n U^{-1} = E_{n+1}$$

An important result here is the existence of a Time operator T :

$$T = \sum_{n=-\infty}^{+\infty} n E_n$$

having the property :

$$U^{-n} T U^n f = (T + n) f$$

which means that if f_n is an eigenfunction of T for the eigenvalue n (i.e. $T f_n = n f_n$) then $U^K f_n$ is an eigenfunction of T for the eigenvalue $n + k$. It suffices now to take Λ as a positive decreasing function of T :

$$\Lambda = \sum_{n=-\infty}^{+\infty} \lambda_n E_n + P_o$$

where P_o is the projector on 1 and $\{\lambda_n\}$ is a decreasing sequence of positive numbers tending to zero as $n \to +\infty$. We have now two theorems for B-systems :

I) If $\varrho(\omega) \geq 0$ then $(\Lambda \varrho)(\omega) > 0$ and $\int_\Omega (\Lambda \varrho)(\omega)\, d\mu = 1$

II) If $\mu_i = \dfrac{\lambda_{i+1}}{\lambda_i}$ is a decreasing sequence then $(\Lambda U \Lambda^{-1} \varrho)(\omega) > 0$ for all $\varrho(\omega) \geq 0$.

We should stress here that this last property does not hold for $t \leq 0$ and $\Lambda U_t \Lambda^{-1} \varrho$ may yield negative probability. This expresses the symmetry breaking with respect to time.

It is possible to show that the time operator describes in fact the approach of the distributions to the uniformity in a precise sense, but we have no place here to describe this and other related results (see [3], [5].

On the quantum level the implications of the irreversibility are worth of some discussion. Here also, we express the irreversibility by the existence of a Liapounov functional. The question therefore is : does quantum mechanics, formulated in the frame of wave functions, admit the existence of such a Liapounov variable. The answer is that the existence of a Liapounov variable is incompatible with the description in terms of the wave functions :

THEOREM : Let $\mathcal{H}$ be the Hilbert space of the wave functions evolving by Schröedinger evolution :

$$\Psi_t = e^{-itH}\Psi \quad , \quad \Psi \in \mathcal{H}$$

where H is the (positive) hamiltonian. If there exists a (bounded) entropy operator $M \geq 0$ such that

$$\Omega\left(\Psi_t\right) = \langle \Psi_t \, , M\Psi_t \rangle$$

is a Liapounov functional i.e. $i\left[H, M\right] = D \geq 0$ and such that entropy operator M and the entropy production operator D are compatible observables (i.e. $\left[D, M\right] = 0$) then $D = 0$

This theorem justifies the need for density matrices in order to incorporate irreversible processes in quantum mechanics. In fact a mixture, ϱ , evolves under the von-Neumann equation :

$$\frac{\partial \varrho_t}{\partial t} = -i L \varrho_t \quad , \quad L\varrho = [H , \varrho]$$

and L is no longer positive. Now, M may exist as a positive decreasing function of a time operator T ; defined as the conjugate of L. Indeed, this operator, which cannot exists in the standard quantum mechanics, may exists for density matrices :

THEOREM : If H has (absolutly) continuous spectrum extending from 0 to $+\infty$, then a time operator T may exists such that $[L , T] = i I$ and entropy operator is a decreasing function of T :

$$M = f(T)$$

Let us end by these two theorems :

I - If M exists then it cannot be factorizable ; i.e. M cannot have the form :

$$M = A_1 \varrho A_2$$

where A_1 and A_2 are operators on $\mathcal{H}$

II - If M preserves the pure states then $D = 0$.

These theorems mean that it is possible to define a new evolution group $\tilde{U}_t = \Lambda e^{-itL} \Lambda^{-1}$ on the mixed states which does not preserve the pure states. The paradox of the reduction of the wave packet is therefore solved as measurement is described by an irreversible process.

ACKNOWLEDGMENTS

I wish to thank Professors Prigogine and Misra for elaboration of ideas summarized in this note.

I acknowledge the financial support of the Actions Concertées of the Belgian Government.

REFERENCES

[1] B. Misra, I. Prigogine and M. Courbage : From deterministic dynamics to probabilistic descriptions. Proc. Natl. Acad. Sci. USA 76, p. 3607-3611, 1979 (see also Physica, to appear).

[2] B. Misra, I. Prigogine and M. Courbage : Liapounov variable, Entropy and Measurement in Quantum Mechanics to appear in Proc. Natl. Acad. Sci.

[3] M. Courbage and B. Misra : On the intrinsic randomness of classical dynamical systems to appear in J. Stat. Phys.

[4] B. Misra, Proc. Natl. Acad. Sci USA 75, 1627-1631 (1978).

[5] M. Courbage, C. Coutsomitras and B. Misra : On the of the correspondance of the deterministic dynamics and Markov processes. To appear.

IRREVERSIBILITY VERSUS HAMILTONIAN DYNAMICS IN STATISTICAL THERMODYNAMICS

Francis FER

Ecole des Mines de Paris
60, Brd St-Michel 75006 PARIS

1.- INCOMPATIBILITY OF THERMODYNAMIC IRREVERSIBILITY AND HAMILTONIAN CHARACTER.

In the frame of that conception of Microphysics which attaches a major importance to transient processes,[1] let us formulate the following problem for the systems of statistical Thermodynamics: is the hamiltonian tool capable of explaining the irreversibility of natural processes?

The answer is quite disputable as long as the word "irreversibility" has not been given a definite content. Letting aside such definitions as "privileged evolution" or some others of the same kind, I shall use until further notice the word irreversibility *in the quantitative sense assigned to it by macroscopic thermodynamics: an isolated system undergoes an irreversible evolution if and only if its entropy increases strictly:* $dS/dt > 0$.*

* for a non isolated system the definition is slightly more elaborate, but we shall not complicate the problem by treating this case

Consider then a hamiltonian *isolated* system, of energy U. Let us call P the phase-point (coordinates q^i, p_i, $i = 1$ to N), Γ as usual the phase space, $H(P) = H(q^i, p_i)$ the hamiltonian function, Σ the hypersurface $H(P) = U$, and Ω the classical measure induced on Σ by the volume measure of the extension-in-phase. As is known, the motion on Σ is measure-preserving.

Now if a quantity S pretends to play the part of the entropy of a system, it must have the three following characteristics:

a) it is a function of the macroscopic state of the system. As one microscopic state P (q^i, p_i), whatever be the correspondence we can imagine between macroscopic and microscopic states, corresponds to one and only one macro-state, then *entropy is also a phase-function* $S(q^i, p_i)$. In addition to this property derived from physical requirements, we require for the entropy to be a measurable - with respect to the measure Ω - function of q^i, p_i. This is a mathematical condition which is not very restrictive, and without which we could not calculate.

b) also because of macroscopic requirements, entropy is a function of the temperature of the system, or of the different temperatures of its macroscopic elements. As such, entropy is an even function of the ensemble of the velocities $\dot{q}^i$, as is easily seen (if, naturally, we trust in the usual interpretation of the temperature). Therefore, if there are no electromagnetic forces between the molecules of the system, entropy is an even function of the momenta p_i. This is still true if the molecules interact electromagnetically - *which is the general physical case* - because the momenta of the form $m\vec{v} + e\vec{A}$ change sign with velocities and the correspondence between the ensemble of the $\dot{q}^i$'s and the ensemble of the p_i's is bijective (except may be for a subset of Σ of measure zero). However this second statement supposes also essentially that we can consider the velocity of propagation of the electromagnetic field as infinite, an approximation we admit for the moment: otherwise the motion is no more hamiltonian (the accelerations at a time t would depend on positions and velocities at times anterior to t). Then, *for all cases of physical interest, entropy must be an even function of the momenta* p_i.

c) lastly entropy must increase (strictly) with the time in all macroscopic natural evolutions, as already stated.

We can now state the following result.[2]

For any hamiltonian system of physical interest, it is impossible to find an integrable function f(P) which has the three properties a, b, c, of an entropy. More precisely:

- given any integrable function $f(P)$, its time-increase averaged over Σ is zero;
- moreover, if the hamiltonian $H(q^i, p_i)$ is an even function of the p_i's - which is the general physical case - and if the function $f(P)$ satisfies to the condition b of an entropy, the measure of the subset of Σ where f is strictly decreasing is equal to the measure of the subset where f is strictly increasing.

Proof.- Although elementary, I give it in detail because of its further inferences.

For the sake of *really needed* generality, we do not suppose that $f(P)$ is derivable. We therefore choose a finite interval of time τ over which we evaluate the increase of the function $f(P)$ by

$$\Delta_\tau f(P) = f(T_\tau P) - f(P) \tag{1}$$

where T_τ denotes the transformation on Σ due to the hamiltonian motion. I first assert that

$$\int_\Sigma \Delta_\tau f(P)\, d\Omega = 0 \tag{2}$$

In effect the left-hand member is

$$\int_\Sigma f(T_\tau P)\, d\Omega - \int_\Sigma f(P)\, d\Omega \tag{3}$$

Now in the first integral $d\Omega$ is the measure of a subset σ surrounding the point P, not $T_\tau P$; but, as T_τ is measure-preserving, $d\Omega$ is also the measure of the subset $T_\tau\,\sigma$ surrounding the point $Q = T_\tau P$. Thus the first integral (3) may be written $\int_\Sigma f(Q)\, d\Omega$ and equals the second integral; hence the equality (2).

Therefore the hypersurface Σ is divided in three parts: Σ_0 where $\Delta_\tau f(P) = 0$, Σ_1 where $\Delta_\tau f(P) > 0$, and Σ_1' where $\Delta_\tau f(P) < 0$; these three subsets are measurable. Consider then the case when $H(q^i,p_i)$ is an even function of the p_i's.

We can obviously call symmetry the mapping $P \to P'$, where P represents a set of coordinates $\{q^i,p_i\}$ and P' the set $\{q^i,\ -p_i\}$ for all i. This symmetry preserves Σ and likewise the measure of any subset of Σ. On the other hand it is easily seen, on the Hamilton equations, that if P_1P_2 is an arc of trajectory described from P_1 to P_2 when the time increases from 0 to τ, the symmetrical arc $P_1'P_2'$ is also an arc of trajectory, but described from P_2' to P_1' when the time increases from 0 to τ. Thus, if $f(P)$ is an even function of the p_i's, the definition (1) leads to the equality $\Delta_\tau f(P_2') = -\,\Delta_\tau f(P_1)$. Hence if σ is a measurable subset of Σ where $\Delta_\tau f(P) > 0$, the symmetrical set $(T_\tau\sigma)'$ of the transformed subset $T_\tau\sigma$ is such that $\Delta_\tau f(Q) < 0$ for any $Q \in (T_\tau\sigma)'$; but, as the measures of σ, $T_\tau\sigma$ and $(T_\tau\sigma)'$ are equal, one sees that to every subset of Σ on which f is strictly increasing there corresponds a subset, of exactly the same measure, on which f is strictly decreasing (and by the same amount). Q.E.D.

Of course the proof would be much easier if $f(P)$ were derivable but, for a reason which will appear hereafter, it is necessary to avoid this restriction.

Lastly to complete the physical signification of the theorem, it must be observed that the hypothesis that $H(q^i,p_i)$ is an even function of the p_i's is the only one which corresponds to the reality: if there are no electromagnetic forces in the system, H depends on the p_i's by the only intermediate of the kinetic energy, which is a quadratic form in p_i; and if electromagnetic forces are present (always with the approximation that c is infinite), the magnetic energy is still a quadratic form of the q^i's, hence of the p_i's.

It is clear that this theorem is a revival of well-known Loschmidt's objection* to Boltzmann, but reinforced by the consideration of the measures on Σ ; and from this point of view the importance of the equality of the measures of the subsets Σ_1 and Σ_1' where any ΔS has opposite signs must be emphasized. Because of this equality it becomes impossible to expect that a macroscopic situation implying $\Delta S < 0$ cannot be practically obtained because its measure would be very small, whereas the macro-states for which $\Delta S > 0$ would be met sufficiently often. *Both areas being equal, the corresponding macro-states occur equally often;* or then we should suppose that any given temperature can be obtained with only the half of the set of all possible p_i's - a case uneasily admissible, and which would oblige to abandon even Maxwell's law of distribution of velocities.

Let us now make a last remark about the subject. Many authors[3] think that it is possible to dodge Loschmidt's objection by the means of the following argument. Owing to the errors of experimental measurements, a given macroscopic state is represented, not by one phase-point of Σ, but by many such points constituting a subset of strictly positive measure, sometimes called a "constellation". Define then a suitable entropy as a function constant over each constellation. Whereas Loschmidt's objection is doubtlessly valid for any single phase-point, one may hope it is no more true for constellations, for the following reason: due to the dispersion of the points of a constellation between many others during the motion, the mathematical expectation of the variation of the entropy could keep the same sign over all constellations.

* that is, the reversal of momenta at any time reverses the course of the trajectory (always under the hypothesis that $H(q^i, p_i)$ is an even function of the p_i's).

But the above theorem is equally valid for functions of constellation, even if the constellations have very distorted forms (provided they are measurable) and despite the fact that entropy, as a function of constellation, is not derivable everywhere; this is why it was necessary to dismiss the restriction of derivability.

2.- SOME PROBLEMS RAISED BY THE PRECEDING RESULT

Let us now discuss the difficulties which arise from the - real or apparent - contradiction between the above theorem and some previous results.

The first of these latter is obviously Boltzmann's *H*-theorem. As is known, the quantity *H* of Boltzmann (which has nothing to do with the hamiltonian, in spite of the identity of notations) is the integral of a suitable function of the density of particles (supposed identical) in the μ-space of one particle, say $h(xyz,p_x p_y p_z)$. It is clear that the datum of the phase-point on the hypersurface Σ determines the position and the momentum of every particle, hence the value of the function h at any point of the μ-space. Therefore H is a phase-function (or a function of constellation, which does not change anything, as already seen) on the hypersurface Σ. Now Boltzmann's theorem asserts that the average value $\overline{(dH/dt)}$ is always negative ($\leqslant 0$) (so that H could represent, save the sign, an entropy). Then there seems to be a contradiction with the above result.

But a careful examination of the proof of Boltzmann's theorem shows that this proof rests not only on the hamiltonian character of the motion, but also on the hypothesis of the so-called "molecular chaos". Now nobody - even Boltzmann - has ever given a precise definition of the molecular chaos: it is an intuitive - and rather physically attractive - notion, but not an axiomatic one. And, on the other hand, Boltzmann concedes that, if one reverses the velocities of the particles of a gas, the hypothesis of the molecular chaos is no more satisfied. Can it be said better that the molecular chaos is an extra-hamiltonian hypothesis, and accordingly that Boltzmann's H-theorem itself is non-hamiltonian? Then there is no contradiction with the theorem of paragraph 1, and Boltzmann's theorem is not to be challenged.

Now, and as if two proofs were better than one, one has tried to supplement Boltzmann'stheorem by the so-called "generalized $\bar{H}$-theorem". Here $\bar{H}$ is the integral, *over all the hypersurface* Σ, of a suitable function of a density of probability which is assigned to the system, and the theorem asserts (on the basis of hamiltonian motion) that $\bar{H}(t_2) \leqslant \bar{H}(t_1)$ for $t_2 > t_1$. This suggests to assimilate $-\bar{H}$ to the entropy, and seems yet to raise a contradiction with the result of paragraph 1.

But, firstly, the quantity $\bar{H}$ is not a function of state*, neither macroscopic nor microscopic, so that it cannot at all represent a true entropy and does not interfere with the result of paragraph 1. Secundly, a careful examination of the demonstration shows that the decrease of $\bar{H}$ is proved only for a short while following the initial time, itself characterized by the possession of a maximum of information about probabilities, and that this decrease proceeds only from a loss of knowledge. $\bar{H}$ is then an entropy of information and not a physical entropy.

I leave aside, as bringing us nothing new, a lot of calculations of the kinetic theory of fluids in irreversible transformation, because they all call for probabilistic hypotheses which cannot be derived - which have at any rate not been derived until now - from the hamiltonian formalism.

3.- IRREVERSIBILITY AND ERGODICITY

We arrive now to a more puzzling, and also more interesting, question: the compatibility or incompatibility between the theorem of paragraph 1 and the property of ergodicity in hamiltonian systems.

* even the integrand of $\bar{H}$ is not a phase-function: it depends on the time.

Let us recall briefly the definition of ergodicity. Given a hamiltonian isolated system and an integrable phase-function $f(P)$, consider any trajectory on the hypersurface Σ of energy U issued from a point P_o. A well-known theorem of Birkhoff asserts that the time-average

$$\hat{f}(P_o) = \lim_{T\to\infty} \frac{1}{T} \int_0^T f(T_t P_o)\, dt \qquad (4)$$

(where T_t is the transformation due to the motion, as above) exists for almost every P_o and depends only on the trajectory, not on P_o. The hamiltonian system is said to be ergodic if, *whatever be the function f*, $\hat{f}$ has the same value for all trajectories of Σ, in which case it is equal to the phase-average $\int_\Sigma f(P)\, d\Omega$.

It has a long time been an open question whether hamiltonian systems are ergodic, or at least some of them. If we refer to the works of the soviet school, those of Sinai[4] in particular, we can to-day consider as a certainty that the motion of the gas of Boltzmann, that is the gas represented by hard spheres colliding elastically in a vessel, is ergodic. There are counter-examples for other sorts of hamiltonian systems; moreover the model of hard spheres implies discontinuities of the velocities at every collision, which is a too schematic picture of the physical reality, and one can wonder whether the ergodicity would hold yet when restoring the continuity of the motion. But the existence of *one* ergodic example is sufficient to raise the problem of the compatibility to the full, and we have then to examine it.

The property that time-averages, *for any function, are independant of the trajectory, therefore of initial conditions,* seems to many people to deserve the name of "irreversibility", in that sense: the system undergoes an evolution which possesses a "privileged tendancy", and this tendancy cannot be reversed (even by such a manipulation as the reversal of velocities: this operation can delay the final outcome, but not modify it).

Then, if this "irreversibility" would be the same that that I treated in paragraph 1, there would be a contradiction.

But it is not the case. The irreversibility defined by the entropy criterion - the only valid from the thermodynamical point of view - does not coincide with, and is not implied by, the "tendency" which is expressed by the property of ergodicity.

According to the importance of the distinction, it is necessary to enlighten it by simple examples.

Let us take for the first a process which, although outside the field of Dynamics, has exactly the same property of ergodicity as above, I mean a Markov process. Let E_j $(j = 1$ to $n)$ a finite number of states of a system and P_k^j the transition matrix from the state E_j to the state E_k in one step. It is known that, under suitable conditions satisfied by the matrix P_k^j, the process is ergodic, that is, the probability of occurrence of E_j tends to a limit $\bar{\omega}_j$ when the number of steps tends to infinity; $\bar{\omega}_j$ is the unique solution of the equations

$$\Sigma_j P_k^j \bar{\omega}_j = \bar{\omega}_k \tag{5}$$

and is independant of the initial state of the system. Moreover the evolution of the probability of each state cannot be reversed.

However there exists no *state* function whose mathematical expectation value of increase at any step has a given sign, and the proof is elementary. Let $g_k = g(E_k)$ a function of state; its average variation for one step issuing from the state E_l is

$$\overline{\Delta g_l} = \Sigma_k P_k^l g_k - g_l$$

By multiplying both members by $\bar{\omega}_l$ and using (5) one gets

$$\Sigma_l \bar{\omega}_l \overline{\Delta g_l} = 0$$

As $\bar{\omega}_l \geqslant 0$ it is impossible that $\overline{\Delta g_l}$ has the same sign for every state E_l(except if $\overline{\Delta g_l} = 0$ $\forall l$, a case without any interest).

We cannot therefore attach to an ergodic Markov process any state function whose behaviour be that of an entropy in an irreversible phenomenon.

It is known however that one can attach to such a process an entropy of information, namely $-\Sigma_j p_j \log p_j$, p_j being the probability of the state E_j at a given stage of the process, and that this sort of entropy is increasing. But it is clear that it is not a state function, but a function of all the possible states of the system; it is exactly the same case as for the function $\overline{H}$ of paragraph 2.

We have then a first example of a type of evolution which is perfectly ergodic and shows an irrepressible tendancy, and which is not irreversible in the thermodynamic sense of the word.

Let us take now a second example which is as instructive, in another respect, the "small planets" of H. Poincaré. Poincaré considers a great number of material points (the planets) describing the same circle, with angular velocities constant in time but differing possibly from one planet to another, and without collisions between them.* This type of motion has a close connection with the motion of particles in a vessel (which makes its interest), and it is easy to see that it is hamiltonian. Then Poincaré shows that, whatever be the initial conditions prescribed to the planets - all of them gathered at the same longitude for $t = 0$, for instance, they will be, for $t = +\infty$, spread uniformly over the circle. This conclusion may appear, at first sight, as a plain illustration of "irreversibility", in the sense of a privileged tendancy of the motion.

But, first of all, this conclusion cannot be accepted, from a physical point of view, because it rests on the hypothesis, made by Poincaré, that the great number of planets authorizes us to handle the spec-

* The fictitious character of this last assumption has no importance for the prupose of the study.

trum of their velocities as if it were continuous. Now the asymptotic behaviour of a set of such planets is essentially different according to the spectrum of the velocities. If it were really continuous, we would have the conclusion of Poincaré; but if it is discrete, *as in reality*, the distribution of the planets tends no more towards a limit as $t \to +\infty$, but remains an almost-periodic function of the time. As for an illustration of irreversibility, the difference speaks by itself.

Nevertheless the case of an almost-periodic distribution has a property which reminds of a privileged tendancy. It is easy to see that the time-average of the density of the planets (calculated as in the formula (4), that is for $T \to \infty$) is independant of the longitude. Therefore, although the motion of the planets is not ergodic in the full sense of the word (because the planets have no interaction; it is a well-known result of the ergodic theory), there is *one* function whose time-average is independant of the position on the circle and then equal to the position-average.

Now, as the motion is hamiltonian, the theorem of paragraph 1 applies fully, and it is impossible to the system a function entropy.

The two examples examined, and it would be possible to find some others, show that *we must distinguish carefully* between the "privileged tendancy" of an evolution*, even when this somewhat vague notion is rendered more precise by the property of ergodicity, and the irreversibility characterized by the entropy criterion. That is why, henceforth and as long as I remain at the macroscopic level, I shall use the word irreversibility only in this thermodynamical meaning.

We can proceed a little further. Let us return to the time-average (4) of a phase-function $f(P)$. It is known that, for hamiltonian systems

* Macroscopic thermodynamics itself often fails to give a more precise qualitative definition of irreversibility, which is however possible and leads, together with the second principle, to the entropy criterion.

such as isolated ones, any trajectory on the hypersurface Σ passes an infinitely many times near any one of its points: this is the recurrence theorem of Poincaré. One knows on the other hand, since the answer of Boltzmann to Zermelo, that the "time of return" is, for a macroscopic system such as 1 cm^3 of matter, immensely long, say $10^{(10^{10})}$ years to fix ideas. Although this value is only an estimate (and, moreover, an upper bound) we can trust in this order of magnitude, which can be corroborated by easier calculations made on almost-periodic functions, such as in the case of the small planets of Poincaré. Therefore the time-averages (4) can be reached only by integrating over intervals of time of the same order of magnitude.

Now the irreversible phenomena we can see every day are on the contrary very short: seconds or minutes for the Joule expansion of a gas, hours for these "slow" phenomena that are diffusion or thermal conductibility.

Therefore, even if the hamiltonian mechanics were adequate to the physical reality in macroscopic irreversible processes as well as in equilibria, it would occur, in the formation of the averages (4) and due to the infinitely short duration of the irreversible stage in comparison with the practical while of integration, *that the influence of the irreversible stage would be completely deleted. In fact the hamiltonian formalism, especially when it uses time-averages and, a fortiori, phase-averages and ergodicity, can represent only stationary regimes - that is, macroscopic equilibria.*

One can presume, moreover, that the subsets of the hypersurface whose phase-points correspond to non-equilibrium macroscopic states have a measure much smaller than the subset which corresponds to equilibrium. If not, of course, a logical inference, this may be taken as a fairly reliable assertion.

4.- THE IRREVERSIBILITY AS A CONSEQUENCE OF THE RETARDATION OF ELECTROMAGNETIC POTENTIALS AND HEREDITARY MECHANICS

If we accept the idea that hamiltonian dynamics can absolutely not account for the macroscopic irreversibility, then we have to look for mechanics which will do it in its place. I set forth here only the principles of this replacement; for mathematical results which derive from these principles, the reader will refer to the paper of D. Fargue in this volume.

Abandoning hamiltonian mechanics, we have no more reason to locate the origin of irreversibility in the great number of particles, and the most natural way is to place it *in the elementary processes, that is at the microscopic level.*

But doing so we lack unfortunately for a precise definition of irreversibility, since until now we used only the entropy criterion, which is a macroscopic attribute. We have therefore to choose a definition of the microscopic irreversibility, for which I propose the following: a (mechanical) motion is irreversible if we cannot reverse it without a *finite change* of external forces. It is easily seen that this definition is only an extension of the macroscopic *qualitative* definition of irreversibility, this one which leads, together with the second principle of Thermodynamics, to the entropy criterion. One sees also that the physically interesting hamiltonian motions do not possess this property.

There are certainly lots of non-hamiltonian dynamics which, on the contrary, possess it (for instance dynamics with limit-cycles). But it is not sufficient to find abstract mathematical models; we must look for microscopic irreversible processes which exist in nature.

From this point of view we know only *one* elementary phenomenon which possesses the irreversibility just defined: it is the emission of a radiation by an accelerated charge, emission which involves in many cases* a damping of the charge.

In particular when a charge q collides with another one, it undergoes a loss of kinetic energy the value of which has been established by H. A. Lorentz and is $(\mu_o\, q^2/6\pi c)\int\gamma^2 dt$ (in M.K.S.A. units, γ acceleration, μ_o permeability of the vacuum, and the integration being performed during the practical time of collision). If the colliding particles are composed with several charges, one must add the energy losses of each component. The kinetic energy so dissipated is converted into an electromagnetic radiant energy which withdraws from the place of the collision and, in the case of a gas enclosed in a vessel for example, is absorbed by the walls or by other molecules. The loss of kinetic energy is indeed an irreversible process because, if one would reverse the colliding trajectory, one should supply to the charge twice the above expression, hence a finite quantity.

Naturally the energy absorbed by the walls modifies the electromagnetic field which they emit, and this supplementary field is capable to re-accelerate the molecules of the gas and thus to send back to them a part or the totality of the energy they lost. This causes that the system of interest - the gas - can be no more considered as perfectly isolated: the energy hypersurface of the hamiltonian motion is transformed in a shell of some thickness, but we can probably, at least for the moment, neglect this effect.

The process thus put forward may seem, at a first sight, so small that it would be practically negligible; and in fact the loss of energy is extraordinarily small: one can take as a rough estimate a value of 10^{-33}J per collision for an atom of argon in the ordinary conditions of pressure and temperature[2]. But there are an enormous number of collisions per molecule (say 10^{10}/s) and an enormous number of molecules in

* at least in these two important cases: closed trajectories and trajectories of collision.

a cm^3 of matter. So that the loss of energy calculated in this manner has exactly the same order of magnitude that the loss, calculated by Stefan's formula, which would be suffered by a gas enclosed in a vessel with perfectly transparent walls.

Now many people - and not of the least* - if they can contemplate the process just exhibited as a possible explanation of irreversibility - because drawing an irreversibility from reversible motions has finally never convinced anybody - can hardly conceive the reverse process, that is, how an unceasing emission of radiation during the collisions can end in an equilibrium.

There is actually a difficulty, at the first sight. One knows quite well since Lorentz the process of emission by an accelerated charge, and it is nearly easy to estimate its intensity and its duration; but the phenomenon of absorption of a very fluctuating field by wandering molecules has been less studied, and it may seem that the lost radiations will be badly recovered.

There is however a current phenomenon of which nobody thinks of giving another explanation, I mean the equilibrium which sets up between two bodies which are only coupled by thermal radiation, for example the system constituted by a container the walls of which are at some temperature and a second body inside it at a different temperature, with the vacuum between them. Nobody doubts that the quicker velocities of the hotter body will transfer themselves to the colder one by the medium of the electromagnetic field, and everyone agrees that the equilibrium is achieved, not when both emissions cease, but when they balance. There is no reason why a mechanism admitted for solids at more or less high temperatures would not be valid for gases at the ordinary temperature.

* Einstein[5] for example considered that "if the elementary processes would be dependant of the time-arrow, then the achievement of thermodynamical equilibrium would be quite incomprehensible".

Furthermore a rough estimate shows that the absorption of the damping radiation by molecules passing in the neighbourhood of a collision is not so unlikely. For electrons going to meet one another on the same line with velocities corresponding to 0°C, the duration of the collision is very small: $\sim 10^{-10}$s, and the emitted wave-packet initially quite short: $\sim$ 2 cm; so that, although the loss of energy is very small (10^{-33}J as already seen), the average electric field just after the emission is yet sufficiently high (10^{-9}V) to accelerate another electron and give it again an energy of comparable value.

Now we cannot stop at this point of merely employing, when the opportunity offers, Lorentz's formula, because its somewhat strange expression calls for some physical deepening.

As shows the original calculation of Lorentz[6], the expression recalled above of the energy loss, or the value of the damping force $(\mu_0\, q^2/6\pi c)(d\vec{\gamma}/dt)$ during the acceleration of a charge, rests upon an expansion in series,of which one keeps only the second term*; the expansion itself is only a manner to calculate the action exerted on the charge, *supposed essentially non punctual*, by the electromagnetic field created by itself; and finally the rigourous value of this action is given by an integral bearing on the retarded potentials created by the different elements of the charge.

Thus one arrives to the conclusion that the present attempt to explain irreversibility at the microscopic level leads to a mechanics which is no more instantaneous as usual, but is an *hereditary mechanics: the motion at the time t is determined, not only by the state of the system at this very instant, but also by the past states of the system.*

One can notice that such mechanics includes in itself by nature the possibility of an explanation of the irreversibility, this word

* the first corresponding to the electromagnetic mass

in the second sense defined above: contrary to hamiltonian (or more generally suitable differential) dynamics, wherethe (conceptual) reversal of velocities is sufficient to reverse the motion, it would be necessary, for hereditary mechanics, to reverse also the whole past, and this is of course impossible, even conceptually (it is this impossibility which implies, in Lorentz's formula, the presence of the square of γ). Naturallythis is not sufficient to explain the macroscopic irreversibility and find again the entropy criterion, and rejoining this latter is perhaps a long and exacting labour; but one can hope the direction is good.

There is furthermore a reason to think that it is not so bad, which is the following: even at the macroscopic level, three *irreversible* phenomena known as transfer phenomena (viscosity, thermal conduction and diffusion), which are theoretically explained by differential equations, depend in reality upon hereditary mechanics. Let us take thermal conduction for example. The usual formulas of the kinetic theory[7] are obtained by introducing in the calculations the thermal kinetic energy attached to an average-value of the free path. Instead, the rigourous calculation would consist in integrating the thermal kinetic energies relative to all the *true* free paths, which would lead then to an integro-differential equation for the heat equation.[8]

5.- CONNECTION BETWEEN HAMILTONIAN AND NON-HAMILTONIAN PICTURES

I allow readily that, contrary to the preceding paragraphs for which some mathematical results are available, here or elsewhere,[9] I can now only outline a working plan. But is it possible to start calculations without first defining a physical aim?

We have seen that the hamiltonian formalism can only represent stationary regimes, that is, macroscopic equilibria, and that the property of ergodicity, far from expressing a true thermodynamic irreversibility, is on the contrary wonderfully adapted to the treatment of these equilibria. On the other hand, it is known that the hamiltonian formalism is experimentally perfectly fitted to this object. It is then clear that this acquisition must be kept carefully.

Consequently, if another, and non-hamiltonian, mechanics pretends to represent the totality of all the natural evolutions, it must firstly rejoin the hamiltonian mechanics in the case of equilibria. This appears to be possible for the hereditary mechanics I proposed above: the equilibrium being reached when a balance has been established between the losses and the recoveries of energy, it ought to be possible to find an equivalence with the hamiltonian dynamics, where these losses and recoveries are simply suppressed.

Now what does happen for irreversible evolutions? We have seen that hamiltonian mechanics cannot account for them; we have seen too that one can reasonably guess that the areas of the hypersurface Σ where the phase-point represents a macroscopic non-equilibrium state are negligible in comparison with the whole surface; and the classical point of view itself agrees with this conclusion since it takes as granted that the non-equilibrium states, though possible, are highly improbable.

But, for the "true" mechanics we are aiming at, and if it is really fitted to the facts, one is lead to think that *the return from the equilibrium state to the non-equilibrium ones is, by the simple work of the equations, impossible*. Thus the effect of the use of the hamiltonian dynamics had been to replace a true impossibility by an improbability. *Perhaps it might be to-day desirable to reverse the step.*

REFERENCES

1.- See the paper of G. LOCHAK in this volume.

2.- FER F.- *L'irréversibilité, fondement de la stabilité du monde physique*. Gauthier-Villars, Paris, 1977.

3.- TOLMAN R.C.- *The principles of statistical mechanics*. Oxford Univ. press, Oxford, 1962.

4.- ARNOLD V.I. & AVEZ A.- *Problèmes ergodiques de la Mécanique classi-*

que. Gauthier-Villars, Paris, 1967, with many references. For more recent works, see SINAI Y. G.- *Russian math. surveys*, 25, 137, 1970 and 28, 49, 1973.

5.- EINSTEIN A.- *A. Einstein et M. Besso, correspondance 1903-1955*. Hermann, Paris, 1972.

6.- LORENTZ H. A.- *The theory of electrons*. Dover, New York, 1952.

7.- CHAPMAN S. and COWLING T. G.- *The mathematical theory of non-uniform gases*. University press, Cambridge, 1952.

8.- FER F.- in *Mélanges*, a book offered to Th. Vogel for his 75th birthday. Editor: Paul Janssens, Université libre de Bruxelles, 1977.

9.- See FARGUE D., in this volume.

IRREVERSIBILITY AND ENVIRONMENTAL FORCES

SALMON Jean
Conservatoire National des Arts et Métiers
292, rue Saint-Martin, 75141 PARIS CEDEX 03

Abstract

Irreversibility in fluids is traditionally introduced by means of Boltzmann's postulate of molecular chaos.

We suggest another way, based on a non-Hamiltonian microdynamics characterized by environmental forces dependent on the velocities.

I – Introduction

The fundamental system of equations of classical statistical mechanics is reversible because it is based on Hamiltonian equations. It brings about the temporal invariance of the entropy of an isolated system which is contrary to the second principle of thermodynamics away from equilibrium.

It was Boltzmann who broke this irreversibility on establishing a kinetic equation ensuring the increase in entropy away from equilibrium and allowing the determination of the transport coefficients [1,2].

We suggest another way, based on the introduction of a non-Hamiltonian micro-dynamics bringing in two opposing environmental forces, one of damping and one of diffusion. A kinetic equation of the Fokker-Planck-Frey-Salmon type is thus obtained and leads to a correct result with respect to the entropy and to a determination of the viscosity coefficient of a dilute gas in good agreement with experiment.

II' – The B.B.G.K.Y. System and Boltzmann's Equation

Let us consider a dilute gas made up of molecules of mass m. t designates the time and $\vec{x}_1, \vec{x}_2, \vec{w}_1, \vec{w}_2$ the position and velocity vectors of the number 1 and 2 particles. $\vec{X}_1$ and $\vec{X}_2$ are the external forces deriving from a potential φ_0, $\vec{X}_{12}$ and $\vec{X}_{21}$ are the interaction forces deriving from a potential φ.

The single and double distribution functions are shown by the symbols F_1 and F_{12}. The low value of the particle density allows us to neglect the triple collisions, and, in these conditions, the system of the first two BBGKY equations is written:

(1) $$\frac{\partial F_1}{\partial t} + \vec{w}_1 \cdot \frac{\partial F_1}{\partial \vec{x}_1} + \frac{\vec{X}_1}{m} \cdot \frac{\partial F_1}{\partial \vec{w}_1} + \int \frac{\vec{X}_{12}}{m} \cdot \frac{\partial F_{12}}{\partial \vec{w}_1} \, \underline{dx}_2 \, \underline{dw}_2 = 0$$

(2) $$\frac{\partial F_{12}}{\partial t} + \vec{w}_1 \cdot \frac{\partial F_{12}}{\partial \vec{x}_1} + \vec{w}_2 \cdot \frac{\partial F_{12}}{\partial \vec{x}_2} + \frac{(\vec{X}_1 + \vec{X}_2)}{m} \cdot \frac{\partial F_{12}}{\partial \vec{w}_1} + \frac{(\vec{X}_2 + \vec{X}_{21})}{m} \cdot \frac{\partial F_{12}}{\partial \vec{w}_2} = 0$$

This reversible system is incompatible with the second principle of thermodynamics. To obtain the transport coefficients, the collisions are considered as local and instantaneous whereas the external forces are considered as weak compared to the interaction forces, whence, on setting:

(3) $$\vec{x}_{12} = \vec{x}_2 - \vec{x}_1$$

we have the following equation in which the integral is only stretched to the interaction domain of the number 1 and 2 particles:

(4) $$\frac{\partial F_1}{\partial t} + \vec{w}_1 \cdot \frac{\partial F_1}{\partial \vec{x}_1} = \int (\vec{w}_2 - \vec{w}_1) \cdot \frac{\partial F_{12}}{\partial \vec{x}_{12}} \, \underline{dx}_{12} \, \underline{dw}_2$$

Let us designate by O the centre of the interaction sphere and by $Oxyz$ a system of axes such that the vector $g_{12} = \vec{w}_2 - \vec{w}_1$ is parallel to, and in the same direction as, Oz. Let us consider a straight line parallel to Oz and lying at a distance b from O. This line cuts the interaction sphere at the points A_E and A_S as well as the plane xOy at the point A of the polar co-ordinates b and ε (Fig. 1):

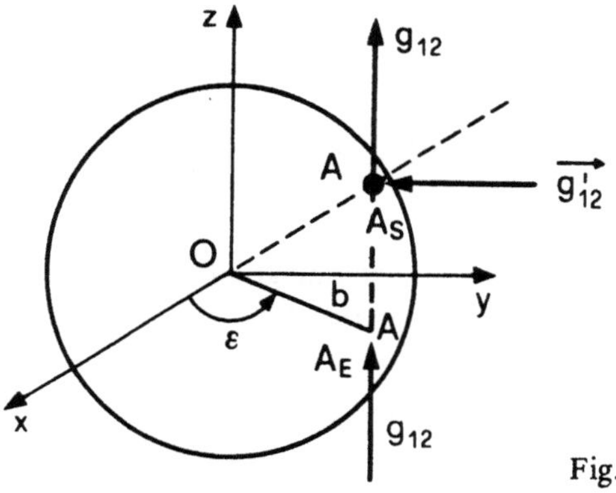

Fig. 1

The point A_E corresponds to the beginning of a collision and the point A_S to the end of a collision, whence:

$$\iiint (\vec{w}_2 - \vec{w}_1) \cdot \frac{\partial F_{12}}{\partial \vec{x}_{12}} \, \underline{dx}_{12} = \iint g_{12} \left[F_{12}(Z_S) - F_{12}(Z_E) \right] b \, db \, d\varepsilon \tag{5}$$

The changing of t in $-t$ and consequently, $\vec{w}_1$ into $\vec{w}_2$ will change the sign of the two sides of the equation (4) if Oz does not change direction. On the other hand, if Oz is made to take the same direction as $\vec{g}_{12}$, only the sign of the left-hand side of (4) will be modified and irreversibility is introduced.

The equation (4) contains both F_1 and F_{12}. The postulate of molecular chaos ensures closure on imposing the loss of information of the particules, thus:

$$Z = Z_E \qquad F_{12} = F_1(\vec{w}_1) \, F_2(\vec{w}_2) \tag{6}$$

$$Z = Z_S \qquad F_{12} = F_1(\vec{w}'_1) \, F_2(\vec{w}'_2) \tag{7}$$

$\vec{w}'_1$ and $\vec{w}'_2$ corresponding to the velocities at the beginning of a collision which lead to the velocities $\vec{w}_1$ and $\vec{w}_2$ at the end of a collision according to the reversible laws of motion of the two bodies. On using (6) and (7), the equation obtained is that of Boltzmann:

$$\frac{\partial F_1}{\partial t} + \vec{w}_1 \cdot \frac{\partial F_1}{\partial \vec{x}_1} = \iiiint\!\!\int g_{12} \left[F_2(\vec{w}'_2) F_1(\vec{w}'_1) - F_2(\vec{w}_2) F_1(\vec{w}_1) \right] b \, db \, d\varepsilon \, \underline{dw}_2 \tag{8}$$

The importance of this equation is considerable. On the one hand, the entropy of an isolated system cannot decrease:

(9) $$S_1 = -K_0 \int \left[\log F_1\right] F_1 \, \underline{dw}_1 \, \underline{dx}_1 \; ; \quad \frac{\partial S_1}{\partial t} \geq 0 \quad (K_0 \text{ Boltzmann's constant})$$

on the other hand, the theoretical determination of the kinetic viscosity coefficient μ is obtained on solving the equation (8).

Let us designate by $-K_0 T_i$ the minimum value of the interparticle potential and by σ the distance which makes the latter vanish. The deviation χ, corresponding to an impact parameter b and to a relative velocity $g = g_{12}$ is:

(10) $$\chi = \pi - 2b \int_{r_d}^{\infty} \left[1 - \frac{b^2}{r^2} - \frac{4\varphi}{mg^2}\right]^{-\frac{1}{2}} \frac{dr}{r^2}$$

(r_d being the root of the quantity in brackets).

From (10) we calculate the integrals:

(11) $$Q_\ell(g) = \frac{2}{\left[1 - \frac{1}{2}\left(\frac{1+(-1)^\ell}{1+\ell}\right)\right]} \int_0^{\infty} \left(1 - \cos^\ell \chi\right) \frac{b\,db}{\sigma^2}$$

(12) $$\Omega^{\ell,s} = \frac{2}{s+1!} \left(\frac{T_i}{T}\right)^{s+2} \int_0^{\infty} \left(\frac{m}{4K_0T_i}\right)^{s+2} Q_\ell(g)\, g^{2s+3} e^{-\frac{mg^2}{4K_0T}} dg .$$

The expression for the viscosity coefficient is then[2]:

(13) $$\mu = \frac{S}{16} \frac{(\pi m K_0 T)^{1/2}}{\pi \sigma^2 \Omega^{2,2}} \left[1 + \frac{3}{196}\left(8 \frac{\Omega^{2,3}}{\Omega^{2,2}} - 7\right)^2\right]$$

An effective potential is that suggested by H.J.M. Hanley and M. Klein[3,4]

(14) $$\varphi = -K_0T_i \left[\frac{6+2\gamma}{m_0-6}\left(\frac{d\sigma}{r}\right)^{m_0} - \frac{[m_0 - \gamma(m_0-8)]}{m_0-6}\left(\frac{d\sigma}{r}\right)^6 - \gamma\left(\frac{d\sigma}{r}\right)^8\right]$$

The agreement with experimental results for Argon is excellent[5]:

$m = 6{,}633\ 10^{-26}$ kg ; $T_i = 152{,}8$ K ; $\sigma = 3{,}297\ 10^{-10}$ m ; $\gamma = 3$

$K_0 = 1{,}3804\ 10^{-23}$; μ in 10^{-6} poise ; $m_0 = 11$.

Table I

T K	μ_{exp}	μ_{th}	$\Delta\mu / \mu$
298,16	227,5	226,3	0,5 %
463,16	232,4	323,1	0,1 %
783,16	471,5	468,4	0,6 %
973,16	546,3	547,7	0,2 %

III – A NON-HAMILTONIAN MICRO-DYNAMICS

Since 1975, we have suggested a hypothesis introducing the irreversibility at the level of the molecular interaction[6,7]. Instead of first of all respecting the postulates of Hamiltonian mechanics and then breaking the past-future symmetry as we have shown in the preceding paragraph, we have chosen to add non-Hamiltonian terms to the fundamental equation of dynamics. The first which seems to impose itself is a damping term $\vec{X}_{12F}$ of the form:

$$\vec{X}_{12F} = -\beta\left[(\vec{w}_1-\vec{v}_1)-(\vec{w}_2-\vec{v}_2)\right] \tag{15}$$

$$n_1 = \int F_1\, d\underline{w}_1 \qquad n_2 = \int F_2\, d\underline{w}_2 \tag{16}$$

$$n_1\vec{v}_1 = \int F_1\, \vec{w}_1\, d\underline{w}_1 \qquad n_2\vec{v}_2 = \int F_2\, \vec{w}_2\, d\underline{w}_2 \tag{17}$$

This damping term must be compensated by a diffusion term in velocity space $\vec{X}_{12d}$ ensuring the continuation of the motion and the velocity at the thermodynamic equilibrium, which leads to the expression:

$$\vec{X}_{12d} = \beta\left[\frac{K_0T}{m}\left(\frac{\partial}{\partial\vec{w}_2}-\frac{\partial}{\partial\vec{w}_1}\right)\log F_{12}\right] \tag{18}$$

The ensemble $\vec{X}_{12F} + \vec{X}_{12d}$ represents an environmental force. The term must vanish for an interparticle Coulomb potential for particles of the same charge in the absence of a neutralizing bottom. Thus β should be proportional to the Laplacien $\Delta\varphi$ of φ. Lastly, the results obtained by means of the hypothesis of linear relaxation show that β must be the product of $\Delta\varphi$ by the average half-duration of the crossing time of the particle of velocity g and of reduced mass $m/2$ through the repulsive part of the interparticle potential of quantum origin.

We thus arrive at an equation of dynamics of the form:

$$m\frac{d\vec{w}_1}{dt} = \vec{X}_{1T} = -\frac{\partial\varphi}{\partial\vec{x}_1} - \tau\Delta\varphi\left[(\vec{w}_1-\vec{v}_1)-(\vec{w}_2-\vec{v}_2)+\frac{K_0T}{m}\left(\frac{\partial}{\partial\vec{w}_1}-\frac{\partial}{\partial\vec{w}_2}\right)\log F_{12}\right] \tag{19}$$

The true Liouville equation is written:

$$\frac{\partial F_{12}}{dt}+\vec{w}_1\cdot\frac{\partial F_{12}}{d\vec{x}_1}+\vec{w}_2\cdot\frac{\partial F_{12}}{d\vec{x}_2}+\frac{\partial}{\partial\vec{w}_1}\cdot\frac{(\vec{X}_{1T}F_{12})}{m}+\frac{\partial}{\partial\vec{w}_2}\cdot\frac{(\vec{X}_{2T}F_{12})}{m}=0 \tag{20}$$

whence, on fully writing out the irreversibility kinetic equation governing F_{12} :

$$\frac{\partial F_{12}}{\partial t}+\vec{w}_1\cdot\frac{\partial F_{12}}{\partial\vec{x}_1}+\vec{w}_2\cdot\frac{\partial F_{12}}{\partial\vec{x}_2}-\frac{1}{m}\frac{\partial\varphi}{\partial\vec{x}_1}\cdot\frac{\partial F_{12}}{\partial\vec{w}_1}-\frac{1}{m}\frac{\partial\varphi}{\partial\vec{x}_2}\cdot\frac{\partial F_{12}}{\partial\vec{w}_2}=$$

$$\frac{\tau\Delta\varphi}{m}\left[\frac{\partial}{\partial\vec{w}_1}\cdot\left[(\vec{w}_1-\vec{v}_1)F_{12}-(\vec{w}_2-\vec{v}_2)F_{12}+\frac{K_0T}{m}\left(\frac{\partial F_{12}}{\partial\vec{w}_1}-\frac{\partial F_{12}}{\partial\vec{w}_2}\right)\right]\right]+$$

$$+\frac{\tau\Delta\varphi}{m}\left[\frac{\partial}{\partial\vec{w}_2}\cdot\left[(\vec{w}_2-\vec{v}_2)F_{12}-(\vec{w}_1-\vec{v}_1)F_{12}+\frac{K_0T}{m}\left(\frac{\partial F_{12}}{\partial\vec{w}_2}-\frac{\partial F_{12}}{\partial\vec{w}_1}\right)\right]\right] \tag{21}$$

If two particles are in presence, the distribution functions are the Dirac type, thus Maxwellian at zero temperature. The irreversible term disappears and the traditional motion of the two bodies is re-established. Let us return to the case of a population of particles and in right-hand side of the equation (21) let us use the approximation:

$$F_{12} = F_1 F_2 e^{-\frac{\varphi}{K_0T}} \tag{22}$$

For a dilute gas, the position of the functions F_1 and F_2 varies little within the interaction sphere. Let us multiply (21) by $\underline{dx_2}$, $\underline{dw_2}$ and integrate. We then have the Frey-Salmon kinetic equation[8,9] :

$$\frac{\partial F_1}{\partial t} + \vec{w}_1 \cdot \frac{\partial F_1}{\partial \vec{x}_1} + \frac{\dot{\vec{X}}_1}{m} \cdot \frac{\partial F_1}{\partial \vec{w}_1} = \frac{n K_0 T Z B}{2m} \frac{\partial}{\partial \vec{w}_1} \cdot \left[(\vec{w}_1 - \vec{v}_1) F_1 + \frac{K_0 T}{m} \frac{\partial F_1}{\partial \vec{w}_1} \right] \quad (23)$$

$$B = \frac{8\pi}{3 K_0^2 T^2} \int_0^{\infty} \left(\frac{d\varphi}{dr} \right)^2 e^{-\frac{\varphi}{K_0 T}} r^2 dr = \frac{8}{3 K_0 T} \int \Delta \varphi e^{-\frac{\varphi}{K_0 T}} r^2 dr \quad (24)$$

The non-decrease is entropy S_1 is ensured. Indeed, on designating the Maxwellian distribution by F_1^M, it follows that:

$$S_1 = -K_0 \int F_1 \log F_1 \underline{dx}_1 \underline{dw}_1 \quad F_1 = F_1^M h \quad \frac{\partial S_1}{\partial t} = \frac{n K_0^3 T^2}{2 m^2} \tau B \int \left(\frac{\partial h_1}{\partial \vec{w}_1} \right)^2 \frac{F_1^M}{h_1} \underline{dx}_1 \underline{dw}_1 \geq 0$$

(25)

The expression for the viscosity coefficient is:

$$\mu = \frac{m}{\tau B} \quad (26)$$

with for τ : (Fig. 2)

$$\tau = \int_0^{\sigma} \int_0^{\infty} \left[\int_0^{\sigma} \left(g^2 \left(1 - \frac{b^2}{r^2} \right) - \frac{4}{m} \varphi(r) \right)^{1/2} dr \right] \left[\frac{2\pi b}{r \sigma^2} \left(\frac{m}{4\pi K_0 T} \right)^{3/2} e^{-\frac{m g^2}{4 K_0 T}} 4\pi g^2 \, db \, dg \right]$$

(27)

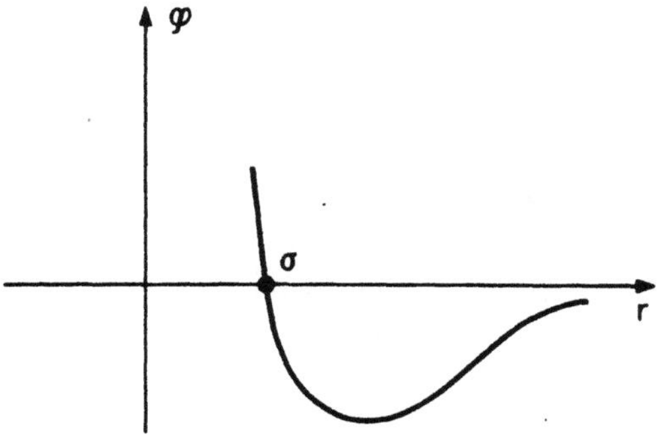

Fig. 2

For a sutherland potential:

(28) $$r < \sigma \quad \varphi = \infty \qquad r > \sigma \quad \varphi = -K_0 T_i f_0 \left(\frac{r}{\sigma}\right) \quad f_0(1) = 1$$

The calculation is made on considering the vertical straight line as the limit of an oblique straight line:

(29) $$r \leq 0 \qquad \varphi = -K_0 T_i \quad \frac{r-a}{\sigma - a} \qquad a \longrightarrow \sigma$$

The final result is:

(30) $$\mu = \frac{g}{32}\left(\frac{mK_0T}{\pi}\right)^{1/2} \frac{1}{\sigma^2} e^{-\frac{T_i}{T}}$$

Agreement with experiment is better than 1% from 300 to 2200 K for Argon[10]

$$\sigma = 2{,}829\ 10^{-10} m \ ; \qquad T_i = 110{,}8\ K \ ; \qquad m = 6{,}633\ 10^{-16} kg.$$

Table 2

T K	$10^5 \mu_{th}$ Pa.s	$10^5 \mu_{exp}$ Pa.s	$\lvert \Delta\mu / \mu \rvert$
250	1,926	1,949	0,012
300	2,272	2,272	0,00
500	3,400	3,365	0,01
700	4,286	4,25	0,008
1000	5,372	5,35	0,004
1600	7,083	7,100	0,002
2200	8,463	8,510	0,006

IV – The Fine Entropy Method

The idea of the fine entropy method has been introduced by J. Fronteau[11,12] and defined as being the logarithm of the distribution function. A molecule of fluid, subjected to an external force $\vec{X}$ and to its close environment, is assumed to obey the following law of motion:

(31) $$m\frac{d\vec{w}}{dt} = \vec{X} - \alpha\left(\vec{w} - \vec{v}\right) - \alpha \frac{K_0 T}{m} \vec{\nabla}_w \log F = \vec{X}_T$$

in which the gradient of the fine entropy is brought in as a generator of disorder to oppose the state of rest towards which the molecule was tending in the absence of $\mathbf{X}$.

The true Liouville equation,

$$\frac{\partial F}{\partial t} + \vec{w} \cdot \frac{\partial F}{\partial \vec{x}} + \frac{\partial}{\partial \vec{w}} \cdot \left(\frac{\vec{X}_T}{m} F \right) = 0 \qquad (32)$$

associated with the equation (30), allows us to recover an equation of the type (23)[12]:

$$\frac{\partial F}{\partial t} + \vec{w} \cdot \frac{\partial F}{\partial \vec{x}} + \frac{\vec{X}}{m} \cdot \frac{\partial F}{\partial \vec{w}} = \frac{\alpha}{m} \left[3F + (\vec{w} - \vec{v}) \cdot \frac{\partial F}{\partial \vec{w}} + \frac{K_0 T}{m} \Delta_w F \right] \qquad (33)$$

P. Combis, J. Fronteau and A. Tellez-Arenas[13] have also studied the motion of a one dimension Brownian quasi-particle at position x and velocity v. On designating the damping coefficient by ν, the equation of motion is written, in a uniform and constant external force $-mg$, as:

$$m \frac{dv}{dt} = -mg - \nu m v - \nu K_0 T \frac{\partial}{\partial v} \log f \qquad (34)$$

and leads, with (32), to the density probability equation f :

$$\frac{\partial f}{\partial t} + v \frac{\partial f}{\partial x} - \frac{\partial}{\partial v} (\nu v + g) f - \frac{\sigma^2}{2} \frac{\partial^2 f}{\partial v^2} = 0 \qquad (35)$$

$$\sigma^2 = 2 \nu \frac{K_0 T}{m} \qquad (36)$$

For an initial position x_0 and an initial velocity v_0 ... i.e. for a f function equal to $\delta(x - x_0)\delta(v - v_0)$ at zero time ... the equation (35) has the explicit solution[14]:

$$f = \frac{1}{2\pi (A)^{1/2}} \exp - \frac{1}{2A} \left[A_{vv}(x - \bar{x})^2 - 2A_{xv}(x - \bar{x})(v - \bar{v}) + A_{xx}(v - \bar{v})^2 \right] \qquad (37)$$

with:

(38) $$A_{xx} = \frac{\sigma^2}{2\nu^3}\left[2\nu t - 3 + 4e^{-\nu t} - e^{-2\nu t}\right]$$

(39) $$A_{xv} = \frac{\sigma^2}{2\nu^2}\left(1 - e^{-\nu t}\right)^2$$

(40) $$A_{vv} = \frac{\sigma^2}{2\nu}\left[1 - e^{-2\nu t}\right]$$

(41) $$A = A_{xx}A_{vv} - \left(A_{xv}\right)^2$$

(42) $$\overline{x} = x_0 + \frac{v_0}{\nu}\left(1 - e^{-\nu t}\right) - \frac{g}{\nu^2}\left(\nu t - 1 + e^{-\nu t}\right)$$

(43) $$\overline{v} = v_0 e^{-\nu t} - \frac{g}{\nu}\left(1 - e^{-\nu t}\right)$$

The quantity $(x - \overline{x})^2$ has, for an average value:

(44) $$\overline{\left(x - \overline{x}\right)^2} = \frac{\sigma^2}{2\nu^3}\left[2\nu t - 3 + 4e^{-\nu t} - e^{-2\nu t}\right]$$

thus, for large t, the following formula which is well verified by experiment:

(45) $$\frac{\overline{\left(x - \overline{x}\right)^2}}{t} = \frac{\sigma^2}{\nu^2} = \frac{2K_0T}{m\nu}$$

V – Conclusion

We have taken up the problem of irreversibility starting from an equation of irreversibility motion. To the usual force deriving from a position potential, we have added two non-Hamiltonian forces or environmental forces. The first is an usual damping force and the second is a force proportional to the gradient in the velocity space of the logarithm of the distribution function.

We thus obtain an irreversible kinetic equation which allows us to take account of the viscosity coefficient of a dilute gas and of the laws of Brownian motion.

Consequently, we have associated non-Hamiltonian forces with dissipative structures.

BIBLIOGRAPHY

[1] Delcroix, J.L., Physique des plasmas, Dunod, Paris 1963, 1966.

[2] Ferziger-Kiper, Mathematical theory of transport processus in gases, North-Holland Publishing Company.

[3] Hanley, H.J.M., Klein, M.J., Phys. Chem., t.76, 1743, 1972.

[4] Hanley, H.J.M., Klein, M.J., Nat.Bur.Stand. (US), Techn. note, 628, 1972.

[5] Kestin, J., Wakeham, W., Watanabe, K., J. Chem. Phys., t.53, 3773, 1970.

[6] Salmon, J., C.R. Acad. Sc. Paris, t.280, Série A, 1559, Juin 1975.

[7] Salmon, J., Ann. Institut Henri Poincaré, vol. 27, n°1, 73, 1977.

[8] Frey, J., Salmon., Nuovo Cimento, t.51 B, X, 437, 1977.

[9] Frey, J., Salmon, J., Modern Developments in Thermodynamics, John Wiley and Sons, 391, 1974.

[10] Thermophysical properties of matter, vol. 11, IFI Plenum.

[11] Fronteau, J., Nuovo Cimento, vol. 2 B, 107, 1971.

[12] Fronteau, J., Ann. Institut Henri Poincaré, t.18, 99, 1973.

[13] Combis, P., Fronteau, J., Tellez-Arenas, A., J. Statistical Physics.

[14] Combis, P., Thèse de 3ème cycle, Université d'Orléans, 1979.

COMMON STRUCTURE OF NON-HAMILTONIAN DYNAMICAL THEORIES OF MACROSCOPIC PHYSICS

Miroslav Grmela
Centre de recherche de mathématiques appliquées
Université de Montréal

Investigation of macroscopic physical systems led to the establishment of several levels of their perception. An excellent historical account is available in Refs. 1,2. A well established level of perception consists of a class of macroscopic systems S (the elements of S are the systems that are investigated), another class of physical systems I (the elements of I are the measurement instruments) and a dynamical theory DT. A dynamical theory DT consists of the phase space H (states of systems in S are completely characterized by elements of H), a set of fundamental, also called phenomenological, quantities Q and a family of time evolution equations

$$\frac{\partial f}{\partial t} = R(f,q) \tag{1}$$

$f \in \mathcal{H}$, $q \in \mathcal{Q}$, parametrized by q. Through the fundamental quantities q the individuality of the systems $s \in \mathcal{S}$ is expressed. The level of perception is well established if there exists an association $s \to q$, $s \in \mathcal{S}$, $q \in \mathcal{Q}$ such that the theoretical consequences of (1) are in good agreement with the measurements, i.e. with the results of interactions of the systems from $\mathcal{S}$ with the systems in $\mathcal{I}$. The association $\mathcal{S} \to \mathcal{Q}$ represents a part of the experience, it cannot be obtained theoretically inside the chosen level.

Examples of the well established levels of perception are the following: (i) the level of Hamiltonian dynamics HD, (ii) the level of the Boltzmann kinetic theory KT, (iii) the level of the Euler-Navier-Stokes-Fourier hydrodynamics H and (iv) the level of thermodynamics TH. The intersection of the classes of systems investigated on the above four levels is certainly not empty (for example a gas in this lecture room belongs to the intersection). On the level HD, a system is considered as composed of particles with no internal degrees of freedom. The positions and velocities of the particles are measured, the phase space is the 6N dimensional space of all positions and velocities (N is the number of particles), $\mathcal{Q}$ consists of interaction potentials. The time evolution equations (1) are the Hamilton equations. On the level KT, a state of a system is characterized by one particle distribution function $f(\underline{r},\underline{v},\underline{t})$, $f(\underline{r},\underline{v},t)d^3\underline{r}d^3\underline{v}$ denotes the number of particles in $d^3\underline{r}$ about $\underline{r}$ with velocities in $d^3\underline{v}$ about $\underline{v}$ at the time t. The time evolution equation (1) is the Boltzmann kinetic equation, $\mathcal{Q}$ consists of cross sections of binary collisions. On the level H, a state is characterized by five functions $N,E,\underline{U}$ of $\underline{r}$

and t, N is the mass, E the internal energy, $\underline{U}$ the velocity. The equation (1) is for example the system of the Navier-Stokes-Fourier equations, $\mathcal{Q}$ consists of kinetic coefficients and the relation between the local pressure and E,N. On the level of thermodynamics only specially prepared systems are investigated. The process of preparation consists of leaving a system isolated from the outside world by boundaries that are specified by the set of measurement instruments belonging to the level TH for sufficiently long time. A state is characterized completely by two parameters (e.g. the temperature and chemical potential), the equation (1) is replaced by the thermodynamic equation of state expressing the third parameter (e.g. the pressure) in terms of the two state variables. The individuality of the system is expressed only by the thermodynamic equation of state. Some authors use the name thermostatics for the theory that we have introduced as thermodynamics.

Two most common attitudes towards the existence of different levels of perception of macroscopic systems are the following: (i) one level is chosen, the others are ignored, (ii) one level is taken as absolutely superior (usually the level HD), the other levels are considered only as consequences arising in the analysis of trajectories corresponding to the chosen superior level. I shall speak about a third attitude in which the levels are considered simultaneously. The objective is to extract a common structure from the dynamical theories on the levels HD, KT, H and formulate an abstract dynamical theory such that the dynamical theories corresponding to the levels mentioned above arise as its different realizations. We observe first that one common property of the levels HD, KT, H must be their compatibility with the level TH. It means intuitively that a study of the long time behaviour of trajectories must reveal somehow the structure of thermodynamics. On all three levels considered there are in fact methods developed to obtain the thermodynamic equation of state corresponding to the levels. It is the Gibbs method on the level HD, the Boltzmann choice of the H-function and its relation to the thermodynamical entropy on the level KT, and the assumption of the local equilibrium on the level H. These methods are not proven to be rigorous consequences of the long time behaviour of trajectories, they however give results that are in good agreement with observations. It is therefore natural to try to express the three methods in a unified language as a structure that allows to speak about the thermodynamic equation of state implied by (1). If the abstract equation (1) is replaced by the dynamical equations corresponding to the levels HD, KT, H, the abstract structure should then become equivalent to the corresponding methods. We shall first present the abstract structure (it has been called the structure DDS - the structure of dissipative dynamical systems - in Ref. 3). Then, we show briefly that in the particular context of the levels HD, KT, H we recover Gibbs' method, Boltzmann's method and the assumption of the local equilibrium respectively. Some applications are considered at the end of this lecture.

[1]. In the phase space H an involution $J: H \to H$, $J \circ J$ = identity map, is defined. In HD the involution J is defined by

$(\underline{r}_1,\ldots\underline{r}_N,\underline{v}_1,\ldots,\underline{v}_n) \overset{J}{\mapsto} (\underline{r}_1,\ldots,\underline{r}_N,-\underline{v}_1,\ldots,-\underline{v}_N)$. This involution J is clearly the involution in H that compensates the involution in time $t \mapsto -t$ so that the flow $H \times \mathbb{R} \to H$ generated by the Hamilton equations remains invariant. On the level of KT the involution J is defined by $f(\underline{r},\underline{v}) \overset{J}{\mapsto} f(\underline{r},-\underline{v})$, on the level H, $(N,E,\underline{U}) \overset{J}{\mapsto} (N,E,-\underline{U})$. The involution J on the level of KT and H is clearly inherited from J on the level HD. In the context of KT and H the flow $H \times \mathbb{R} \to H$ is not however invariant if $\mathbb{R} \to \mathbb{R}$, $t \mapsto -t$, and $H \overset{J}{\to} H$. With the help of J we define a *thermodynamic equilibrium state* as a solution of $R(f,q) = 0$ that is moreover invariant with respect to J. The involution J also splits uniquely $R(f,q)$ into two parts, $R(f,q) = R^+(f,q) + R^-(f,q)$, where $R^{\pm}(f,q) = \frac{1}{2}(R(f,q) \pm JR(Jf,q)$. As a direct consequence of the definition of the thermodynamic equilibrium states, we see that the thermodynamic equilibrium states are solutions to $R^+(f,q) = 0$ and $R^-(f,q) = 0$ that are invariant with respect to J. The solutions to $R^+(f,q) = 0$ that are invariant with respect to J are called *dissipative equilibrium states*.

[2]. In a one component system there are two functionals B_1, B_2 ($B_1: H \to \mathbb{R}$, $B_2: H \to \mathbb{R}$) that are invariant with respect to J (i.e. $B_{1,2}(Jf) = B_{1,2}(f)$) and invariant with respect to the time evolution generated by (1) (i.e. $\frac{dB_1}{dt} = \frac{dB_2}{dt} = 0$)

[3]. There exists one functional S: $H \to R$ invariant with respect to J such that $\frac{dS}{dt} \le 0$. The equality holds if and only if the system considered is in the state of a dissipative equilibrium that was defined in [1].

[4]. We introduce a new functional $W = S + \alpha B_1 + \beta B_2$, where α, β are constants. The thermodynamic equilibrium states defined in [1] are identical with the solutions of $\frac{\delta W}{\delta f} = 0$.

[5]. Let f_0 be an equilibrium state such that $A_0 = D_f^{(2)} W|_{f_0}$, i.e. the second derivative of W with respect to f evaluated at f_0, is a positive definite bounded operation in the tangent space to H at f_0 that is considered as a Hilbert space with a fixed inner product $(.,.)$. Let P^+ resp. P^- denotes the linear parts of R^+ and R^- at f_0. The operator P^+ is formally selfadjoint with respect to $(.A_0.)$ and P^- is formally skewadjoint with respect to the same inner product $(.A_0.)$. Equivalently, but more simply, we can say that P (the linear part of R at f_0) is formally selfadjoint with respect to the indefinite inner product $(.,A_0 J.)$. This property is known in the context of H as the Onsager-Casimir symmetry.

The two parameters α, β introduced in the definition of W, see [4], are the two parameters determining a state of a system in thermodynamics, the thermodynamics equation of state is defined by $\gamma = -\frac{1}{\text{vol } \Omega} W|_E$, where E denotes the set of equilibrium states, vol Ω is the volume of the region Ω in which the system considered is confined. The relation between α, β, γ and the specific thermodynamic quantities as e.g. the temperature T, the chemical potential μ and the pressure p is obtained if for example the thermodynamic meaning of the conserved quantities B_1 and B_2 is known. If, say, $\frac{1}{\text{vol } \Omega} B_1|_E$ is the thermodynamic mass and $\frac{1}{\text{vol } \Omega} B_2|_E$ is the thermodynamic energy then $\beta = \frac{1}{T}$, $\alpha = -\frac{\mu}{T}$ and $\gamma = \frac{p}{T}$.

In the context of KT, $B_1 = \int_\Omega d^3\underline{r} \int d^3\underline{v}\ f(\underline{n},\underline{v},t)$, $B_2 = \frac{1}{2}\int_\Omega d^3\underline{r} \int d^3\underline{v}\ v^2 f(\underline{r},\underline{v},t)$, $S = \int_\Omega d^3\underline{v}\ f(\underline{r},\underline{v},t)\log f(\underline{r},\underline{v},t)$, the dissipative equilibrium states are the so-called local equilibrium states, the thermodynamic equation of state obtained is the thermodynamic equation of state of the ideal gas. In the context of HD the operator $R^+ \equiv 0$. The structure [1] - [5] described above is equivalent to the Gibbs method provided the dissipative equilibrium states are postulated ad hoc. They are the states with the Maxwellian dependence on velocities and with the unspecified dependence on the position coordinates (these states are called thermalized in Ref. 4). The presentation of the Gibbs method that is obtained by following the structure [1] - [5] is very similar to the presentation of equilibrium statistical mechanics by Yvon.[4] More details can be found in Ref. 5. The presence of the structure [1] - [5] in the context of hydrodynamics is probably least obvious. We shall therefore discuss hydrodynamics in more detail.

We denote by $n(\underline{r},t)$, $e(\underline{r},t)$, $\underline{u}(\underline{r},t)$ the field of mass, total energy and impulse. We shall assume that: (a) the total mass $\int_\Omega d^3\underline{r}\ n(\underline{r},t)$, the total energy $\int_\Omega d^3\underline{r}\ e(\underline{r},t)$ and the total impulse $\int_\Omega d^3\underline{r}\ u(\underline{r},t)$ remain invariant under the time evolution, (b) the theory is local, i.e. a change of any quantity at $\underline{r}$ depends only on $n,e,\underline{u}$ evaluated at the same position $\underline{r}$, (c) the fluid considered is isotropic and (d) the time evolution equations do not include the third and higher order derivatives with respect to $\underline{r}$. It follows from (a), (b) that

$$-\frac{\partial n}{\partial t} = \partial_\gamma \overset{\circ}{n}_\gamma$$
$$-\frac{\partial e}{\partial t} = \partial_\gamma \overset{\circ}{e}_\gamma \qquad (2)$$
$$-\frac{\partial u_\alpha}{\partial t} = \partial_\gamma \overset{\circ}{u}_{\alpha_\gamma} ,$$

$\partial_\gamma \equiv \frac{\partial}{\partial r_\gamma}$, the summation convention is used, $\underline{\overset{\circ}{n}}$, $\underline{\overset{\circ}{e}}$, $\underline{\underline{\overset{\circ}{u}}}$ remain to be unspecified. New state variables N,E,$\underline{U}$ are introduced by

$$\begin{aligned} n &= N \\ u_\alpha &= NU_\alpha \\ e &= \tfrac{1}{2}NU_\gamma U_\gamma + NE . \end{aligned} \qquad (3)$$

The constitutive laws for $\underline{\overset{\circ}{n}}$, $\underline{\overset{\circ}{e}}$, $\underline{\underline{\overset{\circ}{u}}}$ follow from (b), (c), (d) and the consideration of physical dimension,

$$\begin{aligned} \overset{\circ}{n}_\alpha &= NU_\alpha \\ \overset{\circ}{u}_{\alpha\gamma} &= NU_\alpha U_\gamma + P_{\alpha\gamma} \\ \overset{\circ}{e}_\alpha &= \tfrac{1}{2}NU_\gamma U_\gamma U_\alpha + NEU_\alpha + P_{\alpha\gamma}U_\gamma + V_\alpha , \end{aligned} \qquad (4)$$

$$\begin{aligned} P_{\alpha\gamma} &= \tilde{p}\, \delta_{\alpha\beta} - \eta_0 \partial_\gamma U_\gamma \delta_{\alpha\beta} - \eta_1 \tfrac{1}{2}(\delta_\alpha U_\beta + \delta_\beta U_\alpha) \\ V_\alpha &= \lambda \delta_\alpha (\tfrac{1}{\tau}) \\ & \tau>0,\ \lambda>0,\ \eta_0>0,\ \eta_1>0 \end{aligned} \qquad (5)$$

The involution J is defined by $(N,E,\underline{U}) \overset{J}{\mapsto} (N,E,-\underline{U})$. The choice of the scalar local functionals $\tilde{p},\tau,\lambda,\eta_0,\eta_1$ is restricted by the inequalities (5), they remain otherwise unspecified. It can be shown now, without any additional assumption that the time evolution equations implied by (2) - (5) possess the structure [1] - [5] provided the choice of $\tilde{p}$ and τ introduced in the constitutive laws (5) and the functional s ($S = \int d^3\underline{r}\; N(\underline{r},t)\; s(\underline{r},t)$, see [3]) is restricted by another condition (called a compatibility condition)

$$\frac{\partial s(\underline{r},t)}{\partial N(\underline{r},t)} = -\frac{1}{(N(\underline{r},t))^2}\frac{\partial s(\underline{r},t)}{\partial E(\underline{r},t)}\tilde{p}(\underline{r},t)$$
$$\tau(\underline{r},t) = -\left(\frac{\partial s(\underline{r},t)}{\partial E(\underline{r},t)}\right)^{-1} . \tag{6}$$

There are no other restrictions on the choice of kinetic coefficients λ,η_0,η_1. We shall briefly indicate the proof of the above statement. It is clear that the two functionals B_1,B_2 (see [2]) that are invariant with respect to the time evolution are $\int_\Omega n(\underline{r},t)d^3\underline{r}$ and $\int_\Omega e(\underline{r},t)d^3\underline{r}$. In order to satisfy [3] we must find a functional $S = \int d^3 r N(\underline{r},t)s(\underline{r},t)$ depending on $N(\underline{r},t)$ and $E(\underline{r},t)$ such that this functional remains invariant if only R^- (i.e., as it is easily found, the part of the time evolution equations obtained by putting $\lambda = \eta_0 = \eta_1 = 0$) is considered. We shall denote by $\frac{\partial^{(-)}}{\partial t}$ the change in time generated by R^-. Thus $\frac{\partial^{(-)}}{\partial t}$ (Ns) must be another conservation law

$$\frac{\partial^{(-)}}{\partial t}(Ns) = -\partial_\alpha S_\alpha , \tag{7}$$

where $\underline{S}$ is an unspecified vector. We can however also calculate $\frac{\partial^{(-)}}{\partial t}$ (Ns) by using R^- part of the time evolution equations for N and E. By comparing we obtain

$$\frac{\partial S_\alpha}{\partial N} = U_\alpha \frac{\partial (Ns)}{\partial N}$$

$$\frac{\partial S_\alpha}{\partial E} = U_\alpha \frac{\partial (Ns)}{\partial E} \qquad (8)$$

$$\frac{\partial S_\alpha}{\partial U_\beta} = N \frac{\partial (Ns)}{\partial N} \delta_{\alpha\beta} + \frac{\partial s}{\partial E} \tilde{p}\, \delta_{\alpha\beta}$$

The integrability condition for (8) is exactly the compatibility condition (6). The rest of the structure [1] - [5] follows from direct calculations. Details can be found in Ref. 5. This ends the proof. The condition (6) is in the conventional introduction of hydrodynamic equations[6] the starting postulate known as the assumption of the local equilibrium. According to this assumption the functional dependence of s on $N(\underline{r},t)$ and $E(\underline{r},t)$ is considered to be the same as the dependence of the thermodynamic entropy on the thermodynamic mass and the thermodynamic energy. The functionals $\tilde{p}$ and τ are introduced in the Navier-Stokes-Fourier constitutive laws as the local pressure and the local temperature. We have thus shown that also in the context of the Euler-Navier-Stokes-Fourier hydrodynamics the structure [1] - [5] is equivalent to the standard method establishing the compatibility of hydrodynamics with thermodynamics.

In the rest of this lecture I mention briefly some application of the structure [1] - [5]. Interesting applications arise in the context of the construction of extensions of the well established dynamical theories. The strategy is to suggest first an extension and then to require that also the extended dynamical theory possesses the structure [1] - [5]. This requirement, that is in fact a hypothesis that any dynamical theory of macroscopic systems having a chance to be of physical importance must possess the structure [1] - [5], then restricts the arbitrariness of the chosen extension. As the first example we mention the extension of the Euler-Navier-Stokes-Fourier hydrodynamics obtained by extending the set of the functions describing completely a state (for example a fourteen state variable hydrodynamics in which pressure tensor and heat flux join $N, E, \underline{U}$ as state variables). In such extended theory the assumption of the local equilibrium cannot be clearly used. The generalization of the presentation of the Euler-Navier -Stokes-Fourier hydrodynamics described above is however straightforward (details can be found in Ref. 5). An example of an extension of the Boltzmann equation that was discussed in detail is the Enskog kinetic theory.[2] The Enskog extension of the Boltzmann equation is known for a long time.[2] The binary collisions become in the Enskog theory non-local and a new fundamental quantity χ (arising by replacing the Boltzmann assumption $f_2(\underline{r}_1,\underline{v}_1,\underline{r}_2,\underline{v}_2 t) = f(\underline{r}_1,\underline{v}_1,t)f(\underline{r}_2,\underline{v}_2,t)$ by $f_2(\underline{r}_1,\underline{v}_1,\underline{r}_2,\underline{v}_2,t) = \chi(n,\underline{r}_1,\underline{r}_2,t)\ f(\underline{r}_1,\underline{v}_2,t)\ f(\underline{r}_2,\underline{v}_2,t);\ n = \int d^3\underline{v} f(\underline{r},\underline{v},t)$ f_2 denotes the two particle particle distribution function) is introduced. The analog of the Boltzmann H-theorem and the thermodynamic equation of state implied by the Enskog equation were not known. The requirement of the structure [1] - [5] leads to the compatibility condition restricting the choice of χ and to the functional S, depending on χ, and thus also to the thermodynamic equation of state implied by the Enskog equation are obtained.[7] Some of the results obtained have been confirmed by an independent study due to P. Résibois[8] of a special case of the Enskog equation. Another extension of the Boltzmann kinetic theory that was considered is the two point extension. In the two point kinetic theory a state of a system is described by a pair, two particle and one particle distribution functions. Two examples of the two point kinetic equations possessing the structure [1] - [5] are now available.[5,9] Consequences of the property [5], that can be used with advantage in the study of the relations between two dynamical theories (e.g. KT and H), were studied systematically in Refs. 3, 10.

References.

1. Carnot, S., Reflection on the motive power of fire, Ed. E. Mendoza, Dover Publ., 1960.
2. Brush, S.G., Kinetic theory, vol. 1,2,3, Pergamon Press, 1966.
3. Grmela, M., Dynamic Stability and Thermodynamics in Kinetic Theory and Fluid Mechanics, J. Math. Phys. 16, 2441, 1975.

   Grmela, M. and Iscoe, I., Reductions in a Class of Dissipative Dynamical Systems of Macroscopic Physics, Ann. Inst. Henri Poincaré. XXVIII, III, 1978.

   Grmela, M., Dissipative Dynamical Systems of Macroscopic Physics in Proceedings of the Seminar on Global Analysis, Lecture Notes in Mathematics (to appear), Eds. M. Grmela and J. Marsden, 1979.
4. Yvon, J., Les corrélations et l'entropie en mécanique statistique classique, Dunod, Paris, 1966.
5. Grmela, M., Compatibility of a Dynamical Theory with Thermodynamics, Rapport CRMA-888, 1979.

6. De Groot, S.R. and Mazur, P., Non-equilibrium thermodynamics, McGraw-Hill, New York, 1969.

7. Grmela, M. and Garcia-Colin, L.S., Compatibility of the Enskog-like Kinetic Theory with Thermodynamics Part I and II, Phys. Rev.

8. Résibois, P., H-theorem for the (modified) Non-linear Enskog Equation, J. Stat. Phys., 19, 593, 1978.

9. Grmela, M. and Salmon, J., in preparation.

10. Grmela, M. and Rosen, R., Some Consequences of the Onsager-Casimir Symmetry, Rapport CRMA - 753, 1979.

SYSTEM THEORETIC FOUNDATIONS FOR MODELLING PHYSICAL SYSTEMS

JAN C. WILLEMS
Mathematics Institute
P.O. Box 800
9700 AV Groningen
The Netherlands

ABSTRACT

In this paper we will outline the main ideas of an axiomatic approach to the modelling of dynamical systems. This approach differs from the usual one in that the input/output, i.e., the cause/effect relationship is not assumed to be given a priori. More details on the theoretical developments and examples from physics may be found in [1]. These ideas are here applied to time-reversibility, where it is shown how one may deduce the existence of a time-reversing involution on the state space.

1. Introduction. In the axiomatic mathematical frameworks used in physics there are in our opinion two disturbing elements. The first one is the absence of 'inputs' and 'outputs' and the second one, related to the first, is the lack of a clear picture of the nature and the definition of the concept of 'state'. Let us illustrate these two points.

In most books on mechanics it is customary to start with Newton's second law (which, incidentally, cannot be interpreted in anything but an input/output framework). However, in an effort to arrive at an Hamiltonian or sortlike framework as soon as possible, something like the following is then claimed: 'in most cases of interest the forces will be due to a potential field'. This assumption, valid for planetary motion and made with the best intentions, cannot stand any serious scrutiny since for practically all mechanical systems which are part of our daily circle of experiences (as cars, bicycles, windwills, etc.), the assumption that the forces are due to a potential field is totally absurd!

Let us now consider a concrete situation in which the absence of an abstract understanding of the concept of state is rather striking. Assume that a kitchen spunge is glued between two plates which are then mounted horizontally such that the lower plate is fixed and such that a vertical force can be applied on the upper plate which thereby acts as a sort of piston. The following physical quantities are assumed to be measured: the position of the upper plate, the force applied on it, the temperature of the spunge (assumed to be uniform), and the heat leaving or being absorbed by the spunge. Now, assume that after an adequate sequence of experiments one has obtained mathematical expressions for the dynamical relations between the force history, and the position, the temperature, and the heat. In other words, the experimentor is satisfied to identify the physical behavior of this device with the obtained mathematical model. Now, assume that one asks for expressions for the internal energy, E, and the internal entropy, S. The problem on the one hand consists in defining the 'state space', i.e., the domain of the functions E and S and on the other hand the values of these functions. If one poses this problem to a physicist then one usually gets the answer that unless the behavior is elastic (which will not be a likely situation for our spunge because of the apparent aftereffect) then one should start the modelling practically at a molecular level in order to come up with a description of the state space and the internal entropy. It is quite remarkable that, even under the assumption of an adequate mathematical model as we have assumed to be the case here, there are no recipes in physics to construct the state space from the experimental 'evidence' and it is hard to take seriously suggestions to start this modelling process over again at a molecular level, taking into consideration the constraints imposed by the observed macroscopic behavior.

An axiomatic framework for describing dynamical systems with inputs and outputs has been developed in mathematical system theory. In physical terms, the inputs model the influence from the environment, while the outputs correspond to the behavior on the 'configuration' space. In addition, the important problem of associating a state space and an input/state/output model with a given 'experimentally' obtained input/output model has been given a great deal of attention in this field. An important text in which this theory is treated is [2]. However, in some applications this framework shows some inadequacies, notably in the starting point that the cause/effect (input/output) signal graph is given a priori. In [1] a number of examples are given of situations in which such an input/output framework is not an appropriate starting point for the mathematical modelling. In [1] we describe in detail an alternative of which we will now give the main ideas.

2. Definitions and Qualitative Properties of Systems. The first definition introduces a very general and yet very usable 'external' definition of a dynamical system. For simplicity we will only consider continuous time systems.

Definition 1: Let W be a set, called the (external) *signal alphabet*, and let the *time axis* $T = \mathbb{R}$. A *dynamical system* Σ (on W) is then simply a subset of W^T, i.e., a family of functions mapping the time axis T into the signal alphabet W. It is said to be *time-invariant* if $\{w(\cdot) \in \Sigma\} \Leftrightarrow \{w(\cdot + \Delta) \in \Sigma\}$ for all $\Delta \in T$; it is said to be *linear* if W is a vector space and if Σ is a linear subspace of W^T. It is said to be *autonomous* if $w_1, w_2 \in \Sigma$ and $w_1(t) = w_2(t)$ for all $t < t'$ or for all $t > t'$ implies $w_1 = w_2$.

The counterpart of the above definition in which the state of the system is explicitely displayed leads to;

Definition 2: Let W be a set called the (external) *signal alphabet*, X a set called the *state space*, and $T = \mathbb{R}$ the *time axis*. A dynamical system on $X \times W$ (i.e., a subset of $(X \times W)^T$) is said to be in *state space form* if it satisfies

(Axiom A) : $\{(x_i, w_i) \in \Sigma,\ i = 1, 2 \text{ and } x_1(t) = x_2(t)\} \Rightarrow \{(x, w) \in \Sigma\}$

with (x,w) defined by

$$(x, w)(\tau) := \begin{cases} (x_1, w_1)(\tau) & \text{for} \quad \tau < t \\ (x_2, w_2)(\tau) & \text{for} \quad \tau \geq t \end{cases}$$

The relationship between systems in external form and those in state space form is given in the following definition:

Definition 3: Let Σ be a state space system. Then $\Sigma_e := \{w \mid \exists x \ni (x,w) \in \Sigma\}$ is called the *external behavior* of Σ. We denote this by $\Sigma \Rightarrow \Sigma_e$. If Σ_e is given and if Σ is a state space system such that $\Sigma \Rightarrow \Sigma_e$ then we call Σ a *(state space) representation* or *realization* of Σ_e.

Example 1: Consider $X \subset \mathbb{R}^n$, $f : X \times U \times T \to \mathbb{R}^n$, $g : X \times U \times T \to W$. Then $\Sigma_e := \{w: T \to W \mid \exists\ x : T \to X$, absolutely continuous, and $u : T \to U$ such that $\dot{x}(t) = f(x(t), u(t), t)$ for almost all t, and $w(t) = g(x(t), u(t), t)$ for all t$\}$ defines a dynamical system. It is time-invariant if f and g donot depend on t explicitely; and it is linear if X, U, W are vector spaces and $f(\cdot,\cdot,t)$, $g(\cdot,\cdot,t)$ are linear maps. An example of an autonomous system in this class is $\{w : R \to \mathbb{R}^p \mid \exists x : \mathbb{R} \to \mathbb{R}^n$ such that $\dot{x}(t) = Ax(t)$; $w(t) = Cx(t)$ for all t$\}$. Here A and C are matrices of appropriate dimension. It is furthermore clear that $\Sigma := \{(x,w) : T \to X \times W \mid$ x is absolutely continuous, and $\exists u : T \to U$ such that $\dot{x}(t) = f(x(t),$

$u(t)$, $t)$ for almost all t, and $w(t) = g(x(t), u(t), t$ for all $t\}$ defines a state space realization of Σ_e.

This example in a very general sense illustrates the problem of state space realization: one starts with a general class of signals (e.g., solutions of Maxwell's equations, an experimentally observed class of signals, a convolution relation, or a system of differential equations) and one attemps to write them as a system of first order differential equations with an input. However, this input may or may not be present. A more general formulation explicitated in [1] is to consider motions governed by $(\dot{x},w) \in f(x)$ with f a point to set map.

In the next section we will discuss a general class of systems for which the realization question may be solved explicitely.

3. Realization Theory. In the problem of state space realization one starts with a given external behavior Σ_e on a signal space W, and we need to construct the state space X and associate with every $w \in \Sigma_e$ at least one $x \in X^T$ such that the pairs (x,w) thus obtained satisfy (Axiom A), In applications, it is important to have as efficient a realization as possible. This may be expressed by asking minimality, which is defined as follows:

Definition 4: Let Σ_i, $i = 1, 2$ be dynamical systems on W_i. Then $\Sigma_1 > \Sigma_2$ if there exists a partial, set to point, surjective map $F : W_1 \to W_2$ such that $\{w_1 \in \Sigma_1, w_1(t) \in A(t), A(t) \in Do(F)\ \forall t\} \Rightarrow \{Fw_1(\cdot) \in \Sigma_2\}$. *Equiv-*

alence of Σ_1 and Σ_2 is defined by the existence of a bijection $F : W_1 \to W_2$ such that $\{w_1 \in \Sigma_1\} \Leftrightarrow \{Fw_1 \in \Sigma_2\}$. A system Σ in a particular class S is said to be *irreducible* or *minimal* if $\{\Sigma' \in S, \Sigma' < \Sigma\} \Rightarrow \{\Sigma, \Sigma' \text{ equivalent}\}$. We will use this terminology in case the class S consists out of all state space realizations of a given external behavior Σ_e. The map F in this case is of the type $F : X_1 \times W \to X_2 \times W$ with $F : (A_1, w) \to (a_2, w)$.

In [1] we describe several canonical procedures for obtaining minimal realizations of a given external behavior. One key problem which arises naturally in this context in whether or not all minimal realizations are equivalent. In general, in fact, this needs not be the case. However, for two important classes it may be shown that all minimal realizations are indeed equivalent:

Proposition 1: *Assume that* Σ_e *is autonomous. Then all its minimal realizations are equivalent.*

An explicit theory for realization may be worked out for the following class of systems which arises very frequently in applications:

Example 2: Let P and Q be matrix polynomials of respective dimensions $(r \times q)$ and $(r \times \ell)$. Consider now the following system $\Sigma_e(P, Q) := \{w : \mathbb{R} \to \mathbb{R}^q \mid w \in L_1(\text{loc})$ and $\exists\, \xi : \mathbb{R} \to \mathbb{R}^\ell$ such that $P(\frac{d}{dt})w = Q(\frac{d}{dt})\xi$ where this equality holds in the sense of distributions$\}$. It is easily seen that $\Sigma_e(P, Q)$ is linear and time-invariant. As a particular case consider $\Sigma(A, B, C, D)$ defined by $\{(x, w) : \mathbb{R} \to \mathbb{R}^n \times \mathbb{R}^m \mid \exists\, u \in L_1(\text{loc}) \ni \dot{x} = Ax + Bu;\ w = Cx + Du\}$. Obviously $\Sigma(A, B, C, D)$ is a linear time-invariant system in state space form with state space $X = \mathbb{R}^n$.

One may prove the following:

Theorem 1: *Consider* $\Sigma_e(P, Q)$. *Then*

(i) *there exists a minimal realization* $\Sigma(A, B, C, D)$ *of* $\Sigma_e(P, Q)$;

(ii) *all minimal realizations of* $\Sigma_e(P,Q)$ *are equivalent;*

(iii) *all minimal realizations* $\Sigma(A, B, C, D)$ *of* $\Sigma_e(P,Q)$ *with* $\mathrm{Ker}\begin{bmatrix} B \\ D \end{bmatrix} = \{0\}$ *are obtainable by applying the transformation group* $\{A, B, C, D\} \xrightarrow[\substack{\det S \neq 0 \\ \det R \neq 0}]{(F,S,R)} \{S(A + BF)S^{-1}, SBR, (C + DF)S^{-1}, DR\}$;

(iv) *there exists a (componentwise) partition of* w *into* $w = (w_1, w_2)$ *such that* Σ *is (minimally) described by* $\dot{x} = Ax + Bw_1$, $w_2 = Cx + Dw_2$, $w = (w_1, w_2)$;

(v) *there exists a nonsingular matrix* R *and a (componentwise) partition of* $Rw := w' = (w_1', w_2')$ *such that* $\Sigma_e(P, Q)$ *is (minimally) described by* $\dot{x} = Ax + Bw_1'$, $w_2' = Cx$, $w' = (w_1', w_2')$.

The interesting interpretation of the result of (iv) in the above theorem is the following: if we are given a smooth linear time-invariant finite-dimensional dynamical system described, as we have assumed here, by a system of differential equations, then there exists a componentwise partition of the 'externally observed' variable w into $w = (w_1, w_2)$ with w_1 the '*input*' and w_2 the '*output*'. This decomposition is not unique but the dimensions of w_1 and w_2 are unique and are thus intrinsically defined by $\Sigma_e(P, Q)$. Thus the construction of an input/output representation may always be carried out for this class of systems which shows in a very appealing and general way that the usual input/

output framework for this class of linear time-invariant system constitutes, as a starting point, no loss of generality! In [1] a number of related representation questions are introduced.

4. Time-Reversibility. A number of physical examples and applications of using this framework in the analysis of physical dynamical systems is given in [1]. In the present paper we will restrict ourselves to time-reversibility which is an important issue in the analysis of physical systems.

Definition 5: Let Σ_e be a dynamical system over W. Its *time-reverse* is defined by $R\Sigma_e := \{w \in W^T | Rw \in \Sigma_e\}$ where $(Rw)(t) := w(-t)$. Σ_e is *time-reversible* if $R\Sigma_e = \Sigma_e$.

Obviously R is an involution ($R = R^{-1}$). The problem is to find a realization Σ of Σ_e such that Σ reflects the time-reversibility of Σ_e. From the usual set ups (e.g. in Hamiltonian mechanics) it is clear that one cannot expect that there exists a state space realization which would also be time reversible (the initial velocities for instance will need be inverted while reversing the time). Consequently, one can at best expect to show the existence of a minimal realization Σ of Σ_e such that there exists an involution S on the state space X of Σ such that $\{(x, w) \in \Sigma\} \Leftrightarrow \{R(Sx, w) \in \Sigma\}$. We have not yet been able to prove this in complete generality:

Theorem 2: *Assume that Σ_e is time-reversible and that all minimal realizations of Σ_e are equivalent (e.g., that Σ_e is autonomous* *or that* $\Sigma_e = \Sigma_e(P, Q)$). *Then* Σ_e *is time-reversible if and only if any minimal realization* Σ *of* Σ_e *admits an involution* S *on its state space* X *such that* $\{(x, w) \in \Sigma\} \Leftrightarrow \{R(Sx, w) \in \Sigma\}$.

The above theorem is easily generalized to the situation in which there exists an involution S_e on W such that $\Sigma_e = RS_e\Sigma_e$, i.e., when one starts with a somewhat more general type of time-reversibility. This theorem sheds some light on the mathematical nature of time-reversibility: even if one starts with time-reversibility one needs an time reversing involution on the state space and in fact one can (under the assumptions of Theorem 2) always deduce the existence of such an involution.

In the linear case, the involution may be made more explicit:

Corollary 1: *Consider* $\Sigma_e(P, Q)$ *and assume that* $\Sigma_e(P, Q) = R\,\Sigma_e(P, Q)$ *(which will be the case e.g., if* $P(s) = P(-s)$ *and* $Q(s) = Q(-s)$). *Then there exists a minimal realization* $\Sigma(A, B, C, D)$ *of* $\Sigma_e(P, Q)$ *for which* x *may be partitioned into* $x = (x_+, x_-)$ *and* $\{(x_+, x_-, w) \in \Sigma(A, B, C, D)\} \Leftrightarrow \{R(x_+, -x_-, w) \in \Sigma(A, B, C, D)\}$.

The above corollary generalizes one of the results in [3] where also a stochastic version of this result may be found.

5. Conclusions. In this paper we have given a general theoretical framework for treating systems which are influenced by their environment. This framework, in contrast with what has been customary in

mathematical system theory, leaves the cause/effect set-up to be deduced from the mathematical model. In addition, we have given some ideas on realization theory. This construction of the state space ought to be a theory which is of much interest in physics.

References

[1] J.C. Willems, "System Theoretic Models for the Analysis of Physical Systems", to appear in *Ricerche di Automatica*, 1980.

[2] R.E. Kalman, P.L. Falb, and M. Arbib, *Topics in Mathematical System Theory*, McGraw-Hill, 1969.

[3] J.C. Willems, "Time Reversibility in Deterministic and Stochastic Dynamical Systems", pp. 318-326 of *Recent Developments in Variable Structure Systems, Economics and Biology* (Eds.: R.R. Mohler and A. Ruberti), Vol. 162 of Springer Lecture Notes in Economics and Mathematical Systems, 1978.

A HAMILTON-JACOBI TREATMENT OF DISSIPATIVE SYSTEMS WITH ONE DEGREE OF FREEDOM

Giacomo Della Riccia

Istituto di Scienze dell'Informazione
Università di Salerno (Italy)
and
Department of Mathematics
Ben-Gurion University of the Negev (Israel)

1. Introduction

Until recently, it had been thought that the equation of motion of classical systems with dissipative forces could not be derived from a Lagrangian function by a pure variational procedure. The dissipative terms of the equation were usually introduced directly in the Lagrange's equation through a Rayleigh dissipative function or by using complex coordinates and velocities in the Lagrangian. Recently,[1,2] a pure Lagrangian description was found for the linearly damped oscillator and certain dynamical systems with velocity-dependent forces and a Hamilton-Jacobi treatment was developed[3] for these systems. We extend these results to the general class of classical systems with one degree of freedom governed by a Lienard equation of motion which is of the form

$$\ddot{x} + f(x)\dot{x} + g(x) = 0 \ .$$

We integrate the Lienard equation with a Hamilton-Jacobi formalism and then we turn our attention to the construction of time-independent Hamiltonian for more general systems. As we shall see, time-independent Hamiltonians can be useful for predicting the existence of limit cycles in the time evolution of a system. We also discuss the broader question of finding all possible equivalent Lagrangians and q-equivalent Hamiltonians for the general equation of motion $\ddot{x} = f(x,\dot{x},t)$. It turns out that this problem is equivalent to the Last Multiplier problem introduced by Jacobi in 1844.

As an application of these developments, we give a detailed treatment of a nonlinear oscillator, described by Smith,[4] which is a system with a limit cycle which can be solved analytically.

Classical dissipative systems with limit cycles can be considered as models for quantum systems with discrete eigenstates. Since the Hamilton-Jacobi formalism is an important link between classical and quantum mechanics, the development of a Hamilton-Jacobi treatment of such dissipative systems may be useful in attempting to build a dynamical description of quantum transitions.

2. Solution of the Hamilton-Jacobi equation for the Lienard equation

The Lagrangian that was found for the linearly damped oscillator

$$\ddot{x} + \varepsilon \dot{x} + g(x) = 0 \tag{1}$$

is $$L = e^{\varepsilon t}[\tfrac{1}{2} v^2 - V(x)] \tag{2}$$

where $v = \dot{x} = dx/dt$ and $V(x) = \int g(x)dx$. We now consider the Lienard equation

$$\ddot{x} + f(x) + g(x) = 0 \tag{3}$$

which in the phase plane (x,v) becomes

$$v dv/dx + f(x)v + g(x) = 0 \tag{4}$$

We introduce new variables τ and X defined by $d\tau = f(x)dt$ and $X = \int f(x)dx = F(x)$. In terms of these variables, the Lienard equation becomes

$$\ddot{X} + \dot{X} + \tilde{g}(X) = 0 \tag{5}$$

where $\dot{X} = dX/d\tau = dx/dt = v$, $\ddot{X} = d\dot{X}/d\tau$ and $\tilde{g}(X) = \tilde{g}[F(x)] = g(x)/f(x)$. On the basis of the similarity between equations (5) and (1), we immediately find, in agreement with (2), that a Lagrangian for (5) is

$$L = e^{\tau}[\tfrac{1}{2} v^2 - \tilde{V}(X)] \tag{6}$$

where $\tilde{V}(X) = \int \tilde{g}(X)dX = \int g(x)dx = V(x)$. The momentum conjugate to X is

$$p = \partial L/\partial v = ve^{\tau} , \tag{7}$$

and, thus, the Hamiltonian corresponding to L is

$$H(X,p;\tau) = \tfrac{1}{2} p^2 e^{-\tau} + e^{\tau}\tilde{V}(X) \tag{8}$$

Let us introduce a Hamilton-Jacobi function $\tilde{S}$ of the form $\tilde{S}(X,\tau) = \tilde{Q}(X,\alpha)e^{\tau}$ where α is a parameter. By definition, we have

$$p = \partial\tilde{S}/\partial X = \partial\tilde{Q}/\partial X \; e^{\tau} \tag{9}$$

and the following Hamilton-Jacobi equation

$$\tfrac{1}{2} (\partial S/\partial X)^2 \; e^{-\tau} + e^{\tau} \; \tilde{V}(X) + \partial\tilde{S}/\partial\tau = 0$$

which, after division by $e^{\tau} \neq 0$, reduces to

$$\tfrac{1}{2} (\partial\tilde{Q}/\partial X)^2 + \tilde{V}(X) + \tilde{Q} = 0 \; . \tag{10}$$

From (7) and (9) we obtain $\partial\tilde{Q}/\partial X = v$, therefore after returning to the original variable x, equation (10) becomes

$$\tfrac{1}{2} v^2 + V(x) + Q(x,\alpha) = 0 \tag{11}$$

where $Q(x,\alpha) = \tilde{Q}(X,\alpha)$. This determines $Q(x,\alpha)$ provided we express v

in terms of x. To do this we differentiate (11) with respect to x. We obtain $vdv/dx + g(x) + f(x)v = 0$ (12)

after using $\partial Q(x,\alpha)/\partial x = \partial\tilde{Q}(X,\alpha)/\partial X \, dX/dx = vf(x)$. Equation (12) is nothing else than the Lienard equation in the phase plane (x,v) already mentioned in (4). We assume that an integrating factor $I(x,v)$ for equation (12) can be found. Then

$$I(x,v)\{vdv + [f(x)v + g(x)]dx\} = dC(x,v) = 0$$

and we obtain a solution to (12) in the form

$$C(x,v) = \text{constant} = \alpha \,. \qquad (13)$$

The function $Q(x,\alpha)$ is determined by the elimination of v between (11) and (13). In practice it may be difficult to perform this elimination, but we shall not neet it explicitly. Proceeding according to the Hamilton-Jacobi theory we write

$$\partial\tilde{S}(X,\alpha)/\partial\alpha = e^{\tau} \, \partial Q(x,\alpha)/\partial\alpha = \text{constant} = \beta \qquad (14)$$

Differentiating (11) with respect to α we find

$$v\partial v/\partial\alpha + \partial Q(x,\alpha)/\partial\alpha = 0 \qquad (15)$$

since $V(x)$ does not depend on α. According to (13), we have

$$d\alpha = dC(x,v) = I(x,v)\{vdv + [f(x)v + g(x)]dx\} \,,$$

hence $\partial\alpha/\partial v = vI(x,v)$ (16)

Combining (14), (15) and (16) we get

$$I(x,v) = \partial\alpha/v\partial v = -\,1/\beta e^{\tau}$$

or $I(x,v)e^{-\tau} = I(x_0,v_0)e^{-\tau_0} = -\,1/\beta$ (17)

where x_0, v_0, τ_0 represent arbitrary initial conditions at time $t = t_0$. From equations (13) and (17) we obtain $x = x(\tau-\tau_0;\alpha,\beta)$ and $v = v(\tau-\tau_0;\alpha,\beta)$ and finally since $d\tau = f(x)dt$, we get

$$t = t_0 = \int_0^{\tau-\tau_0} \frac{ds}{f[x(s;\alpha,\beta)]}$$

In other words, we have found a parametric representation, with the variable τ as parameter, of the solution of Lienard equation on the basis of a pure Hamilton-Jacobi formalism.

3. The Last Multiplier problem and the Lagrangian formulation

The concept of the Last Multiplier is due to Jacobi who showed its importance in the theory of systems of differential equations. Let

$$\frac{dx_1}{X_1} = \frac{dx_2}{X_2} = \ldots\ldots = \frac{dx_n}{X_n} = \frac{dx}{X}$$

where $(X_1,\ldots,X_n,X)$ are given functions of the variables $(x_1,\ldots,x_n,x)$, be a given system of equations. The last multiplier $M(x_1,\ldots,x_n,x)$ of the system is any solution of the partial differential equation (p.d.e)

$$\frac{\partial(MX_1)}{\partial x_1} + \ldots\ldots + \frac{\partial(MX_n)}{\partial x_n} + \frac{\partial(MX)}{\partial x} = 0$$

An interesting property is that if M_1 and M_2 are two different last multipliers, then a first-integral of the system is simply $M_1/M_2 = \eta(x_1,\ldots,x_n,x) = \text{constant}$.

For a complete presentation of the last multiplier theory and applications we refer the interested reader to the classic treatise of Whittaker.[5] For our purpose, we only consider the system of equations

$$\frac{dx}{v} = \frac{dv}{f(x,v,t)} = dt \tag{18}$$

which corresponds to the most general equation of motion $\ddot{x} = f(x,\dot{x},t)$

of a classical system with one degree of freedom. In that case, an additional property of the last multiplier is that if $L(x,v,t)$ is a Lagrangian function for $\ddot{x} = f(x,\dot{x},t)$ then $\partial^2 L/\partial v^2 = M$ is a last multiplier. Therefore, the determination of a Lagrangian function L reduces to the determination of a last multiplier M given by

$$\frac{\partial (Mv)}{\partial x} + \frac{\partial (Mf)}{\partial v} + \frac{\partial M}{\partial t} = 0 \tag{19}$$

Furthermore two different last multipliers M_1 and M_2 will necessarily generate equivalent Lagrangians L_1 and L_2 with corresponding Hamiltonians $H_1(x,p_1,t)$ and $H_2(x,p_2,t)$ which are not related by a canonical transformation; however, they are q-equivalent in the sense of Currie and Saletan[6] with the "fouling" constant $M_1/M_2 = \eta(x,v,t) =$ constant.

From now on we restrict our attention to the important class of autonomous dynamical systems, that is systems with an equation of motion $\ddot{x} = f(x,\dot{x})$ invariant under time translation.

In this case the last multipliers $M(x,v)$ which don't explicitly depend on time are given by

$$\frac{\partial (Mv)}{\partial x} + \frac{\partial (Mf)}{\partial v} = 0$$

But this is precisely the p.d.e defining an integrating factor $I(x,v)$ of the equation $vdv/dx = f(x,v)$ in the phase plane (x,v) of the system. Hence $M(x,v) = I(x,v) = \partial^2 L/\partial v^2$. (20)

Let $C(x,v) = \alpha$ be the integral of

$$I(x,v)\,[vdv - f(x,v)dx] = dC(x,v) = 0 \tag{21}$$

The constant of the motion $C(x,v) = \alpha$, implied by the time-invariance of the equation of motion, can be transformed into a Hamiltonian $H(x,p)$

defined by $H(x,p) = C(x,v)$ provided we define the conjugate momentum p in accordance with (20), by

$$p = \partial L/\partial v = \int I(x,v)\partial v \ . \tag{22}$$

Using these definitions and (21) one can easily check that the Hamilton equation $dx/dt = v = \partial H/\partial p = \partial C/\partial v \ \partial v/\partial p = Iv \ \partial v/\partial p$ is verified with $I\partial v/\partial p = 1$ in agreement with (22).

It is well-known that if $I(x,v)$ is an integrating factor, then $I_1(x,v) = C(x,v)I$ is also an integrating factor. Each trajectory of the system in the phase plane is characterized by choosing a particular value of α in the expression

$$I_1/I = C(x,v) = \alpha \ .$$

If the system has a limit cycle which is approached by several trajectories, corresponding to different values of α, it is clear that as $t \to \infty$ the expression I_1/I must become of the form $0/0$ or ∞/∞ in order to remain consistent with the different values of α. Hence on a limit cycle, $C(x,v)$ must become indeterminate; hence if a limit cycle exists, it is necessarily a subset of the set of points where $(C(x,v)$ ceases to be defined (or, equivalently, where $I(x,v)$ vanishes or becomes equal to $\pm\infty$). We shall find an example of this behavior in the case of Smith's oscillator treated in the following section.

Remark: For the Lienard equation, written in terms of X and τ, we have the system of equations

$$\frac{dX}{v} = \frac{dv}{-v-\tilde{g}(X)} = d\tau$$

and $\frac{\partial(Mv)}{\partial X} - \frac{\partial[M(v+\tilde{g}(X))]}{\partial v} + \frac{\partial M}{\partial \tau} = 0$

A last multiplier is obviously $M = e^{\tau} = \partial^2 L/\partial v^2$ leading immediately to (6).

4. The Smith's nonlinear oscillator

The equation of motion of Smith's oscillator is

$$\ddot{x} + [2A(x) + xA'(x)]\dot{x} + x[\omega^2 + A^2(x)] = 0 \quad (23)$$

with $A = bx^n - a$, $a > 0$, and its solution was found to be

$$x = \cos(p+\omega t)[qe^{-nat} + nbe^{-nat} \int_0^t e^{na\theta} \cos^n(p+\omega\theta)d\theta]^{-1/n} \quad (24)$$

where p, q are arbitrary constant.

It is not difficult to see that as $t \to \infty$, $x(t)$ tends towards a periodic function and that there is exactly a limit cycle in the (x,v) plane attracting all trajectories.

Let us find the constant of the motion implied by the time-translation invariance of (23). This is the solution of

$$vdv/dx + [2A(x) + xA'(x)]v + x[\omega^2 + A^2(x)] = 0$$

We found an integrating factor of this equation, namely

$$I(x,v) = \frac{1}{x^2(1+U^2)h(x,U)}$$

where $U = \frac{1}{\omega}[v/x + A(x)]$

and $h(x,U) = bx^n[(U - \frac{a}{\omega})^n + a_2(U - \frac{a}{\omega})^{n-2} + a_3(U - \frac{a}{\omega})^{n-3} + \ldots + a_n] + c$

where the coefficients are given by the relations

$$k\omega^2 a_k + 2(k-1)\omega a\, a_{k-1} - (\omega^2 + a^2)(n-k+2)a_{k-2} = 0$$

for $k = 2,3,\ldots,n$, with $a_0 = 1$, $a_1 = 0$ and $2n\omega a\, a_n - (\omega^2+a^2)a_{n-1} = - n\omega^2 c$. Using this integral factor, we get the solution

$$C(x,v) = \ell n[x^2(1+U^2)]^n - \ell n[h(x,U)]^2 + 2n\,\frac{a}{\omega}\,\text{arctg}\,U = \alpha \qquad (25)$$

which, as we know, can be transformed into a time-independent Hamiltonian. We observe that $I(x,v)$ becomes infinite when $h(x,U) = 0$; this is the equation of the limit cycle and we see that it is an algebraic curve of degree n, a result which would have been difficult to obtain on the basis of (24) alone. The integrating factor becomes infinite also when $x^2(1 + U^2) = 0$, that is when $x = 0$ <u>and</u> $v = 0$; in fact the origin of the phase plane is a fixed point which can be considered as a (degenerate) limit cycle as $t \to -\infty$.

If we introduce U in equation (23) we find

$$dU/\omega dt + 1 + U^2 = 0$$

whose solution is $U + tg(\omega t + \theta) = 0$, representing a first-integral of the system. Replacing $\text{arctg}\,U$ by $-(\omega t + \theta)$ in (25), we can see that when $t \to \infty$ and $h(x,U) \to 0$ as the trajectories approach the limit cycle, the relation $C(x,v) = \alpha$ tends to the undeterminate form $(\infty - \infty) = \alpha$ for every α.

<u>Remark</u>: If in the Smith's oscillator we choose $b = 0$ and $a = -\varepsilon$, $\varepsilon > 0$, we find the linearly damped oscillator, and (25) reduces to the results given in the literature.[3]

5. Conclusion

It was possible to solve the Hamilton-Jacobi equation associated with a special time-dependent Hamiltonian $H(X,p,\tau)$ which is easily obtained for any Lienard equation of motion. This Hamiltonian, when expressed in terms of the original variables (x,v), or in terms of the "energy" $E = \frac{1}{2} v^2 + V(x)$ of the system, takes the following interesting forms:

$$H(X,p,\tau) = [\frac{1}{2} v^2 + V(x)e^{\tau} = E e^{\tau} = e^{\tau_0} EI(x,v)/I(x_0,v_0)$$

We can also notice that the space component $Q(x,\alpha)$ of the Hamilton-Jacobi function we used, is proportional to the energy E. The Hamilton-Jacobi equation can be solved provided an integrating factor $I(x,v)$ of the equation in the phase plane can be found. The integrating factor is also useful for generating a time-independent Hamiltonian when the equation of motion is invariant under time translation. The analytic behaviour of this Hamiltonian and of the integrating factor give useful information on the existence of limit cycles.

References

1. Havas, P., *Phys. Rev.*, 83, 224, 1951; *Nuovo Cimento Suppl.*, 5, 363, 1957.
2. Denman, H.H., On linear friction in Lagrange's equation, *Am. J. Phys.*, 34, 1147, 1966.
3. Denman, H.H., and Buch, L.H., Solution of the Hamilton-Jacobi equation for certain dissipative classical mechanical systems, *J. Math. Phys.*, 14, 326, 1973.
4. Smith, R.A., A simple non-linear oscillation, *J. London Math. Soc.*, 36, 33, 1961.
5. Whittaker, E.T., *A Treatise on the Analytical Dynamics of Particles and Rigid Bodies*, Fourth edition, Cambridge University Press, 1965, 276.
6. Currie, D.G., and Saletan, E.J., q-Equivalent particle Hamiltonians I. The classical one-dimensional case, *J. Math. Phys.*, 7, 967, 1966.

SYMPLECTIC GROUP, QUANTUM MECHANICS AND ANOSOV'S SYSTEMS.

by
André AVEZ
Département de Mécanique, Université Paris VI
2 Place Jussieu, 75005 Paris, France

Abstract.

Dirac's canonical quantization is studied in the general framework of symplectic Mechanics. Existence and uniqueness of such a quantization are investigated. Particular attention is paid to R^{2n}: in particular it is proved that there is only one canonical quantization which recovers the usual spectrum of the harmonic oscillator.

Introduction.

My lectures are devoted to canonical quantization as introduced by Dirac[10] in 1925. Unfortunately it is an old-fashioned topic ! Mainly because people thought the procedure was impossible as soon as polynomials of degree greater than two are involved (see[13], p.41-51). Impossibility arises from the following requirement: *any operator* which commutes with position and momentum operators q and p must be proportional to the identity operator I. This is by no means necessary, we need just to require that any operator, *which is the image of an observable in the correspondence principle* and which commutes with q and p, be proportional to I.

It turns out that canonical quantization is always possible under this milder condition.

This paper is divided into five parts. Parts 1 and 2 are a brief survey of symplectic Mechanics, quantum Mechanics and canonical quantization. Except for notations, they can be skipped by experts. In Part 3 I give a canonical quantization of compact symplectic manifolds and I prove its irreducibility. Results seem new, but meaningless for Physics: symplectic manifolds of Physics are cotangent bundles and they are never compact. However this shows the method for Part 4.

Physicists may go directly to Parts 4 and 5, which deal with the quantization of $R^{2n} = \{(q,p)\}$ under the usual Poisson bracket. In Part 4 I exhibit the canonical quantization of Van Hove and Souriau. In Part 5 I investigate the uniqueness under algebraic and unitary equivalences. The key role is devoted to the harmonic repulsor (repulsion proportional to the distance): its Hamiltonian generates a one-parameter group which looks like the geodesic flow of the Poincaré upper half-plane. A second observation is crucial: the Lie algebra of the polynomials in (q,p) under Poisson bracket is generated, as a Lie algebra, by two elements. This means that we know how to quantify any polynomial as soon as we know how to do it for the two generators.

A few more words: I give (or sketch) the proofs if they do not exist in the literature, otherwise I give a reference.

Part 1. SYMPLECTIC MECHANICS

(A good reference is,[1] where proofs can be found)

1.1. Symplectic manifolds.

Let ω be a 2-form on a smooth manifold M. We say that ω is non-degenerate if the linear map

$$T_m M \longrightarrow T_m^* M$$
$$X \rightsquigarrow \omega(X,\cdot) = i_X \omega$$

of the tangent space at $m \in M$ into its dual is an isomorphism for any m. If ω is closed and non-degenerate, we say that (M,ω) is a symplectic manifold. Example (the harmonic oscillator): $M = R^2 = \{(q,p)\}$, $\omega = dq \wedge dp$.

We recall some fundamental facts about symplectic manifolds:

a) M has even dimension, say $2n$;

b) $\omega^n = \omega \wedge \ldots \wedge \omega \neq 0$ (it is the Liouville volume element of M);

c) Darboux' theorem: around each $m \in M$ there exist local coordinates, the so-called canonical coordinates $(q_1,\ldots,q_n, p_1,\ldots,p_n)$. such that $\omega = \Sigma dq_i \wedge dp_i$, $i = 1,\ldots,n)$.

1.2. Automorphisms.

An automorphism of a symplectic manifold (M,ω) is a diffeomorphism $f: M \to M$ of M onto M which preserves ω : $\omega = f_* \omega$ (inverse image of ω under f). Automorphisms form a group under composition of maps: *the symplectic group* $S(\omega)$.

The generator of a one-parameter subgroup of $S(\omega)$ is a vector X on M such that the Lie derivative $L_X\omega$ of ω with respect to X vanishes. The set of the vector fields Y such that $L_Y\omega = 0$ is a Lie algebra under the Lie bracket $[\ ,\ ]$: it is called the Lie algebra of the Hamiltonian vector fields or, somewhat improperly, the Lie algebra $s(\omega)$ of $S(\omega)$. I say "improperly" because $Y \in s(\omega)$ does not necessarily generates a one-parameter subgroup of $S(\omega)$, except, for instance, if Y has compact support.

Now, let me explain why we call $s(\omega)$ the Lie algebra of *Hamiltonian* vector fields. Take $X \in s(\omega)$; since ω is closed $d(i_X\omega) = d(i_X\omega) + i_X d\omega = L_X\omega = 0$ and $i_X\omega$ is a closed 1-form. By the Poincaré lemma there exists (at least locally) some $H \in C^\infty(M, \mathbf{R}) = F$ *(vector space of the real-valued* C^∞ *functions on* M*)* such that $i_X\omega = -dH$. Write this relation in canonical coordinates; the components of X are

$$X_i = -\frac{\partial H}{\partial p_i}, \quad X_{i+n} = \frac{\partial H}{\partial q_i}, \quad i = 1,\ldots,n,$$

and therefore the integral curves of X satisfy the familiar Hamiltonian equations

$$\frac{dq_i}{dt} = -\frac{\partial H}{\partial p_i} , \qquad \frac{dp_i}{dt} = \frac{\partial H}{\partial q_i} .$$

To summarize: it is equivalent to give Hamiltonian equations with Hamiltonian H or to give $X \in s(\omega)$ such that $i_X\omega = - dH$.

Finally, since the map $X \rightsquigarrow i_X\omega$ is an isomorphism, $i_X\omega = - dH$ defines X unambiguously as a function of H: we shall write $X = X_H$. Example (the harmonic oscillator): $H = -(p^2/m + k^2 \cdot q^2)/2$ and $X_H = (p/m, k^2 \cdot q)$.

1.3. Observables and Poisson bracket.

An observable is a function $f \in F$. Let X_H be an Hamiltonian vector field and look to the evolution of f along an integrable curve of X_H :

$$X_H(f) = \frac{df}{dt} = \sum \left(\frac{\partial f}{\partial q_i} \cdot \frac{dq_i}{dt} + \frac{\partial f}{\partial p_i} \cdot \frac{dp_i}{dt}\right) =$$

$$- \sum \left(\frac{\partial f}{\partial q_i} \cdot \frac{\partial H}{\partial p_i} - \frac{\partial f}{\partial p_i} \cdot \frac{\partial H}{\partial q_i}\right) .$$

Now, define the Poisson bracket of $u, v \in F$ by

$$\{u,v\} = \sum \left(\frac{\partial u}{\partial q_i} \cdot \frac{\partial v}{\partial p_i} - \frac{\partial u}{\partial p_i} \cdot \frac{\partial v}{\partial q_i}\right)$$

We may write

$$X_H(f) = \frac{df}{dt} = \{H,f\} \tag{1.4}$$

Canonical coordinates play an auxiliary role and $\{u,v\}$ can be defined intrinsically :

$$\{u,v\} \cdot \omega^n = n \cdot du \wedge dv \wedge \omega^{n-1} \tag{1.5}$$

The Poisson bracket is obviously bilinear and alterned. With (1.5) one proves easily the Jacobi identity. Thus F is a Lie algebra under Poisson bracket. Now compute the Lie bracket of X_u and X_v (use (1.4) and the Jacobi identity):

$$[X_u, X_v] = X_{\{u,v\}} .$$

Therefore $f \in F \rightsquigarrow X_f \in s(\omega)$ is an homomorphism of Lie algebras. Its kernel is the set of the u such that $0 = X_u(g) = \{u,g\}$ for any $g \in F$, i.e. $du = 0$. *If* M *is connected* the kernel reduces to R (the set of the constant functions) and $F/R \to s(\omega)$ is an isomorphism of Lie algebras. Thus a function f (modulo an additive constant) can be identified with $X_f \in s(\omega)$. This led Heslot[12] to the following definition: *a classical observable* f *must no longer be regarded as a function, but as a generator of the automorphism group.*

Part 2. QUANTUM MECHANICS AND CANONICAL QUANTIZATION

2.1. The phase space.

Let E be a complex separable Hilbert space, $< , >$ its inner product and $\| \ \|$ its norm. The phase space of quantum Mechanics is the projective space $\hat{M}$ of E; it is the quotient of $E - \{o\}$ by an equivalence relation; $u, v \in E - \{o\}$ are identified if $v = k \cdot u$ for some $k \in C$. The following expression is invariant under this equivalence

$$\frac{|<u,v>|}{\|u\| \cdot \|v\|}, \qquad u, v \in E - \{o\} \quad , \tag{2.2}$$

it is the structure of $\hat{M}$, just as ω is the structure of the symplectic manifold (M, ω).

2.3. Automorphisms.

The automorphisms of $\hat{M}$ are the bijections of $\hat{M}$ into itself which preserve its structure. *Wigner's theorem* tells what they are: they are the restrictions to $\hat{M}$ of the unitary and antiunitary operators of M. Two such operators U and V being identified if $V = e^{ir} \cdot U$ for some $r \in R$ (see for a proof: V.S. Varadarajan. Geometry of quantum theory. Vol.1. Van Nostrand (1968), Chap. 7).

These automorphisms form a group $\mathrm{Aut}(\hat{M})$ under composition of maps. The component of the identity of $\mathrm{Aut}(\hat{M})$ arises from the unitary group.

Let us look for the Lie algebra of $\mathrm{Aut}(\hat{M})$. First, the "Lie algebra" of the unitary group is *the space* A *of the skew-adjoint operators* since (Stone's theorem): if U_t , $t \in R$, is a continuous one-parameter group of unitary operators, there exists a skew-adjoint operator A such that $U_t = e^{tA}$. To get the "Lie algebra" of $\mathrm{Aut}(\hat{M})$, recall that two unitary operators must be identified if they differ from a factor e^{ir}, $r \in R$. Therefore *the "Lie algebra" of* $\mathrm{Aut}(\hat{M})$ *is* A/iR: two skew-adjoint operators must be identified if they differ from some $i.r.I$, $r \in R$, I = identity. Exactly as in classical Mechanics *the observables of quantum Mechanics are the elements of the Lie algebra of the automorphisms of the phase space* $\hat{M}$.

The analogue of Hamilton's equations is the Schrödinger's equation $\frac{\partial f}{\partial t} = H(f)$, if $f \in E$ belongs to the domain of $H \in A$.

2.4. Canonical quantization.

Let H be an Hamiltonian on a symplectic manifold (M, ω). How to get a quantum picture of this system? Dirac's rule says we must associate quantum observables (i.e. skew-adjoint operators modulo $i.R$) to classical observables in a way which makes correspond bracket operator to Poisson bracket. In other words we must look for an homomorphism $F/R \to A/iR$ of Lie algebras. This means that we must look for an homomorphism $F \xrightarrow{\wedge} A$ of Lie algebras *(called a Dirac representation)* which is compatible with the equivalences in F and A: $f, g \in F$ and $g - f \in R$ must imply $\hat{g} - \hat{f} \in iR.I$. To be precise:

1) $F \xrightarrow{\wedge} A$ is linear ;

2) $\widehat{\{u, v\}} = [\hat{u}, \hat{v}] = \hat{u}\hat{v} - \hat{v}\hat{u}$ for any $u, v \in F$;

3) $\hat{1} = i \frac{h}{2\pi} I$, where $h \in R$ may be thought as the constant of Planck (we shall take $h = 2\pi$!!).

In fact one may ask for less: one replaces F by a subalgebra which contains the given Hamiltonian H and enough generators to generate a group of automorphisms acting transitively on M. Example (the harmonic oscillator): the subalgebra is spanned by the Hamiltonian and q,p : the translation group of $R^2 = \{(q,p)\}$ acts transitively on R^2 and its Lie algebra is spanned by q,p .

Part 3. QUANTIZATION OF THE COMPACT SYMPLECTIC MANIFOLDS.

3.1. Background.

Let (M,ω) be a compact connected symplectic manifold of dimension 2n . The volume element ω^n defines a Lebesgue measure. Therefore one may define an inner product on $C^\infty(M,C)$ and extend it to the Hilbert space L^2 consisting of all complex-valued ω^n-measurable functions on M, such that $|f|^2$ is ω^n-integrable: $< f,g > = \int f\cdot g^*\cdot\omega^n$, where $*$ is the conjugation. It is useful to introduce the mean-value $\overline{f}$ of f : $\overline{f} = < f,1 > / < 1,1 >$.

I shall need two elementary facts:

$$< \{f,g\},1 > = 0 \ ; \qquad f,g \in C^1(M,R). \tag{3.2}$$

Proof: Since ω is closed, (1.5) implies $\{f,g\}\cdot\omega^n = d[n\cdot f\cdot dg_\wedge\omega^{n-1}]$. Now use boundary $(M) = \phi$ and Stoke's theorem.

$$< \{f,g\},h > = - < g, \{f,h\} > \ ; \quad f,g,h \in C^1(M,R). \tag{3.3}$$

Proof: Use $\{f,g\cdot h\} = \{f,g\}\cdot h + \{f,h\}\cdot g$ and (3.2).
Now, let us construct a Dirac representation of $F = C^\infty(M,R)$ into the Lie algebra of the skew-adjoint operators of L^2.

3.4. Existence.

The map $H \rightsquigarrow \hat{H}$, where $\hat{H}$ is the closure of the operator defined on $C^1(M,R)$ by $f \rightsquigarrow \{H,f\} + i\,\overline{H}\cdot f$, is a Dirac representation.

Proof: (3.3) shows that $X_H = \{H,\cdot\}$ is a skew-adjoint operator on $C^\infty(M,R)$. Since $C^\infty(M,C)$ is dense in L^2 and invariant under X_H, the closure of X_H is a skew-adjoint operator on L^2. Of course the same result holds for the closure $\hat{H}$ of $X_H + i\cdot\overline{H}\cdot I$.

Observe that the $\hat{H}$'s have a common dense domain $D = C^\infty(M,\mathbb{C})$ and that $\hat{H}D \subset D$. Now, a routine computation shows that $\widehat{\{f,g\}}(h) = [\hat{f},\hat{g}](h)$ for $h \in D$ (use (3.2) and the Jacobi identity). A density argument proves that $\widehat{\{f,g\}} = [\hat{f},\hat{g}]$.

Finally $\hat{1} = i\cdot I$ is trivial.

Do there exist other Dirac representations ^, where $\hat{H}$ is the closure of an operator defined on $C^\infty(M,\mathbb{C})$ by

$$\hat{H}(f) = \{H,f\} + i\cdot D(H)\cdot f, \tag{3.5}$$

where $D: F \to F$ is linear ? An easy computation shows that D must satisfy $D(1) = 1$ and $D\{u,v\} = \{Du,v\} + \{u,Dv\}$; $u,v \in F$. Thus D must be a derivation of F into F. All these derivations were found in[9] which was in fact taylored for this purpose. Using this result and $D(1) = 1$ we get:

3.6. Theorem.

A Dirac representation, such that $\hat{H}$ has the form (3.5) on $C^\infty(M,\mathbb{C})$ is $\{H,\cdot\} + i[\{g,H\} + \overline{H}]\cdot I$, for some $g \in F$. (3.4 is the particular case $g = 0$).

Let us prove that the previous Dirac representations are irreducible.

3.7. Theorem.

The representation ^ of theorem 3.4 is irreducible: the only invariant subspaces of L^2 are $\{o\}$, $\mathbb{C}$ and L^2.

Proof: Since $\mathbb{C}$ is obviously invariant we may rephrase 3.7: the only invariant subspaces of the orthocomplement of $\mathbb{C}$ (we call it again L^2) are $\{o\}$ and L^2.

Let us express first the invariance under $\hat{H}$, *where* $H \in F$, $\overline{H} = 0$. We have $e^{t\hat{H}}f = f \circ \varphi_t$ where φ_t is the one-parameter group of $\mathrm{Aut}(\omega)$ generated by $X_H = \{H,\cdot\}$. Hence the real and imaginary parts of L^2 are both invariant under $e^{\hat{H}}$ if $\overline{H} = 0$ and *we may assume first that* L^2 *is a real Hilbert space*.

I introduced once (see[5, 6]) a symplectic Laplacian: since M is compact, there exists a finite set of $f_r \in F$, even with $\overline{f_r} = 0$, such that the vectors $\{f_r,\cdot\}$ span the tangent space $T_m M$ for any $m \in M$. I proved that $\Delta = -\Sigma\{f_r, \{f_r,\cdot\}\}$ is a Laplacian (linear second-order operator, elliptic and self-adjoint). Since we took for L^2 the orthocomplement of the constants, Δ^{-1} is well-defined and it is a compact operator.

Now, let L be an invariant subspace and $p: L^2 \to L$ the orthogonal projection. Since L is invariant under $\hat{H}$ if $\overline{H}=0$ and since $\overline{f_r} = 0$ and $\Delta = -\Sigma(\hat{f}_r)^2$, it follows that $\Delta p = p\Delta$. But Δ and Δ^{-1} are self-adjoint and Δ^{-1} is compact, then $\Delta^{-1}p = p\,\Delta^{-1}$ everywhere, and L is invariant under Δ^{-1}. This implies that $p\,\Delta^{-1} = \Delta^{-1}p$ is a compact self-adjoint operator of L and, therefore, there exists a basis a_n of eigenfunctions in L: $\Delta^{-1}p(a_n) = r_n \cdot a_n$, $r_n \in R$. Apply Δ: $a_n = r_n \cdot \Delta a_n$, and the a_n are eigenfunctions of Δ. It is classical that the a_n are C^∞. Thus we proved that $L \cap F$ is dense in L. Similarly $L^\perp \cap F$ is dense in the orthocomplement $L^\perp$ of L.

Now take $u \in L \cap F$, $v \in L^\perp \cap F$. The invariance of L implies $\{H,u\} \in L \cap F$ for any $H \in F$. Use (3.3): $0 = <\{H,u\},v> = <\{u,v\},H>$. Since H is arbitrary, $\{u,v\} = 0$. We deduce $\{u,v^2\} = 0$, and a new application of (3.3) gives $0 = <\{v^2,u\},f_r> = <u, \{f_r,v^2\}>$. Since $L \cap F$ is dense in L and u arbitrary we get $\{f_r, v^2\} \in L^\perp \cap F$. Repeat the argument: $\{f_r, \{f_r,v^2\}\} \in L^\perp \cap F$. Sum over r: $\Delta(v^2) \in L^\perp \cap F$. Since $L^\perp$ is invariant under Δ^{-1} we obtain $v^2 - \overline{v^2} \in L^\perp \cap F$. Similarly $u^2 - \overline{u^2} \in L \cap F$ if $u \in L \cap F$. Therefore $< u^2 - \overline{u^2}, v^2 - \overline{v^2} > = 0$, i.e. $< u^2, v^2 > = \overline{u^2} \cdot \overline{v^2}$, for any $u \in L \cap F$ and any $v \in L^\perp \cap F$. Now assume $L \neq \{o\}$ and take $u \in L \cap F$, $u \neq 0$. A standard argument shows that $v = 0$ if $v \in L^\perp \cap F$. A density argument proves that $L^\perp = \{o\}$, $L = L^2$.

Let us go back to the complex Hilbert space. Actually we proved that its real and imaginary parts have no non-trivial invariant subspaces under $\hat{H}$ if $\overline{H} = 0$. *Let us express, now, the invariance under* $\hat{H}$ *if* $H \in R - \{o\}$. We have $e^{\hat{H}} f = e^{iH} \cdot f$ and we see that $e^{\hat{H}}$ mix up real and imaginary parts. Then L^2 is the only one non-zero invariant subspace.

A similar argument proves that the representations of Theorem 3.6 are irreducible.

3.8. Problems.

Is any Dirac representation of $C^\infty(M,R)$ completely reducible ?

Are the irreducible Dirac representations of $C^\infty(M,R)$ unitarily equivalent ?

I next turn to the canonical quantization of non-compact manifolds. Let us consider a mechanical system with n degrees of freedom $q = (q_1, \ldots, q_n)$. The state space is $R^n = \{q\}$ and the phase space is $R^{2n} = \{(q,p)\}$, $p = \{p_1, \ldots, p_n\}$. Assume the potentials are polynomial in q (e.g. the harmonic oscillator). The Hamiltonian is a polynomial in (q,p), and this leads to the quantization of the Lie algebra P_n of the polynomials in (q,p) under Poisson bracket.

Part 4. EXISTENCE OF CANONICAL QUANTIZATION OF P_n

4.1. The construction.

Let L^2 be the Hilbert space of the complex square-integrable functions defined on $R^{2n} = \{(q,p)\}$; A is the Lie algebra of the skew-adjoint operators of L^2. We try to imitate the construction of Part 3. Let K be the space of the C^∞ complex functions defined on R^{2n}, with compact supports. We try to get a map $P_n \hat{\rightarrow} A$ such that, for any $H \in P_n$, $\hat{H}$ is the closure of the operator defined on K by

$$f \rightsquigarrow \{H,f\} + i \cdot D(H) \cdot f , \qquad f \in K \tag{4.2}$$

where $D: P_n \to P_n$ is linear.

It is easy to see, as in Part 3, that such a $\hat{H}$ is a linear skew-operator. On the other hand, as in 3.4, $\hat{1} = i\cdot I$ and $\{f,g\}^{\wedge} = [\hat{f},\hat{g}]$, for any $f,g \in P_n$, mean that D must be a derivation of P_n into itself such that $D(1) = 1$.

4.3. Derivations. (see[4,8]).

Let D be a derivation of P_n into itself. Then, there exist $P \in P_n$, $k \in \mathbf{R}$ such that $D = \{P,\cdot\} + k\cdot N$, where $N = I - \Sigma q_i \cdot \frac{\partial}{\partial q_i}$.

Proof: $T(1) = 0$ for an inner derivation T. *Let us prove the converse.* The relations

$$\{q_i,p_j\} - \delta_{ij} = \{q_i,q_j\} = \{p_i,p_j\} = 0$$

imply

$$\{T(q_i),p_j\} + \{q_i,T(p_j)\} = 0, \text{ i.e.}$$

$\frac{\partial}{\partial q_j} T(q_i) = -\frac{\partial}{\partial p_i} T(p_j)\ldots$ etc. Thus the 1-form

$\Sigma(T(p_i)dq_i - T(q_i)dp_i)$ is closed and there exists $P \in F$ such that

$T(q) = -\frac{\partial P}{\partial p}$, $T(p) = \frac{\partial P}{\partial q}$ (in fact $P \in P_n$ since $T(q),T(p) \in P_n$).

From its very construction the derivation $I = T - \{P,\cdot\}$ satisfies $I(1) = I(q) = I(p) = 0$. Therefore I commutes with $\partial/\partial q$ and $\partial/\partial p$; for instance $\frac{\partial}{\partial q} If = \{If,p\} = I\{f,p\} = I\frac{\partial}{\partial q} f$. If $Q \in P_n$, $\deg(Q) = 2$, this implies

$$\frac{\partial}{\partial q}[I(Q)] = I(\frac{\partial Q}{\partial q}) = 0, \quad \frac{\partial}{\partial p}[I(Q)] = 0, \quad \text{and } I(Q) \in \mathbf{R}.$$

But any monomial of degree 2 is the Poisson bracket of two similar monomials; therefore $I(Q) = 0$. Now let us show that I commutes with the product by q_i or p_i (this is a Heslot's idea). Any $R \in P_n$ can be written

$R = \frac{\partial X}{\partial p} = \{q,X\}$, $X \in P_n$. Then

$$2q\cdot I(R) = 2q\cdot I(\frac{\partial}{\partial p} X) = 2q\frac{\partial}{\partial p} I(X) =$$

$$2q\cdot\{q,I(X)\} = \{q^2,I(X)\} = I\{q^2,X\} =$$

$$2I(q\cdot\{q,X\}) = 2\cdot I(q\cdot R).$$

This implies readily $I=0$; therefore $T = \{P,\cdot\}$.

On the other hand one checks easily that N is a derivation and that $N(1) = 1$. Now, if D is a derivation, $T = D - D(1)\cdot N$ satisfies $T(1) = 0$ and therefore $T = \{P,\cdot\}$, $P \in \mathcal{P}_n$. The theorem follows.

Using 4.2, 4.3 and $D(1) = 1$, we get:

4.4. Theorem.

A Dirac representation such that $\hat{H}$ has the form 4.2 on K is

$$\{H,\cdot\} + i[\{P,H\} + H - \Sigma q_i \frac{\partial H}{\partial q_i}] \quad \text{for some } P \in \mathcal{P}_n.$$

The case $P = 0$ was known of Van Hove[16] and Souriau.[14] Streater[15] obtained the general case by a different method. A similar result holds for other non compact manifolds (use theorems 3 and 4 of Ref[9]).

Part 5. UNIQUENESS OF THE CANONICAL QUANTIZATION OF $\mathcal{P}_n$.

5.1. Introduction (see Ref. 1, p.435).

For simplicity I shall take $n=1$, but everything works in the general case. Here $\mathcal{P} = \mathcal{P}_1$ and $\mathcal{H}$ is an *abstract* complex separable Hilbert space; $\mathcal{A}$ is the set of the skew-adjoint operators of $\mathcal{H}$. A Dirac representation is a map $\mathcal{P} \to \mathcal{A}$ satisfying:

a) For each finite subset E of $\mathcal{P}$ there is a dense subspace S(E) of $\mathcal{H}$, invariant under $\hat{f}$ and contained in the domain of $\hat{f}$ for all $f \in E$ (if $E = \{a,b,\dots\}$ we write $S(a,b,\dots)$);

b) $(f+g)\hat{} = \hat{f}+\hat{g}$ on $S(f;g)$ and
$(k\cdot f)\hat{} = k\cdot\hat{f}$ if $k \in R$

c) $\{f,g\}\hat{} = [\hat{f},\hat{g}]$ on $S(f;g)$;

d) $\hat{1} = i\cdot I$.

A Dirac representation is *irreducible* if no proper closed subspace in H is carried into itself by all the $e^{\hat{f}}$, $f \in P$.

Two Dirac representations ^ and - in A are *unitarily equivalent* if there exists a unitary operator U of H such that $\bar{f} = U\hat{f}U^{-1}$ for all $f \in P$.

Do there exist many Dirac representations ? Are two irreducible Dirac representations unitarily equivalent ? The Lie algebra P presents strong rigidity features which imply there are "few" Dirac representations.

5.2. Algebraic uniqueness.

Let us prove (see Ref. 4) that any Dirac representation is faithful. Consequence: two Dirac representations are algebraically isomorphic.

Theorem (Ref. 4, 8).

The only ideals of the Lie algebra P are $\{o\}$ R and P. In particular P/R is a simple Lie algebra.

Proof: An ideal $I \neq \{o\}$, R, contains a polynomial P of positive degree. The derivatives of P belong to I since $\frac{\partial P}{\partial p} = \{q,P\}$, $\frac{\partial P}{\partial q} = \{P,p\}$. Therefore we may assume degree (P) = 1 and P = p + ... or q + ... Then $2q = \{q^2,P\} \in I$ or $2p = \{P,p^2\} \in I$. Since any polynomial f may be written $f = \{q,X\}$ or $\{X,p\}$, $X \in P$, we see that $f \in I$.

Isomorphism follows: $\hat{1} = i\cdot I$ implies Ker(^) $\neq$ R, P and the ideal Ker(^) reduces to $\{o\}$, viz. $\hat{P}$ is isomorphic to P.

Remark: Let G be the smallest closed subgroup of the unitary group $U(H)$ which contains all the $e^{\hat{f}}$, $f \in P$. The simple Lie algebra P/R is, formally, the Lie algebra of G/S^1. One can compute the formal curvature of G/S^1 for a left-invariant Riemannian metric. It turns out that this curvature is $\leqslant 0$. In a subsequent paper I shall use this fact to prove a metric rigidity theorem in the spirit of Selberg et al. (e.g.: G.D. Mostow. Strong rigidity of locally symmetric spaces. Annals of Math. Studies, n°78. Princeton Univ. Press, 1973).

5.3. Very few is needed to know a Dirac map.

Let us prove that ^ is known once we know $\hat{f}$ and $\hat{g}$ for some suitable polynomials f and g .

We say that a Lie algebra L is generated by a subset E of L if L is the smallest Lie subalgebra which contains E (any element of L is obtained from elements of E by repeated linear combinations and Poisson bracket

Theorem (Ref. 4, 8).

P is generated by q and $p^2 + q^4$ (see Ref. 8 for a proof and a generalization to P_n).

Corollary.

P is generated by qp and $a = q + p^2 + q^3$.

Proof:

$\{qp,a\} = -3q^3 + 2p^2 - q = b,$

$3a + b = 5p^2 + 2q = c,$

$\{qp,c\} = 10p^2 - 2q = d,$

$15p^2 = c + d, \qquad 6q = -d + 2c, \qquad q^3 = a - p^2 - q,$

$18q^4 = \{q^3, \{q^3,p^2\}\}$.

Hence q and $p^2 + q^4$ are generated by qp and a.

This implies that, once we know $\hat{q}$ and $(p^2+q^4)\hat{}$ [resp. $\widehat{qp}$ and $\hat{a}$], we know everything. In particular let ^ and - be two Dirac maps; assume $\overline{qp} = U \widehat{qp} U^{-1}$, $\bar{a} = U \hat{a} U^{-1}$ for some unitary operator U for H. Then ^ and - are unitarily equivalent. Of course there exist some relations between $\widehat{qp}$ and $\hat{a}$ (or $\hat{q}$ and $(p^2+q^4)\hat{}$) which show that they cannot be chosen arbitrarily; for instance $\{q, \{q, p^2 + q^4\}\} = 2$ implies $[\hat{q}, [\hat{q}, (p^2 + q^4)\hat{}\,]] = 2iI$. These relations bring constraints to $\hat{a}$ once $\widehat{qp}$ is known and may help to reduce, or even determinate, $\hat{a}$ once $\widehat{qp}$ is known. On the other hand these relations are *obstructions*, as it is transparent from the usual counter-example to the existence of a Dirac map in which $\hat{q}$ and $\hat{p}$ act irreducibly on H (e.g. Ref. 1, p.435 and Ref. 13, p.41). It would be nice to know that P is *finitely presented* (any relation is a consequence of a finite number of relations $r = 0$). Such a result would be true, it would suffice to check that conjectural $\widehat{qp}$, $\hat{a}$ fullfil the conditions $r = 0$.

In the following I prove that $\widehat{qp}$ is determined up to a unitary equivalence, therefore there is only "one degree of freedom" $\hat{a}$ for a Dirac map $\hat{}$.

5.4. The Anosov's structure of P.

Let us call P^+, P^-, P^o the subspaces of P generated by $q^r \cdot p^s$, $s > r$ (resp. $s < r$, resp. $s = r$). They are subalgebras and P is their direct sum as a vector space. Let us look to the action of the one-parameter group $e^{t\cdot\widehat{qp}}$. The relation

$$\{qp, q^r p^s\} = (s - r)\cdot q^r p^s \quad \text{implies}$$
$$[X, A] = (s - r)\cdot A, \text{ where } \widehat{qp} = X, \ (q^r \cdot p^s)\hat{} = A$$

Take $u \in S(qp; q^r p^s)$ and write $\varphi_t(\hat{f}) = e^{tX}\cdot\hat{f}\cdot e^{-tX}$. Then

$$\frac{d}{dt}\varphi_t(A)(u) = \varphi_t\ [X,A](u) = (s - r)\cdot\varphi_t(A)(u) \text{ and}$$

$\varphi_t(A)(u) = e^{(s-r)t}\cdot A(u)$. Therefore, if $t > 0$, $u \in S(f;qp)$, we have

$$\|\varphi_t\ \hat{f}(u)\| \geqslant e^t\cdot\|\hat{f}(u)\| \qquad \text{if} \quad f \in P^+$$
$$\|\qquad\| \leqslant e^{-t}\cdot\|\qquad\| \qquad \text{if} \quad f \in P^-$$
$$\|\qquad\| = \|\hat{f}(u)\| \qquad \text{if} \quad f \in P^o .$$

Hence:

Theorem.

$\hat{P}$ splits into three subalgebras which are respectively dilating, contracting and stable with respect to the flow φ_t: $\hat{f} \rightsquigarrow e^{tX}.\hat{f}.e^{-tX}$ generated by $X = \widehat{qp}$: φ_t looks like an Anosov's flow (see Ref. 3, 7, for this notion).

Spectral properties of Anosov's flows are well-known and we shall use this theorem as a heuristic guide to prove spectral properties of $\widehat{qp}$.

Physical interpretation of $\widehat{qp}$. The linear canonical map $(q,p) \rightsquigarrow (q+p, (-q+p)/2)$ brings qp to the Hamiltonian $(p^2-q^2)/2$ of the "harmonic" repulsor $\ddot{q} = q$.

5.5. $e^{\widehat{qp}}$ has denumerable Lebesgue spectrum.

Definition (see [3]).

A unitary operator U of H has Lebesgue spectrum if there exists an orthogonal basis $e_{k,n}$ $(k \in K, n \in \mathbf{Z})$ such that $Ue_{k,n} = e_{k,n+1}$. The cardinality of K is called the multiplicity: it is finite or denumerable.

A continuous one-parameter group U_t of unitary operators has Lebesgue spectrum if each U_t , $t \neq 0$, has Lebesgue spectrum.

We need two facts:

a) Two unitary operators U and U' with Lebesgue spectrum of the same multiplicity are unitarily equivalent (A: $e_{k,n} \rightsquigarrow e'_{k,n}$ defines a unitary operator A and $U' = A^{-1} U A$).

b) A unitary operator U of H has Lebesgue spectrum if there exists a proper subspace L of H such that $UL \subset L$, $\bigcap_n U^n L = \{o\}$, closure of $[\bigcup_n U^n L] = H$. The multiplicity is $\dim(L \ominus UL)$ [$\ominus$ means orthocomplementation]

Theorem.

$e^{\widehat{qp}}$ has denumerable Lebesgue spectrum and, therefore, $\widehat{qp}$ is determined up to a unitary equivalence.

First proof: We need a *lemma* (Stone, Von Neumann): $e^{\hat{p}}$ *is ergodic.* Under the assumptions of 5.1 the Weyl relation holds:

$$e^{s\hat{q}} \cdot e^{t\cdot\hat{p}} \cdot e^{-s\hat{q}} = e^{ist} \cdot e^{t\hat{p}} \quad \text{for} \quad t,s \in \mathbf{R}$$

Therefore $e^{t\hat{p}}(u) = u$ implies $e^{t\hat{p}}(e^{-s\hat{q}}u) = e^{ist}(e^{-s\hat{q}}u)$. This shows that $u \neq 0$ would imply that any s would be an eigenvalue of $\hat{p}$. Since the set of these eigenvalues is at most denumerable, this is impossible.

Corollary: $-(\hat{p})^2$ *is positive definite.*

If $u \in S(p)$ then $\hat{p}(u) \in S(p)$ and $(\hat{p})^2 u$ exists. Since $\hat{p}$ is skew-adjoint we have $< -(\hat{p})^2 u, u > = \| \hat{p}\, u \|^2 > 0$ if $u \neq 0$.

Now let us go back to the theorem. Write $X = \widehat{qp}$, $B = -(\hat{p})^2$. The relation $\{qp,p\} = p$ implies $\widehat{qp}\cdot\hat{p} - \hat{p}\cdot\widehat{qp} = \hat{p}$ and, as a consequence, $XB - BX = 2B$ on $S(p\,;qp)$. After integration we obtain

$$e^{tX} \cdot B \cdot e^{-tX} = e^{2t} \cdot B \tag{5.6}$$

Let $\{E_\lambda\}$ be the spectral resolution of the identity for the positive self-adjoint operator B:

$$B = \int_{0^+}^{+\infty} \lambda\cdot dE_\lambda \,. \quad \text{We deduce}$$

$$e^{tX} \cdot B \cdot e^{-tX} = \int_{0^+}^{+\infty} \lambda\cdot d(e^{tX} \cdot E_\lambda \cdot e^{-tX}), \quad \text{and}$$

$$e^{2t} \cdot B = \int_{0^+}^{+\infty} \lambda \cdot e^{2t} \cdot dE_\lambda = \int_{0^+}^{+\infty} \lambda \cdot dE_{\lambda\cdot e^{-2t}} \,.$$

Let us use (5.6) and the uniqueness of the spectral resolution:

$$e^{tX} \cdot E_\lambda \cdot e^{-tX} = E_{\lambda\cdot e^{-2t}} \tag{5.7}$$

But since $\{E_\lambda\}$ is a spectral family of projections, it has the familiar properties:

a) The range of E_λ is an increasing function of λ;

b) As $\lambda \to -\infty$ range (E_λ) shrinks to $\{o\}$;

c) As $\lambda \to +\infty$ range (E_λ) tends to H.

Write $L = E_1 H$ and choose $t > 0$, then (5.7) implies:

1) $e^{tX} L = E_{e^{-2t}} H \subset E_1 H = L$;

2) $(e^{-tX})^n L = E_{e^{2tn}} H \to H$ as $n \to +\infty$,

and closure $[\bigcup_n (e^{tX})^n L] = H$;

3) $(e^{tX})^n L = E_{e^{-2tn}} H \to E_o H = \{o\}$ as $n \to +\infty$, and $\bigcap_n (e^{tX})^n L = \{o\}$.

Hence $e^{tX}, t > 0$, has Lebesgue spectrum of multiplicity $\dim(L \ominus e^{tX} L) = \dim(E_1 H \ominus E_{e^{-2t}} H = m(t)$. But $m(t)$ is an increasing function of t, $m(0) = 0$ and $m(+\infty) = \dim E_1 H = \dim(e^{-sX} E_1 H) = \dim(E_{e^{2s}} H) \to \dim H = \infty$ as $s \to +\infty$. Therefore $m(t) = +\infty$ if $t > 0$.

Second proof : (shorter, but less elementary):

Since $[\widehat{qp}, \hat{p}] = \hat{p}$, $\widehat{qp}$ and $\hat{p}$ span a Lie algebra, isomorphic to the Lie algebra of the affine group S_1 of $\mathbb{R}$. The only connected normal abelian subgroup of S_1 is:

$P = \{e^{t\hat{p}} : t \in \mathbb{R}\}$. Now, Mautner's lemma (Ref. 2, p.13) says: the spectrum of $e^{\widehat{qp}}$, restricted to the orthocomplement of $\{u: A(u) = u$ for all $A \in P\}$, is Lebesgue. Since $e^{t\hat{p}}$, $t \neq 0$, is ergodic, this orthocomplement is H.

Remark : Less can be proved in a simple way: $e^{\widehat{qp}}$ is strongly mixing. Just use the ergodicity of $e^{\hat{p}}$ and L.W. Green's technique (Ref. 11).

5.8. Corollary.

Let $F: \mathbb{R} \to \mathbb{R}$ be an arbitrary polynomial and k a positive integer. Let us write $X = \widehat{qp}$, $H = [p^k \cdot F(qp)]^\wedge$. Then

1) The point spectrum of H is $\{o\}$ or ϕ ;

2) e^{X+H} has denumerable Lebesgue spectrum.

Proof: $\{qp, p^k \cdot F(qp)\} = k \cdot p^k \cdot F(qp)$ implies $[X,H] = k \cdot H$ on $S\{qp; p^k \cdot F(qp)\}$. By integration we obtain

$$e^{-tX} \cdot e^{sH} \cdot e^{tX} = e^{s \cdot e^{-kt} \cdot H} \qquad \text{for } t, s \in \mathbf{R}$$

(compare to I. Segal and J. von Neumann. Annals of Math., 52, 509, 1950)

Assume u is an eigenvector of H, with eigenvalue $\Theta \in \mathbf{R}$. We deduce at once that $e^{tX}(u)$ is an eigenvector with eigenvalue $\Theta \cdot e^{-kt}$. Therefore $\Theta = 0$, since the point spectrum of H is at most denumerable. Let us call M the closed subspace spanned by $\{e^{tX}(u)\colon t \in \mathbf{R}\}$ and $M^{\perp}$ its orthocomplement. Both of them are invariant under e^{tX} and e^{sH}, and the restriction of e^{sH} to $M^{\perp}$ is ergodic. Now take the restriction to $M^{\perp}$ of $[X+H,H] = k \cdot H$. Since H is ergodic, we may use the argument of 5.5: the restriction to $M^{\perp}$ of e^{X+H} has Lebesgue spectrum. On the other hand $H+X = X$ on M, and, therefore, the restrictions of e^{H+X} and e^{X} to M coincide. But e^{X} has Lebesgue spectrum on $\mathcal{H}$ and it leaves M invariant; therefore e^{H} (and consequently e^{H+X}) has Lebesgue spectrum on M. This proves that e^{H+X} has Lebesgue spectrum on $\mathcal{H} = M \oplus M^{\perp}$.

Physical implication. The only one possible energy level of a Hamiltonian of the form $p^2/2 + p^2 \cdot F(qp)$ is zero.

5.9. Remark.

Let $A\colon \mathbf{R}^2 \to \mathbf{R}^2$ be a linear canonical mapping. It is clear that $f \rightsquigarrow (f_{\circ} A)\hat{}$ is a Dirac representation and that $(f_{\circ} A)\hat{}$ and $\hat{f}$ share the same spectral properties. Example: $(ap^2 - q^2/a)\hat{}$ has denumerable Lebesgue spectrum for any $a \in \mathbf{R}$.

Now we study the complex counterpart $(ap^2 + q^2/a)\hat{}$ of $(a \cdot p^2 - q^2/a)\hat{}$, $a \in \mathbf{R} - \{o\}$.

5.10. Harmonic oscillator and canonical quantization.

The Hamiltonian of the harmonic oscillator is $(p^2+m^2\omega^2q^2)/2m$, where m is the mass of the particle and $\nu = \omega/2\pi$ is the frequency. According to the usual quantum recipes, the corresponding operator on $L^2(R,dq)$ is

$$-\frac{h^2}{8\pi^2 m}\cdot\frac{d^2}{dq^2}+\frac{1}{2}m\,\omega^2 q^2\cdot I\,.$$

This operator has pure point spectrum, with simple eigenvalues $(n+1/2)\cdot h\cdot\nu$, $n = 0,1,\ldots$ Let us take this fact for granted and *let us look for Dirac representations* $\hat{}$ *in which* $\frac{1}{2m}(p^2+m^2\omega^2q^2)\hat{}$ *has pure point spectrum, with simple eigenvalues* $i(n+c)\cdot h\cdot\nu$, $n \geqslant 0$, $c =$ *constant.* It is convenient to introduce $B = (p^2/m\omega + m\omega\cdot q^2)\hat{}$, which must have simple eigenvalues $i(n+c)h/\pi$, $n \geqslant 0$, with corresponding normed eigenfunctions e_n.

First step. $\hat{q}$ *and* $\hat{p}$ *are determined.*

Let us introduce $\lambda = (m\omega\pi/h)^{1/2}$, $\mu = (\pi/hm\omega)^{1/2}$, $a = \lambda\hat{q}+i\mu\hat{p}$ and its adjoint a^*. From

$$\{p^2/m\omega + m\omega q^2,\ \lambda q + i\mu p\} = 2i(\lambda q + i\mu p)$$

we deduce

$$[B,a] = 2ia,\quad [B,a^*] = -2ia^*$$

The action on e_n gives

$$B\,a(e_n) = i\,[\,(n+c)(h/\pi) + 2\,]\cdot a(e_n)$$

$$B\,a^*(e_n) = i\,[\,(n+c)(h/\pi) - 2\,]\cdot a^*(e_n).$$

Thus $(n+c)(h/\pi)\pm 2$ must be an eigenvalue of B. This forces the Planck's constant h to be of the form $h = 2\pi/N$, N = positive integer. *We shall fix the unit of action by taking* $h = 2\pi$.

Now, the simplicity of the eigenvalues of B implies $a(e_n) = a_n\cdot e_{n+1}$, $a^*(e_n) = b_n\cdot e_{n-1}$, $b_0 = 0$. We have

$$a_n = \langle a(e_n), e_{n+1}\rangle = \langle e_n, a^*(e_{n+1})\rangle = b^*_{n+1}$$

and $[a,a^*] = -I$ gives us $|b_{n+1}|^2 - |b_n|^2 = 1$.

Finally, we obtain

$$a(e_n) = b^*_{n+1} \cdot e_{n+1},$$
$$a^*(e_n) = b_n \cdot e_{n-1},$$
$$|b_n| = n^{1/2}.$$

Let us show that the b_n are fixed up to a unitary transformation $U(e_n) = A_n \cdot e_n$, $|A_n| = 1$. Since $b_n = n^{1/2} \cdot \exp(i\theta_n)$, $\theta_n \in R$, we get

$$a(Ue_n) = A_n \cdot A^*_{n+1} \cdot \exp(-i\theta_{n+1}) \cdot (n+1)^{1/2} \cdot U(e_{n+1})$$
$$a^*(Ue_n) = A_n \cdot A^*_{n-1} \cdot \exp(i\theta_n) \cdot n^{1/2} \cdot U(e_{n-1}).$$

Take $A_o = 1$, $A_n = \exp(-i\theta_n) \cdot A_{n-1}$ and write again e_n instead of $U(e_n)$:

$$B(e_n) = 2i\,(n+c) \cdot e_n,$$
$$a(e_n) = (n+1)^{1/2} \cdot e_{n+1},$$
$$a^*(e_n) = n^{1/2} \cdot e_{n-1},$$

which determinate $\hat{q}$ and $\hat{p}$.

Second step: $\widehat{q^2}$, $\widehat{qp}$, $\widehat{p^2}$ *are determined.*

Let us introduce $X = \widehat{qp}$, $A = (p^2/m\omega - m\omega q^2)\hat{}$, $s = A + 2iX$ and its adjoint $s^* = -A + 2iX$. One checks easily $[B,s] = -4is$, $[B,s^*] = 4is^*$. The action on e_n gives

$$Bs(e_n) = 2i\,(n-2+c) \cdot s(e_n),$$
$$Bs^*(e_n) = 2i\,(n+2+c) \cdot s^*(e_n).$$

In view of the simplicity of the eigenvalues of H, we obtain $s(e_n) = u_n \cdot e_{n-2}$, $s^*(e_n) = v_n \cdot e_{n+2}$, $u_o = u_1 = 0$. We have

$$v_n = \langle s^*(e_n), e_{n+2} \rangle = \langle e_n, s(e_{n+2}) \rangle = u^*_{n+2},$$

and $[a^*,s] = 0$, $[s^*,a^*] = 4ia$ give

$$u_n \cdot (n-2)^{1/2} = u_{n-1} \cdot n^{1/2}, \quad n = 3,4\ldots$$
$$u_{n+2} \cdot (n+2)^{1/2} - u_{n+1} \cdot n^{1/2} = 4i\,(n+1)^{1/2}$$

We find easily $u_n = 2i(n^2-n)^{1/2}$, thus

$$s(e_n) = 2i\,[n(n-1)]^{1/2} \cdot e_{n-2}$$

$$s^*(e_n) = -2i\,[(n+1)(n+2)]^{1/2} \cdot e_{n+2}\,.$$

On the other hand $[s,s^*] = -8iB$ implies $|u_{n+2}|^2 - |u_n|^2 = 16(n+c)$. *Therefore* $c = 1/2$: *the vacuum is given necessarily the energy* $h\nu/2$.

A posteriori we find $s = 2i(a^*)^2$, $s^* = -2ia^2$.
This gives readily

$$\widehat{q^2} = -i(\hat{q})^2, \quad \widehat{p^2} = -i(\hat{p})^2, \tag{5.11}$$

$$\widehat{qp} = -i(\hat{q}\hat{p}+\hat{p}\hat{q})/2\,.$$

Third step: $\widehat{q^3}$ *is determined.*

$\{q^n, qp\} = n\cdot q^n$, n = positive integer, implies $[\widehat{q^n},\widehat{qp}] = n\cdot\widehat{q^n}$. Now, use (5.11) and $[\hat{q},\hat{p}] = iI$, we find $\widehat{q^n} = (-i)^{n+1}\cdot(\hat{q})^n$. Similarly $\widehat{p^n} = (-i)^{n+1}\cdot(\hat{p})^n$.

Fourth step: everything is determined.

In view of the corollary of 5.3, since the generators $\widehat{qp}$ and $(q+p^2+q^3)\hat{}$ are determined and are polynomials in $\hat{q}$, $\hat{p}$, any $\hat{f}$, $f\in P$, is determined and is a polynomial in $\hat{q}$, $\hat{p}$. Therefore there is *at most* one Dirac representation ^, up to a unitary equivalence. In fact:

Theorem.

There is *no* Dirac representation in which the harmonic oscillator has pure point spectrum with simple eigenvalues $i(n+c)\cdot h\cdot\nu$, $n = 0,1..$ c = constant.

Proof. We imitate the proof of Ref. 1, p.437. Using (5.11), $\widehat{q^3} = (\hat{q})^3$, $\widehat{p^3} = (\hat{p})^3$ and $\{q^3, p^2\} = 6q^2p$, $\{q^2, p^3\} = 6qp^2$ we find

$$(q^2p)\hat{} = i(\hat{q} - \widehat{q^2}\cdot\hat{p}),$$
$$(qp^2)\hat{} = i(\hat{p} - \hat{q}\cdot\widehat{p^2}).$$

Using $\{q^2p, qp^2\} = 3q^2p^2$ and the two previous formulas, we find

$$3(q^2p^2)\hat{} = -iI + 6\hat{q}\hat{p} + 3i(\hat{q})^2(\hat{p})^2. \tag{5.12}$$

On the other hand $\{q^3,p^3\} = 9q^2p^2$ and $\widehat{q^3} = (\hat{q})^3$, $\widehat{p^3} = (\hat{p})^3$ imply

$$9(q^2p^2)\hat{} = -6iI + 6\hat{q}\hat{p} - 3i(\hat{q})^2(\hat{p})^2 \tag{5.13}$$

Comparison of (5.12) and (5.13) gives the desired contradiction.

Remarks. The theorem is not trivial since *we did not assume that* $\hat{q}$ *and* $\hat{p}$ *act irreducibly.*

Dirac representation is possible and unique up to polynomials of degree 2. Can we extend this representation ? In other words: What are the subalgebras of P, which contain q, p, q^2, p^2, qp, and where the Dirac representation of the theorem is possible ?

References.

1. Abraham, R. and Marsden, J., *Foundations of Mechanics*, Benjamin. (second edition), 1978
2. Auslander, L., Green L., Hahn, F., Flows on homogeneous spaces, in *Annals of Math. Studies*, n°53, Princeton University Press, 1963.
3 Arnold, V. and Avez, A., *Ergodic problems of classical Mechanics*, Benjamin, 1968.
4. Avez, A., *Cours de Mécanique hamiltonienne* (roneotyped). Conservatoire National des Arts et Métiers, Paris, 1973-1974.
5. Avez, A., Représentation de l'algèbre de Lie des symplectomorphismes par des opérateurs bornés. *C.R. Acad. Sci. Paris*, t.279, 785, 1974.
6. Avez, A. Remarques sur les automorphismes infinitésimaux des variétés symplectiques compactes. *Rend. Sem. Mat. Univ. di Torino*, vol.33, 5, 1975.
7. Avez, A., Sistemi dinamici, *Cahiers des groupes de recherche du C.N.R. Pisa*, 1976.

8. Avez, A. et Heslot, A. L'algèbre de Lie des polynômes en les coordonnées canoniques munie du crochet de Poisson. *C.R. Acad. Sci. Paris*, t.288, 831, 1979.

9. Avez, A. et Lichnérowicz, A., Dérivations et premier groupe de cohomologie pour des algèbres de Lie attachées à une variété symplectique. *C.R. Acad. Sci. Paris*, t.275, 113, 1972.

10. Dirac, P.A.M., The Fundamental equations of quantum Mechanics. *Proc. Roy. Soc.*, A 1909, 642, 1925 (Re-edited in : B.L. van der Waerden, *Sources of quantum Mechanics*, Dover Publ., 1968).

11. Green, L.W., Ergodic flows and functional analysis, *Proceedings of a conference on Functional Analysis*, UC. Irvine, 181, 1966.

12. Heslot, A., *Remarques sur la notion d'observables en mécanique classique et en mécanique quantique.* Thèse de 3ème Cycle, Univ. Paris VI, 1979.

13. Rosen, G., *Formulations of classical and quantum dynamical systems.* Academic Press, 1969.

14. Souriau, J.M., Quantification canonique. *Comm. Math. Phys.*, 1, 374, 1966

15. Streater, R.F., Canonical quantization,*Comm. Math. Phys.*, 2, 354, 1966.

16. Van Hove, L., Sur certaines représentations unitaires d'un groupe infini de transformations. *Acad. Roy. Belg. (classe Sci.)*, t. 26, 61, 1951.

-:-:-:-:-:-

FREE EUCLIDEAN MARKOV FIELDS OF INTEGER SPIN IN THE LANDAU GAUGE

Carlos Delfino GALLES
Mathématiques, Université P. et M. Curie

Ceferino RUIZ GARRIDO
Mathématiques, Université Paris VII

The generalization of the Markov property proposed by Nelson[1] as a means of reconstructing a scalar quantum field theory in Minkowski space starting from a Euclidean one has been extended by Yao[2] and Lim[3] to electromagnetic and gravitational fields. It is to be hoped that the changes resulting from the use of Euclidean space might throw some light on the difficulties attendant on higher-spin theories. In this note we shall consider the case of massless particles of arbitrary spin showing that in the Landau gauge the field is Markovian.

The Landau gauge method was initially devised for the electromagnétic case[4] with the propagator constructed in such a way as to be divergenceless. The generalization to fields of arbitrary integer spin s was carried out in Ref. (5) where a symetric, traceless and divergenceless tensor $A_{\mu_1 \dots \mu_s}(x)$ was considered and the respective propagator, in Minkowski space, was determined.

The corresponding Euclidean propagator can be found using the matrix with values $A_{i\mu} = 1$ if $i = \mu = 1, 2, 3$, $A_{40} = i$ and $A_{i\mu} = 0$ otherwise, as is standard practice. In this way the following Euclidean propagator, obviously non local, is obtained :

$$P^E_{i_1 \dots i_s, j_1 \dots j_s}(p) =$$

$$= \frac{1}{p^2} \sum_{n=o}^{[s/2]} \frac{(-1)^n \, s! \, (2s-2n-1)!!}{2^n \, n! \, (s-n)! \, (2s-1)!!} \Big[\theta_{i_1 i_2} \theta_{j_1 j_2} \dots \theta_{i_{2n-1} i_{2n}} \theta_{j_{2n-1} j_{2n}} \theta_{i_{2n-1} j_{2n+1}} \dots \theta_{i_s j_s} \Big]_{\substack{sym\ (i) \\ sym\ (j)}}$$

where $[s/2]$ is the largest integer smaller than or equal to $s/2$. The operator θ_{ij} is defined by

$$\theta_{ij} = \delta_{ij} - p_i \, p_j /_{p^2} \, .$$

Let us now consider the space S^s of the symetric tensors $A_{i_1 \dots i_s}(x)$ over $\mathbb{R}^4$ whose components are C^∞ functions with compact support. This enables us to identify S^s with the Sobolev space $\mathbb{D}^m$, where $m = \binom{s+3}{3}$.

Considering the inner product,

$$<A,B>_{K_1} = \sum_{i,j} \tilde{A}_{i_1 \dots i_s}(p) \, P^E_{i_1 \dots i_s, j_1 \dots j_s}(p) \, \tilde{B}_{j_1 \dots j_s}(p) \, d^4p$$

and the space K_1 of tensors belonging to S^s such that

$$\| A \|^2_{K_1} = <\tilde{A},\tilde{A}> \, < \infty$$

where the tilde denotes the Fourier transformation.

The fact that the scalar product given by our propagator is positive semi-definite leads us to consider the pre-Hilbert quotient space

$K_2 = K_1 / \text{Kern } \|\ldots\|_{K_1}$ whose completion we call K_3.

Finally we look at K, a subspace of K_3, formed by the functions that satisfy the following conditions,

$$p^i \tilde{A}_{i\, i_2 \ldots i_s} = 0$$

$$\tilde{A}^i_{i\, i_3 \ldots i_s} = 0.$$

In K the scalar product is reduced to,

$$\langle A,B\rangle_K = \sum_i \int \tilde{A}_{i_1 \ldots i_s} p^{-2} \tilde{A}_{i_1 \ldots i_s} d^4p.$$

The rest of the proof is similar to Nelson's. Let Ω be an open set in $\mathbb{R}^4$, Ω' its complement, $\mathring{\Omega}'$ the interior of Ω' and $\partial\Omega$ the boundary of Ω. Let us consider a tensor $f_{i_1 \ldots i_s} \in K$ such as $\text{supp}\, f_{i_1 \ldots i_s} \subset \Omega$, and calling $h_{i_1 \ldots i_s}$ its orthogonal projection onto the set of functions with support in Ω'. Since the propagator in the functional space K is a local operator we have $h = f$ as distributions on $\mathring{\Omega}'$. As by definition $f = 0$ on $\mathring{\Omega}'$ we have $\text{supp}\, h \subset \Omega' - \mathring{\Omega}' = \partial\Omega$, which means that the field considered is a Markov one.

Furthermore, considering a tensorial field defined only on the hyperplane $x_4 = 0$, it is easy to see that the field does not satisfy the reflection property. This failing prevents the theory from fulfilling the requirements of Wightman axioms.

Références.

(1) E. Nelson, J. Func. Anal. 12, 211 (1973).

(2) T.H.Yao, J. Math. Phys. 17, 241 (1976).

(3) S.C. Lim, J. Phys. A 9, 1743 (1976).

(4) L. Landau, E. Lifschitz, Théorie Quantique Relativistique (Première partie). Edition Mir (1972), p. 354.

(5) C.D. Galles, Prog. Theor. Phys. 60, 313 (1978).

WKB-EXPANSIONS FOR PROPAGATORS

F.Langouche [+]

D.Roekaerts [++]

Instituut voor Theoretische Fysica,
Universiteit Leuven, Belgium

E.Tirapegui

Institut de Physique Théorique,
Université de Louvain,Belgium

Given the differential equation ($\underset{\sim}{Q} = (Q^1, Q^2, \ldots, Q^M)$)

$$\frac{\partial}{\partial t} p(\underset{\sim}{Q},t) = L\left(\frac{\partial}{\partial \underset{\sim}{Q}}\right) p(\underset{\sim}{Q},t), \tag{1}$$

where $L\left(\frac{\partial}{\partial \underset{\sim}{Q}}\right)$ contains at most second derivatives, one is interested in the propagator of (1), i.e. the solution $I(\underset{\sim}{Q},t;\underset{\sim}{Q}_o,t_o)$ such that

$$I(\underset{\sim}{Q},t_o;\underset{\sim}{Q}_o,t_o) = \delta(\underset{\sim}{Q}-\underset{\sim}{Q}_o). \tag{2}$$

The general form of L is $\left(\partial_\alpha \equiv \frac{\partial}{\partial Q^\alpha}\right)$

$$L = \frac{1}{2}\eta c \; \Sigma \; A(\underset{\sim}{Q})\partial_\alpha B(\underset{\sim}{Q})\partial_\beta C(\underset{\sim}{Q}) + \Sigma \; D(\underset{\sim}{Q})\partial_\alpha E(\underset{\sim}{Q}) + \frac{1}{\eta c}F(\underset{\sim}{Q}) \tag{3}$$

where Σ indicates a sum of terms of the same form. The WKB-type expansion we are looking for is an expansion of the propagator of the form

+ Onderzoeker I.I.K.W.,Belgium

++ Aspirant N.F.W.O.,Belgium

$$I(Q,t;Q_o,t_o) = I_{WKB}(Q,t;Q_o,t_o).(1+\sum_{k=1}^{\infty} I_{(k)}\eta^k) \tag{4}$$

where the WKB-approximation I_{WKB} is such that

$$I_{WKB}^{-1}\left(\frac{\partial}{\partial t} - L\left(\frac{\partial}{\partial Q}\right)\right) I_{WKB} = O(\eta). \tag{5}$$

Let us first give two examples. In quantum mechanical applications $\eta=\hbar$, $c=i$.[1.A-C] For the quantum system corresponding to the classical Lagrangian

$$L_{cl}(\dot{q},q) = \frac{1}{2}g_{\alpha\beta}(q)\dot{q}^\alpha\dot{q}^\beta + \dot{q}^\alpha A_\alpha(q) - V(q) \tag{6}$$

the Schrödinger equation for the propagator is $(g = \det g_{\alpha\beta})$

$$\frac{\partial}{\partial t}I = \frac{1}{2}i\hbar\frac{1}{\sqrt{g}}\partial_\alpha(\sqrt{g}\, g^{\alpha\beta}\partial_\beta)\, I + \frac{1}{2}[A^\alpha\partial_\alpha I + \frac{1}{\sqrt{g}}\partial_\alpha\sqrt{g}\, A^\alpha I] - \frac{i}{\hbar}[\frac{1}{2}A^\alpha A_\alpha + V]I \tag{7}$$

In the case of Fokker-Planck dynamics $c=1$ and η is a parameter measuring the strength of the fluctuations.[1.D] Let $P(Q,t;Q_o,t_o)$ be the transition probability density of the Markovian process $q(t)$ determined by the Langevin equation

$$\dot{q}^\alpha(t) + a^\alpha(q(t)) = \sqrt{\eta}\,\sigma_i^{\ \alpha}(q(t))\, f_i(t)\ , \quad \alpha = 1,\dots,M \tag{8}$$

where $f_i(t)$, $i=1,\dots,M'\geq M$, is a Gaussian white noise, and set

$$g^{\alpha\beta}(q) = \sum_i \sigma_i^{\ \alpha}(q)\sigma_i^{\ \beta}(q)\ , \tag{9}$$

then the invariant probability density[1.D]

$$I(Q,t;Q_o,t_o) = [g(Q)/g(Q_o)]^{1/2}P(Q,t;Q_o,t_o) \tag{10}$$

satisfies

$$\frac{\partial}{\partial t}I = \frac{1}{2}\eta\frac{1}{\sqrt{g}}\partial_\alpha\sigma_i^{\ \alpha}\partial_\beta\sqrt{g}\sigma_i^{\ \beta}I + \frac{1}{\sqrt{g}}\partial_\alpha\sqrt{g}\, a^\alpha I. \tag{11}$$

Both (7) and (11) are of the form (1,3).

We now turn to the computation of the expansion (4). Equations of the type (1) admit an operator formalism : $\hat{Q}\to Q$ and $\hat{P}\to -i\frac{\partial}{\partial Q}$, hence

$[\hat{Q}^{\alpha}, \hat{P}_{\beta}] = i\delta^{\alpha}{}_{\beta}$. Then L determines the operator (we will call it the "Hamiltonian", also in the non-quantum case)

$$\hat{H}(\hat{P},\hat{Q}) = -i\tfrac{1}{2}\eta c \; \Sigma \; A(\hat{Q})\hat{P}_{\alpha}B(\hat{Q})\hat{P}_{\beta}C(\hat{Q}) - \Sigma \; D(\hat{Q})\hat{P}_{\alpha}E(\hat{Q}) + \frac{i}{\eta c}F(\hat{Q}), \qquad (12)$$

and the propagator $I(Q,t;Q_o,t_o)$ can be written as

$$I(Q,t;Q_o,t_o) = \langle Q|U(t,t_o)|Q_o\rangle \, , \qquad (13)$$

where $U(t,t_o)$ is determined by

$$i\frac{\partial}{\partial t'}U(t',t) = \hat{H}\; U(t',t) \; ; \; U(t,t) = 1. \qquad (14)$$

Owing to our choice of commutation relations (independent of η) all η-dependence of the Hamiltonian is explicit.

The standard procedure to determine the WKB-approximation consists in imposing the condition (5) on a trial solution. It is of the form

$$I_{WKB} = \eta^{-M/2} \; \overline{N}(Q,Q_o) \; \exp[\tfrac{i}{\eta} A_{cl}], \qquad (15)$$

where A_{cl} is some classical action and $\overline{N}(Q,Q_o)$ essentially contains the Van Vleck determinant. The correspondence between the operator $\hat{H}(\hat{P},\hat{Q})$ and the classical action A_{cl} should be fixed by some correspondence rule between $\hat{H}(\hat{P},\hat{Q})$ and a classical Hamiltonian $H_{cl}(P,Q)$. Mizrahi studied the question for which class of correspondence rules the WKB-approximation is of the form (15) with $\overline{N}(Q,Q_o)$ identical to the Van Vleck determinant.[2] The correspondence we need in the general case studied here is not contained in this class, consequently $\overline{N}(Q,Q_o)$ will differ from the Van Vleck determinant by a non-constant factor. We remark that Elworthy and Truman have studied the WKB-approximation in a rigorous way.[4]

The only method, that has been used up to now, to compute explicitly the higher order corrections $I_{(k)}$ is by using the functional integral representations of the propagator I. The first attempts in this direction can be found in refs.2,3. The first explicit calculations of all corrections was done by the present authors.[1.A-D] Here we will only give an outline of the method and indicate the most important stumbling-blocks.

Starting from (13) one can construct phase space functional integral representations for I of the form

$$I(\underset{\sim}{Q},t;\underset{\sim}{Q}_o,t_o) = \int_\gamma \mathcal{D}\underset{\sim}{Q}\mathcal{D}\underset{\sim}{P} \exp i\int_{t_o}^{t} d\tau[\underset{\sim}{P}.\dot{\underset{\sim}{Q}} - H^\gamma(\underset{\sim}{P}(\tau),\underset{\sim}{Q}(\tau))]\delta(\underset{\sim}{Q}(t)-\underset{\sim}{Q})\delta(\underset{\sim}{Q}(t_o)-\underset{\sim}{Q}_o) \tag{16}$$

where γ stands for the discretization by which this expression is defined and H^γ depends on it.[1.E]

An equivalence class of discretizations characterizes a correspondence rule.[1.E] Also in the context of path integrals defined without limiting procedure these rules can be considered.[5] In fact, in ref.1.E we showed that both approaches, with or without limiting procedure, meet the same ambiguities. We also showed that it is the knowledge of the discretization that makes all terms in a formal series expansion unambiguously defined.

For the case we study here the appropriate discretization $\bar{\gamma}$ is such that

$$H^{\bar{\gamma}}(\underset{\sim}{P},\underset{\sim}{Q}) = -i\tfrac{1}{2}\eta c\, \Sigma\, P_\alpha P_\beta A(\underset{\sim}{Q})B(\underset{\sim}{Q})C(\underset{\sim}{Q}) - \Sigma\, P_\alpha D(\underset{\sim}{Q})E(\underset{\sim}{Q}) + \frac{i}{c}F(\underset{\sim}{Q}). \tag{17}$$

Indeed, let $H_{cl}(\underset{\sim}{P},\underset{\sim}{Q}) \equiv H^{\bar{\gamma}}(\underset{\sim}{P},\underset{\sim}{Q})\big|_{\eta=1}$, and let $(\underset{\sim}{y}(\tau),\underset{\sim}{x}(\tau))$ be the solution of the Hamilton's equations of H_{cl} for $\underset{\sim}{x}(t_o)=\underset{\sim}{Q}_o$, $\underset{\sim}{x}(t)=\underset{\sim}{Q}$. Then one can make in (the discretized version of) (16) the change of variables $\underset{\sim}{P} = \underset{\sim}{y}/\eta + \underset{\sim}{p}/\sqrt{\eta}$, $\underset{\sim}{Q} = \underset{\sim}{x} + \sqrt{\eta}\underset{\sim}{q}$ and expand the action around $(\underset{\sim}{y},\underset{\sim}{x})$. Doing so one arrives at an expansion of the form

$$I = \eta^{-M/2} \exp \frac{i}{\eta} \int_{t_o}^{t} d\tau [\underset{\sim}{y}\dot{\underset{\sim}{x}} - H_{cl}(\underset{\sim}{y},\underset{\sim}{x})] \cdot \int_{\bar{\gamma}'} \mathcal{D}\underset{\sim}{q}\mathcal{D}\underset{\sim}{p} \exp i\int_{t_o}^{t} d\tau [\underset{\sim}{p}\dot{\underset{\sim}{q}} - \bar{H}_o{}' - \bar{H}_1{}'] \qquad (18)$$

where $\bar{\gamma}'$ is a known discretization, H_o is quadratic in $\underset{\sim}{p}$ and $\underset{\sim}{q}$ and independent of η and H_1 is a series of the from $\sum_{k>2} \sqrt{\eta}^{k-1} z_1 \ldots z_k$, where each z_i, $i=1,\ldots,k$ is a component of either $\underset{\sim}{p}$ or $\underset{\sim}{q}$. One sees then that, since all η-dependence is made explicit, that the series generated when $\exp -i\int H_1 d\tau$ acts on the free generating functional constructed with H_o, is the expansion in powers of η we are looking for.(It turns out that the terms in odd powers of $\sqrt{\eta}$ vanish.) All terms of the expansion are well-defined because $\bar{\gamma}'$ is known. More details can be found in refs.1, here we only want to point out that the reason why explicit calculation of higher order corrections was never done before, has its origin in the severe ordering problem of the operator (12) and in fact that in general the appropriate classical Hamiltonians $H^{\bar{\gamma}}$ and H_{cl} are not related to $\hat{H}$ by any of the usual correspondence rules. Therefore the construction of the discretization $\bar{\gamma}$ is an essential part of the present method. (When B(Q) is symmetric in $(\alpha\beta)$, $\bar{\gamma}$ is equivalent to the simpler γ_3.)

<u>References</u>

1. Langouche,F.,Roekaerts,D.,Tirapegui,E.,A.*Lett.Nuov.Cim.*25,307,1979 ; B.*Phys.Lett.*72A,413,1979 ; C.*Preprint* KUL-TF-79/028 ; D.*Preprint* KUL-TF-79/027 ; E.*Phys.Rev.*D20,419,1979.
2. Mizrahi,M.M.,*J.Math.Phys.*18,786,1977 ,*J.Math.Phys.*19,298,1978
3. De Witt-Morette,C.,*Ann.Phys.*97,367,1976
4. Elworthy,K.D.,Truman,A., *Preprint* Edinburgh
5. Mizrahi,M.M., in *Feynman Path Integrals*, Lecture Notes in Physics 106, Ed. : S.Albeverio et al.; Springer-Verlag, Berlin, 1979.

HAMILTONIAN AND HEREDITARY PROPERTIES
OF CONTINUOUS ELECTRICAL MEDIA

FARGUE Daniel, Ecole des Mines,
60, boulevard Saint-Michel
75006 Paris, France

Hamiltonian mechanics of matter, which are almost exclusively used in microphysics, are nevertheless unable to give account of some of the essential features of matter: the privileged part of quantum mechanical stationary states and their stability, the sudden changes between these states[1,2] or a sufficiently convincing explanation of macroscopic irreversibility[2,3] .

We are thus led to look for a non-hamiltonian description whose basis fortunately exists already in physics. The Lorentz equation[4] ,which is experimentally well confirmed, is indeed such an evolution equation. Its main characteristics, which will be kept up in the following are

- a stretched, non point-like particle
- self-acting electromagnetic forces in the particle.

In the application to microphysics, the rigidity of the Lorentz electron will be replaced, following F. Fer[5,2], by a more refined description using the quantum potential of L. de Broglie[6], and a cohesive force, similar to the Poincaré forces. This will allow us not only to recover the main results of quantum mechanics, but also to study some new problems dodged by usual quantum mechanics, such as the stability of the states or the details of a quantum jump and its corresponding emission of radiation.

These possibilities come from the fact that we consider matter, and the retarded electromagnetic field generated by it, but here without the usual difficulties of infinite self energy encountered in quantum electrodynamics, because our description is based on a *continuous* mediun

The first step is thus to study the dynamics of electrical continuous medium, which, as we will see below, will be governed by integro-differential non-linear equations; then we shall see some associated results in microphysics.

1 Mechanics of electrical continuous media.

Definitions. For simplicity we consider here only the case of newtonian dynamics, but the relativistic generalization has the same qualitative properties. The mass density is $\mu(x^j)$, the velocities are $v^i(x^j)$. We suppose that the charge density is $\rho = e\,\mu\,/\,m$, where e is the total electrical charge and m the mass. The electromagnetic field generated by the medium is defined by its potentials V, A^i w i $E^i = -\partial^i V - \partial_t A^i$, $\mu_o H^i = \varepsilon^{ijk}\,\partial_j A_k$ (ε^{ijk} : completely antisymmetric unity tensor). The external potentials are ϕ, α^i, and the fields ε^i, β^i.

The forces acting on the medium are the electromagnetic forces, those caused by a stress tensor T^{ij} derived from a free energy f or a quantum potential Q , and an internal cohesive force whose densit derives from the potential

$$G(x^i, t) = \int \mu(x^j,t)\, g(|x^i-y^i|)\, d^3y \qquad (1)$$

The dynamic equation is ($d/dt = \partial_t + v^i \partial_i$)

$$\mu \frac{dv^i}{dt} = -\mu\, \partial^i G + \partial_j T^{ij} + \rho\,(E^i + \varepsilon^i) + \rho\, \varepsilon^{ijk} v_j\,(B_k + \beta_k) \qquad (2)$$

which needs to be completed by the Maxwell-Ampère equations

$$\partial_t D^i = -\rho v^i + \varepsilon^{ijk} \partial_j H_k \quad , \qquad \partial_i (\varepsilon_o E^i) = \rho \qquad (3)$$

and by the mass conservation equation

$$\frac{d\mu}{dt} + \mu\, \partial_i v^i = 0 \qquad (4)$$

Hamiltonian formalism. The evolution equations (2) , (3), of the whole system, matter and electromagnetic field, can be showed [8] to derive from a hamiltonian formalism, with

$$H = \int \Big(\frac{1}{2\mu} \{ p^i - \rho\,(A^i + \alpha^i) \}^2 + \frac{1}{2}\mu G + \mu f + \rho\,(V+\phi) \Big)\, d^3x$$
$$+ \int \Big(\frac{(D^i)^2}{2\,\varepsilon_o} + \frac{(\varepsilon^{ijk}\partial_j A_k)^2}{2\,\mu_o} + D^i \partial_i V \Big)\, d^3x \qquad (5)$$

where the canonical variables are

$$x^i \text{ and } p^i = \mu v^i + \rho\,(A^i + \alpha^i)$$
$$A^i \text{ and } P^i = -D^i = -\varepsilon_o E^i \quad ; \quad \phi \qquad (6)$$

The variational principle is then

$$\delta H = \int \{ \delta(p_i\, d^3x)\, v^i - \frac{d\,(p_i\, d^3x)}{dt}\, \delta x^i \} + \int \{ -\partial D_i \frac{\partial A^i}{\partial t} + \frac{\partial D^i}{\partial t}\, \partial A_i \}\, d^3x \qquad (7)$$

where ∂ is the variation with constant x^i and $\delta = \partial + \delta x^i \partial_i$. The corresponding invariant 2-form is

$$\omega = \int \mu \, \delta v_i \wedge \delta x^i \, d^3x + \int \rho \, \delta(A_i + \alpha_i) \wedge \delta x^i \, d^3x - \int \partial D_i \wedge \partial A^i \, d^3x \quad (8)$$

But all this is allowed only if we consider V and A^i as independent variables, exactly as the positions x^i of the different points of matter, and correspondingly the Maxwell equations (3) as evolution equations. If, for practical or observational reasons, we consider only the matter variables as independent, the motion is projected on the phase space of matter alone and is generally no more hamiltonian, because there is an interaction term in (5) or (8) between matter and field.

This is true, however, only if we consider every possible motion of the system. In the class of permanent motions in an external static field we have $\partial_t \phi = \partial_t A^i = 0$ and (8) shows that the last term is $= 0$. Thus, this class of permanent motions is hamiltonian with

$$H = \int \{ \mu \left(\frac{v^2}{2} + \frac{G}{2} - f \right) + \rho \phi + \frac{1}{2} \rho \left(V + v^i A_i \right) \} \, d^3x \quad (9)$$

The same physical system is then hamiltonian or not, depending on the evolutions considered: stationary states or transitions for instance. Moreover, we see that, for permanent motions, the more general dynamics of matter considered here reduce to ordinary hamiltonian dynamics.

Retarded actions and hereditary dynamics. In the general, non stationary state, it is interesting to know what kind of dynamics governs the evolution of matter considered as only independent (or controllable). We have only to solve the Maxwell equations (3) and eliminate the fields from (2).

The question then often arises whether to use the retarded or the advanced potentials, or a mixed form. This not a matter of philosophical taste (principle of causality) but of resolution of a hyperbolic equation. The unique solution of Cauchy's problem in that case is the following, as was already shown by Lorentz[4]

$$H^i = H^i_1 + H^i_2 \quad , \quad E^i = E^i_1 + E^i_2 \tag{10}$$

where H^i_1 and E^i_1 are terms depending only on the initial values H^i_o and E^i_o at time $t = t_o$, and H^i_2 , E^i_2 are the "forced terms" depending on the material sources ρ and $\rho\, v^i$.

The forced terms have a purely retarded expression if $t > t_o$ (our only'philosophical' assumption is that we know what this inequality means, and this is done as soon as we have a clock: as a matter of fact, we have already assumed it when we wrote (2) or (3) !)

$$H^i_2(x^j,t) = \frac{1}{4\pi} \int_{r \leqslant c(t-t_o)} \frac{\varepsilon^{ijk}\, \partial_j \{\rho v_k\}\, (y^j, t-r/c)}{r}\, d^3y \tag{11}$$

($r = |x^k - y^k|$) , and a similar expression for E^i_2 . If $t < t_o$, only the advanced expression with $t + r/c$ is to be used.

As we usually look for solutions in the future, this is enough to prove that the dynamics of electrical matter is neither hamiltonian nor differential but *hereditary* because, through $t - r/c$, equation (11) involves an integration over the past of the system.

Under the unavoidable assumption that the initial electromagnetic energy is finite, it is possible to show[9] some new results. First, the behavior of the field is ergodic, that is it "forgets" asymptotically its initial values, at least in every finite domain. The precise theorem is: *if* H_o *and* $D_o \in L^2$ *, then* H_1 *and* D_1 *converge strongly to* 0 *with* $1/(t-t_o)$ *on every finite domain.* The same is true of the part of the field created by the matter (assumed enclosed in a finite domain) between t_o and t_1 . It can be said that the electromagnetic field generated by the matter irreversibly goes away to infinity.

The heredity is of a forgetting type and this is necessary to explain the success of differential dynamics, as an approximation. Moreover, the impossibility to control the infinite number of degrees of freedom of initial electromagnetic fields, as weak as their actions may be after some time, could be the source of the probabilistic appearance of matter behaviour.

Asymptotic behaviour. A possible approach is the following. As we have seen, the electromagnetic field generated by the medium goes away to infinity, and carries some of the energy of the whole system, which is then 'lost'. For an isolated system, the energy is constant; if we assume it finite, so is the radiation loss. In the same way as for a radio source one is then tempted to say that the system will tend to equilibrium as t grows indefinitely.

In the case of continuous media, particularly if one tries to obtain mathematically exact results, it is not so simple. First of all, what are the admissible asymptotic motions, i. e. those which certainly radiate only a finite amount of energy? As soon as we assume the energy of matter alone to have a lower bound, it implies that the electromagnetic energy remains finite. Among the terms of H^i or E^i given by (10), only two may grow indefinitely as t grows indefinitely: their Fourier transforms are of the form (k^i is the wave number) :

$$\int_{t_o}^{t} \varepsilon_{ijk} \frac{k^j}{|k|} \partial_\tau(\rho v^k)(k^i,\tau) \frac{\begin{smallmatrix}\cos\\ \sin\end{smallmatrix}\{|k|\ (t-\tau)\}}{|k|} d\tau$$

A sufficient condition for the energy to remain finite is thus that $\partial_\tau \operatorname{curl}(\rho\vec{v}) = 0$. Are these motions, including the permanent ones, asymptotically the only allowed ones? Equivalently, is the following proposition true: *a necessary condition for* $\left\| \int_{t_o}^{t} f(k^i,t) \frac{\sin|k|\ (t-\tau)}{|k|} d\tau \right\|$.

to remain finite when $t \to \infty$ *is that, in a certain way* $f(k^i,t) \to 0$ *as* $t \to \infty$? The general answer is unknown, but some results may be obtained by answering the inverse question. Motions which radiate an infinite energy are, for instance the motions confined ina finite domain and which depend from t through a trigonometric polynomial, or else the finite superpositions of the stationary states of the hydrogen atom. This could at least partially, explain the much more important part played by the eigenstates in comparison with their superpositions.

2 Results in microphysics.

We intend to give in this paragraph a brief sketch of the description of a spinless electron[2,5], in the light of the preceding reflexions. The only change is the use of the quantum potential of L. de Broglie

$$Q = \frac{\hbar^2}{2\,m}\,\frac{\Delta\sqrt{\mu}}{\sqrt{\mu}} \tag{12}$$

instead of the free energy f ; this will account for the undulatory properties of matter.

Permanent motions. First, let us consider the case of equilibrium of an electron alone, far from any exterior action. The solution has spherical symmetry, and defining $s(r)$ by $s^2(r) = r^2\mu(r)$, we are led to the variational problem (remember that permanent motions are hamiltonian) of finding $s_0(r)$ defined on $(0,\infty)$ and such that $||s'||<\infty$, $||s||= S$, which minimizes the non convex functional

$$H(s) = \int_0^\infty s'^2(r)\,dr + \int_0^\infty\int_0^\infty s^2(r)\,K(r,r')\,s^2(r')\,dr\,dr' \tag{13}$$

$K(r,r')$ is a kernel which summarizes the action of the electrical and cohesive potentials, and is such that H is allowed to take negative values :

$$K(r,r') = \frac{2\pi}{rr'} \int_{|r-r'|}^{r+r'} c(\rho)\, d\rho$$

This condition is mathematically and physically necessary to ensure the possibility of non zero solutions, that is the possibility of existence of a particle through condensation by cohesive forces greater than the electrical repulsion.

Alternatively, when s_o is of class C^2, the corresponding Euler equation is the equilibrium condition:

$$s''(r) - 2\, s(r) \int_o^\infty K(r,r')\, s^2(r')\, dr' - \gamma\, s(r) = 0$$

We have proven[10] that *for a broad class of cohesive potentials, such that* $|c(r)| < C < \infty$ *, a solution* s_o *exists,* satisfying the preceding conditions. Thus we have a description of the point-like character of free electrons.

What happens for bounded electrons? The following results were proven by F. Fer[5] in 1964. Let us consider permanent irrotational motions such that there exists a $S(x^i,t)$:

$$v^i + \frac{e}{m} (A^i + \alpha^i) = \partial^i S \tag{14}$$

Defining the wave function as

$$u = \sqrt{\mu}\; e^{\frac{i\, m}{\hbar} S} \tag{15}$$

so that, conversely we have the relations

$$\mu = u\, u^\dagger \qquad\qquad \mu\, \partial^i S = \frac{i\hbar}{2m} (u\, \partial^i u^\dagger - u^\dagger\, \partial^i u) \tag{16}$$

usually interpreted in quantum mechanics as the probability density and the probability current. By Madelung's method[11] (2) and (4) reduce to the wave equation:

$$i \hbar \partial_t u = L(u)\, u \qquad (17)$$

$$L(u) = \frac{1}{2m} \{ i \hbar \partial_i + e (A^i + \alpha^i) \}^2 + e \phi + (e V + m G) \qquad (18)$$

$L(u)$ depends actually on u through the internal electromagnetic terms and the cohesive term, given by Maxwell's equation and (1).

Even if this wave equation is formally similar to Schrödinger's and eventually reduces to it in first approximation, two essential features are different: *it is always non-linear and integro-differential* owing to the introduction of the self acting terms A^i, V, G.

Assuming then, that the external electromagnetic field ϕ, α^i is static, for instance created by the nucleus of an atom, the permanent motions of the electron are such that $\partial_t \mu = \partial_t v^i = \partial_t V = \partial_t A^i = 0$. (14) proves that $\partial_t S$ is a function of t alone, and one finds :

$$m \partial_t S = - \int \{ \mu \frac{(v^i)^2}{2} + \rho (V + \phi) + \mu G - \mu Q \} d^3x = - F \qquad (19)$$

where F is a constant since the integral is independent of t. *The permanent motions correspond then to the "stationary" solutions* of (17):

$$u = u_o(x^i)\, e^{-iFt / \hbar} \qquad (20)$$

with $$(L - F)\, u_o(x^i) = 0 \qquad (21)$$

(20) is an eigenvalue equation similar to Schrödinger's, but non linear (although no more hereditary). The eigenvalue F differs slightly from the constant energy of the electron (9), and there is no loss of energy through electromagnetic radiation, a property which was forbidden to

the Lorentz electron, owing to its rigidity, involving the impossibility of permanent motions.

Thus far the calculus was rigourous, but as equation (21) is integrodifferential and non-linear, a very little studied type, we are forced to make some approximations. The first is that A^i is negligible, i.e. that the magnetic forces are small. The second is that $eV + mG << \phi$, i.e. the cohesive force balances almost exactly the electric repulsion so that the electron neither explodes nor collapses. Equation (21) is then Schrödinger's modified by terms small enough to be treated by a perturbation method. *A first order calculation shows that there exists solutions of (20) which are quantized* and that their energy values are just slightly different from the usual quantum states.

Transient motions. The preceding results were hamiltonian and thus not very different from the usual theory. But we will now take account of the propagation delay of electromagnetic actions, i.e. the hereditary character of our dynamics: this will allow a detailed study of the quantum jumps and of the stability of the stationary states.

Let a small external perturbation, which can be very brief, slightly disturb the permanent motion whose eigenvalue is F_i .A transient motion starts which we assume as follows.

The medium begins to oscillate about its permanent motion, as does a perturbed hamiltonian system, but the the propagation delay breaks in and amplifies the vibration. After some time, the amplitude is large enough for the medium to be drawn back to the new permanent motion $F_j \neq F_i$. The vibration goes on but this time the propagation delay damps the motion which tends to the j state.

As strange as it may seem, the idea that the propagation delay first amplifies, then damps the motion is however confirmed by the calculation which may be done of the beginning and of the end of the process i.e. by the study of the small movements about a permanent motion.

Setting $\delta u = \chi \exp-(i F_i t)$, and approximating the retarded potentials, one finds an equation in $\partial\chi/\partial t$ which is linear. By looking for a solution of the form

$$\chi = e^{\sigma t} \{ \xi(x^i) \cos \omega t + \zeta(x^i) \sin \omega t \} \tag{22}$$

by a perturbation method, one finds

$$\omega = \frac{|F_i - F_j|}{\hbar} + \delta\omega \tag{23}$$

where $\delta\omega$ is a small frequency shift. The value of σ depends on the part of the equation containing the retarded terms (the one which $\to 0$ when $c \to \infty$), and $\sigma > 0$ if $F_k < F_i$, $\sigma < 0$ if $F_j > F_i$.

This shows the semi-stability of the stationary states (except perhaps for the ground level) and rejoins the ideas exposed by G. Lochak[1]. We thus see that this theory gives, at least by a perturbation method a means to study these quantum jumps whose irreversibility prevents the explanation by usual quantum mechanics.

Another situation where the propagation delay of electromagnetic actions plays a hidden, but probably important part,[2,3] is the impact of particles in statistical thermodynamics. This is easy to study in the mathematical scheme given here: it suffices to remark that the mass density $\mu(x^i)$ is then nearly zero everywhere, except in the neighbourhood of the particles. In the case of one particle the first approximation of this hereditary mechanics is naturally the Lorentz equation. But more generally in the case of many particles, one can show[8], by using the invariant form (8), that the volume element of the phase space is no more invariant, and evaluate its rate of change. A comforting result is that the energy loss caused by such radiation is of the order of magnitude of the energy loss given by the macroscopic law of radiation from a gas.

References

1. LOCHAK G. Quantization as a stability problem, in this volume.
2. FER F., L'irréversibilité, Gauthier-Villars, Paris 1977.
3. FER F., Irreversibility versus hamiltonian dynamics in statistical thermodynamics, in this volume.
4. LORENTZ H. A., The theory of electrons, reprint, Dover, New-York, 1952.
5. FER F., C.R.Ac.Sc.Paris, 258, 2983, 3215, 3435, 1964; ibid., 262, 1417, 1966, ibid., 263, 103, 1966.
6. De BROGLIE L., La réinterprétation de la mécanique ondulatoire, Gauthier-Villars, Paris, 1971.
7. FARGUE D., FER F., Annales de la Fondation L. de Broglie, 1, 30, 1976.
8. FARGUE D., Thèse, Université de Provence, 1974.
9. FARGUE D., C.R.Ac.Sc.Paris, 283 B, 21, 1976.
10. FARGUE D., FER F., C.R.Ac.Sc.Paris, 288 A, 232, 1979.
11. MADELUNG E., Zs.f.Phys., 40, 322, 1926.

SOLITONS OF A NONLINEAR SCHRÖDINGER EQUATION

T.A. Minelli and A. Pascolini
Istituto di Fisica dell'Università
Sezione INFN, Padova

1. Introduction.

In recent years, there has been a considerable development in the study of the soliton solutions of the class of nonlinear Schrödinger equations

$$i\hbar\frac{\partial}{\partial t}\psi(\underline{r},t)=\left[-\frac{\hbar^2}{2m}\nabla^2+U(\underline{r})+F(\rho)\right]\psi(\underline{r},t); \qquad \rho=|\psi|^2 \tag{1}$$

for several forms of the nonlinear term $F(\rho)$.[1]

Upon such an equation have Bialynicki-Birula and Mycielski based their approach towards a nonlinear wave mechanics,[2] choosing $F(\rho)=-b\ln(a^n\rho)$, where n is the dimension of the space, a is a dimensionless constant and the nonlinearity parameter b has the dimension of an energy.

Their construction is made possible by the fact that the logarithmic nonlinear Schrödinger equation (LSE) possesses solutions to which particle-like properties can be attributed consistently with quantum theory: 1) The LSE with U=0 is invariant with respect to the full Galilei group, its solutions transform properly, and conservation laws can be derived. 2) Energy, momentum and angular momentum can be defined in a consistent way and retain all their usual properties. 3) For stationary states - i.e. for solutions of the form $\psi(\underline{r},t)=\exp(-i\omega t)\psi(\underline{r})$ - the Planck's relation $E=\hbar\omega$ is satisfied. 4) The norm of the solutions is preserved in time and can be chosen equal to 1. 5) A weak form of the superposition principle holds: the sum of two solutions whose overlap is negligible is also a solution. 6) A system made up of two non-interacting subsystems is described by a solution of LSE which is the direct product of the solutions of the subsystems (separability of non-interacting subsystems). 7) If $b>0$ the energy is bounded from below, for a constant potential U_0.

In this case the lower bound is attained only for the gaussons, soliton-like solutions of gaussian shape which exist in any dimensions:

$$\psi(\underline{r},t)=A\exp\left[i(\underline{k}\cdot\underline{r}-\omega t)\right]\exp\left[-\frac{mb}{\hbar^2}(\underline{r}-\underline{v}t-\underline{r}_0)^2\right]; \qquad \underline{k}=\frac{m}{\hbar}\underline{v} \tag{2}$$

where $\underline{v}$ is the envelope velocity and $\underline{k}$ the wave number. Gaussons satisfy the nonlinear dispersion relation $\hbar\omega=(\hbar^2/2m)k^2+U_0+nb\left[1-\ln(aA^2)\right]$. The radius of the gausson depends only on the mass m and on the nonlinearity parameter b, which plays the role of a universal constant.

The dynamical behaviour of gaussons has also been considered in few cases: analytical solutions have been obtained for LSE with electric and magnetic fields, and a numerical study of collisions of free gaussons in one and two dimensions has proved their stability in scattering processes.[3]

The present contribution deals with the study of the onedimensional LSE with simple quantum potentials like a square well and a square barrier, and considers in particular the tunnel effect for gaussons.[4]

2. Dynamics of Gaussons.

Conservation laws are the most successful and direct analytical instrument in studying the soliton dynamics.[5] In terms of the charge densi ty ρ and the current density $\underline{j}=\mathrm{Re}\left[\psi^*(\hbar/im)\underline{\nabla\psi}\right]$, equation (1) can be rewritten as a couple of fluid-dynamical equations

$$\frac{\partial}{\partial t}\rho+\underline{\nabla}\cdot\underline{j}=0 \quad , \quad \frac{\partial}{\partial t}j_k+\sum_i\nabla_i T_{ik}+\rho\nabla_k U=0 \tag{3}$$

where the stress tensor T is defined by

$$T_{ij}=-\rho\left[(\hbar^2/4m)\nabla_i\nabla_j\ln\rho+\delta_{ij}(G-F)\right]+\frac{m}{\rho}j_i j_j \quad , \tag{4}$$

where $G(\rho)=(1/\rho)\int d\rho' F(\rho')$.

For U=0 equations (3) may be viewed as differential conservation laws, to which the integrated quantities $Q\equiv\langle\psi|\psi\rangle=\int\rho d^n\underline{r}$, $\underline{P}\equiv\langle\psi|\underline{p}|\psi\rangle=m\int\underline{j}d^n\underline{r}$ correspond. Q and $\underline{P}$ may be interpreted as total charge and momentum. If $U\neq0$, the momentum conservation is broken and $\underline{P}$ changes in time according to the Ehrenfest theorem:

$$\frac{d}{dt}\underline{P}=-\langle\psi|\underline{\nabla U}|\psi\rangle=-\int\rho\underline{\nabla}Ud^n\underline{r} \quad . \tag{5}$$

As already stated, also the energy

$$E=\frac{\hbar^2}{2m}\langle\nabla\psi|\nabla\psi\rangle+\langle\psi|\left[U+G(\rho)+\text{const.}\right]|\psi\rangle \tag{6}$$

is always conserved, and the angular momentum when U=0.

For a gausson of amplitude A, frequency ω and wave vector $\underline{k}$, the invariants charge, momentum and energy are $Q=A^2(\hbar\sqrt{\pi/2mb})^n$, $\underline{P}=\hbar\underline{k}Q$, $E=\hbar\omega Q$. In the case of the cubic nonlinearity, analytical treatments exist, taking advantage of the existence of additional infinite conservation laws,[6] and of the applicability of the inverse scattering method. In the LSE case, on the contrary, no additional invariance laws are given and this prevents a full analytical treatment and makes it necessary the use of numerical calculations even for the simplest dynamical problems, such as the scattering of a particle by a force concentrated at one point.

An analytical treatment of the same process is possible in the linear case for plane waves and for wave packets - at least as long as they do not spread - in terms of the tunnelling through a potential jump taking advantage of the superposition principle.

For the LSE this principle holds only in a weak form, but after the interaction of a gausson with a potential barrier a solution composed by two spatially separated parts is produced, such that each of the two parts tends to an equilibrium configuration of solitary shape (Fig. 1a). So, after a proper time, one can apply the weak superposition principle and describe the process as the decay of a gausson in two smaller ones.

From the Ehrenfest theorem the current variation after the interaction of a gausson (incident from the left) with the potential barrier $U=U_0, x>x_0$, is given by

$$P_F - P_I = -U_0 \int_{-\infty}^{+\infty} \rho(x_0, t)\, dt \tag{7}$$

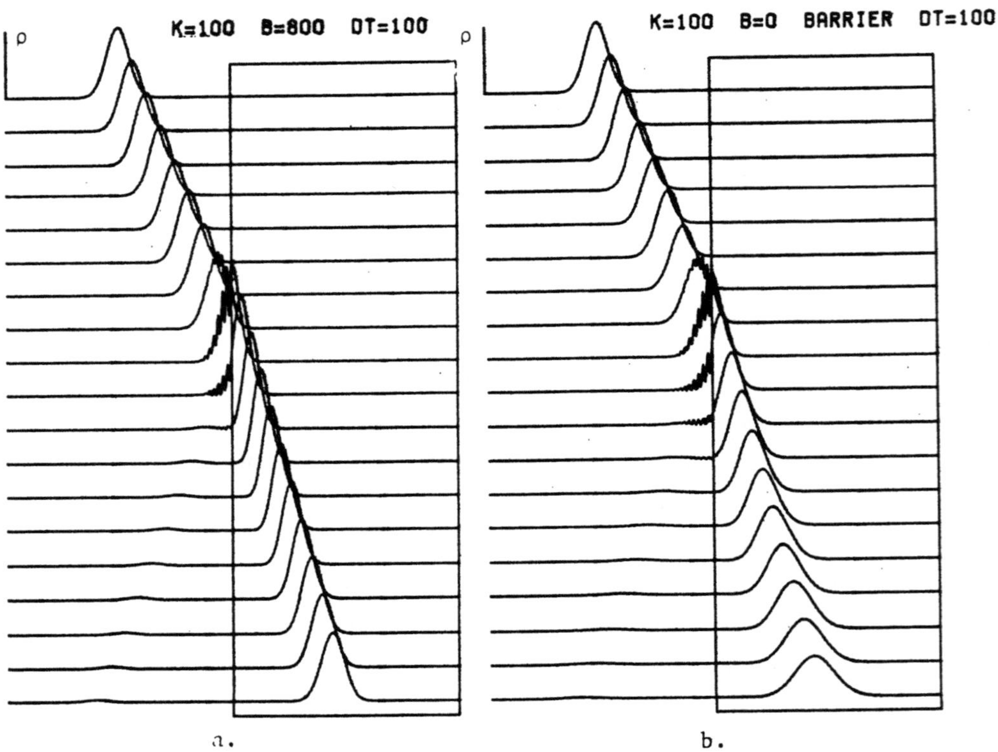

Fig. 1.

In the region $x > x_0$ the equilibrium condition is reached without any interference of the "incoming" and "reflected" solutions. So it is possible to assume that the charge density of the solution for $x > x_0$ is a function of x and t of the form $\rho = \rho(x-vt)\,(\rho(x_0,t) = \rho(x_0^+,t) = \rho(x_0-vt))$ for every time and the integral (7) becomes

$$P_F - P_I \simeq -U_0 Q/v \quad , \tag{8}$$

being Q the transmitted charge.

Note that the linear case can be considered as the limit of the nonlinear one when the parameter b tends to 0: then the stablest solution is a gausson of infinite radius, to be reached in infinite time, as happens for linear packets that spread (Fig. 1b).

3. Numerical Calculations.

Equation (1) has been solved numerically with positive and negative potential steps and a square well and a barrier in one dimension. The kinetic energy of the incoming normalized gausson was chosen equal double and half the potential strength U_0. In one case different values of the nonlinearity parameter b have been considered. The parameter a was set equal to 1.

The method of solution was taken from Goldberg, Schey and Schwartz,[7] with additional iterations to include the nonlinear term. U_0 was chosen $2(50\pi)^2$ and 400 was the standard value for b in a system of units in which $\hbar = 1$, $m = \frac{1}{2}$. A comparison with exact solution in the case U=0 gave a precision of 10^{-4} in the computed energy and momentum in the worst case. The validity of the choice of a free gausson as initial solution was also tested: for b large enough in the first stages the potential jump is sufficiently far not to affect the solution.

A sample of the results is given in Figs. 1-5 which show the charge density of the solutions, or a function λ of its logarithm to enhance the smallest effects.

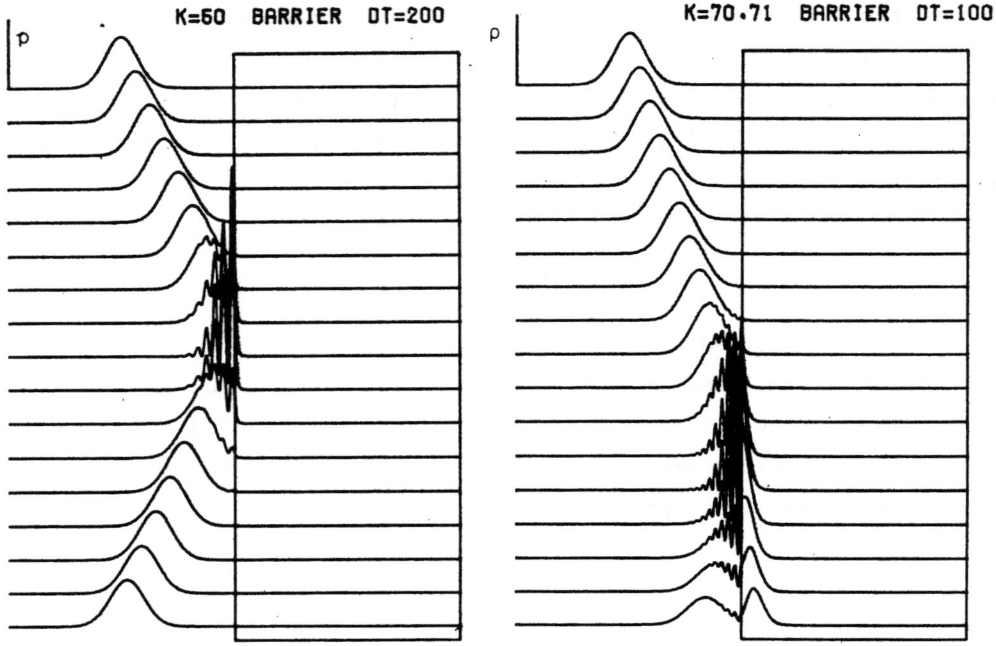
K=60 BARRIER DT=200
p
K=70.71 BARRIER DT=100
ρ

Fig. 2.

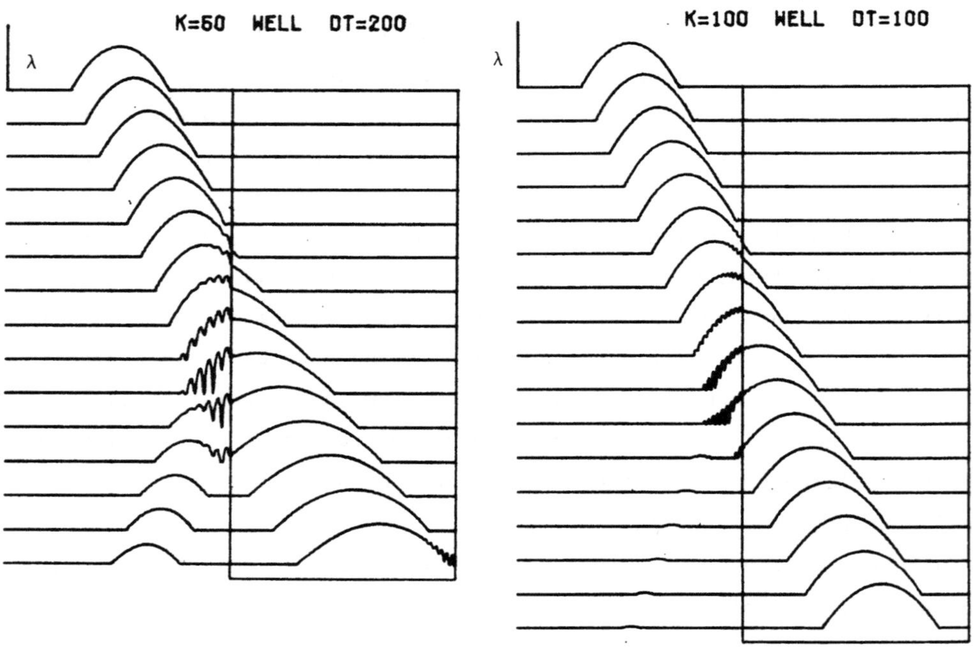
K=60 WELL DT=200
λ
K=100 WELL DT=100
λ

Fig. 3.

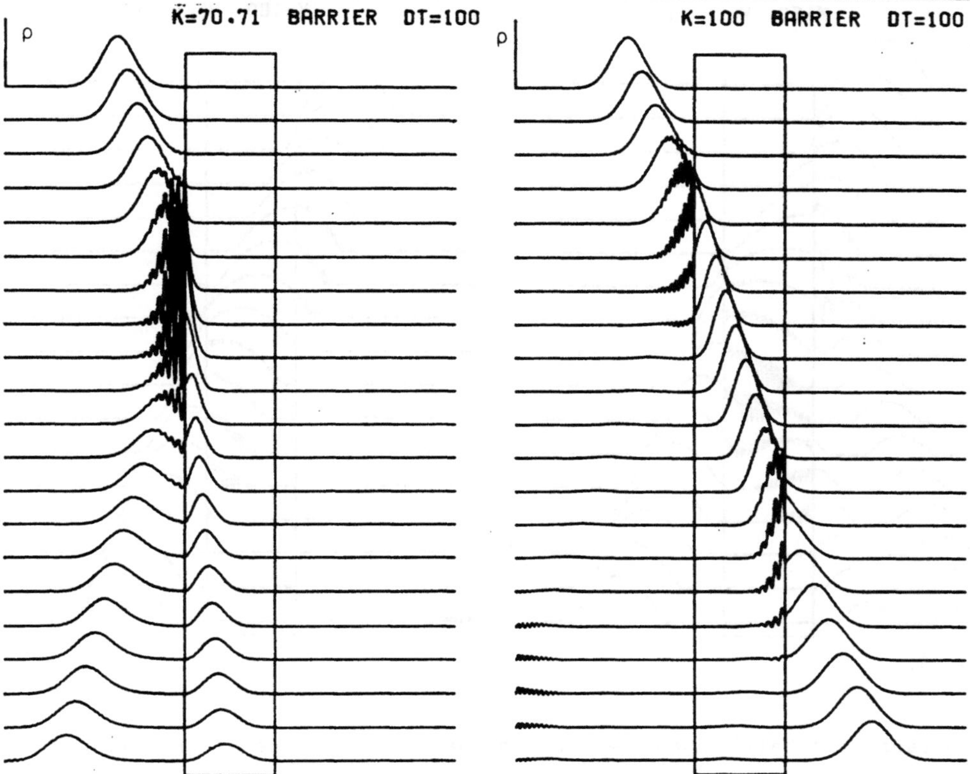

Fig. 4.

Fig. 1 exhibits the different behaviour of a gausson of LSE and a packet of the linear equation, identical at t=0. In the linear case the packets spread, while both components of the nonlinear solution evolve towards gausson solutions, as has been confirmed by best-fit calculations.

In Fig. 2 comparison is done among the processes induced by gaussons of different kinetic energy and same b scattered by a positive potential step. When the energy E_k is half U_0 we have an elastic reflection of the gausson. When $E_k=U_0$ a third of the total charge is transmitted, and for $E_k=2U_0$ only a small fraction is reflected. The difference from the linear tunnel effect is very small, presumably within the computational errors and formula (8) holds better than 1% in this last case.

The main feature exhibited by finite barrier potential is the trapping of part of the charge within the barrier for $E_k \geq U_0$. It is also evident for $E_k=2U_0$ that the velocity of the envelope solutions decreases as

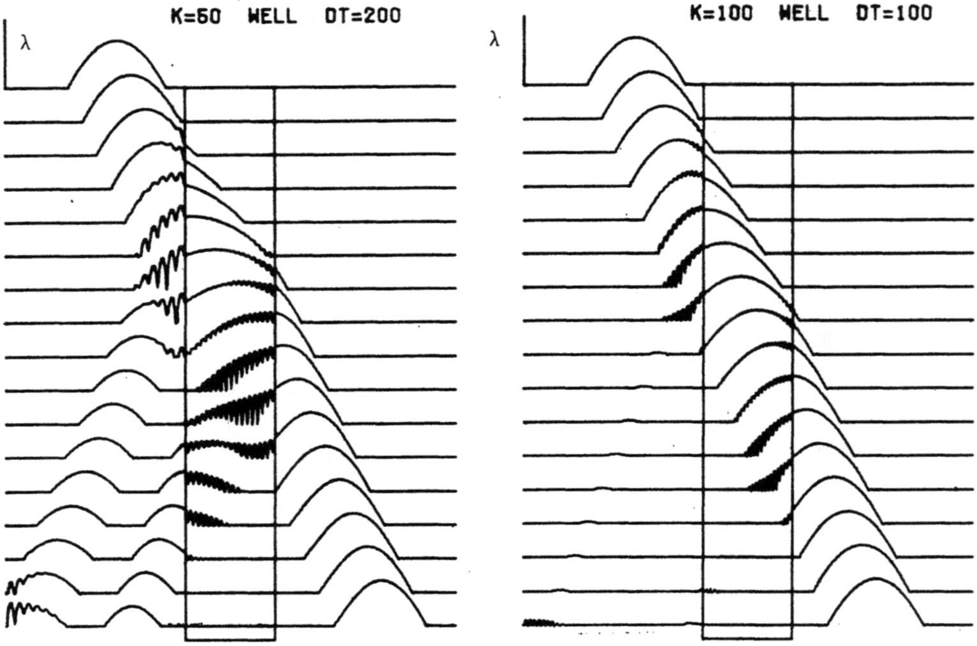

Fig. 5.

it passes the barrier and returns to the initial value after coming out (Fig. 4).

In the case of the negative potential step (Fig. 3) formula (8) holds for all E_k's at the level of few 10^{-3}. A new feature of the interaction of gaussons with a square well is the production of two reflected waves: one reflected in the initial interaction, the second transmitted back after a reflection against the second wall (Fig. 5).

In all cases the total charge and the total energy are conserved within the accuracy of the computation.

In conclusion, a wave mechanics that associates a gausson wave solution with a particle can describe with LSE the interaction of particles with simple forms of the potential, with the new feature of producing many particle solutions, consistently with the LSE conservation laws.

4. Physical Interpretation.

The proponents of the wave mechanics based on LSE move in the realm of the Copenhagen interpretation of quantum mechanics - they find out consistency with the Born interpretation of the wave function - and consider their theory useful to describe phenomena in an intermediate region (mesoworld), which separates the microworld where linearity is a natural law from the macroworld, where linearity is only an approximation to its intrinsic nonlinearity. An experimental test to measure the nonlinearity parameter b has been recently proposed by Shimony.[8]

An alternative interpretation of the theory can be found within de Broglie's theory of double solution,[9] which considers the microworld at smallest distances of nonlinear nature, and suggests that its best linear approximation is given by a couple of solutions, one regular and one singular, properly matched via a phase identity condition.[10]

A possible approach to connect the two theories is to solve both for the same dynamical problem and find out their equivalence, and that is the program the authors intend to pursue. In particular, the peculiar quantum effect of creation of particles in interaction also in the non relativistic case - and so the non conservation of particle number - must be fully explained in this context.

References.

1. Lonngren K. and Scott A.C. Eds., *Solitons in Action* ,Academic,N.Y.,1978. Scott A.C., Chu F.Y.F. and McLaughlin D.W., The Soliton etc., *Proc. IEEE,* 61, 1443, 1973.

2. Bialynicki-Birula I. and Mycielski J., Wave equations with logarithmic nonlinearities, *Bull. Acad. Polon. Sci.* ClIII, 23, 461, 1975. Bialynicki-Birula I. and Mycielski J., Nonlinear wave mechanics, *Ann. Phys.* 100, 62, 1976. Bialynicki-Birula I. and Mycielski J., Gaussons, *Physica Scripta* (topical issue *Solitons*), 1979. Ventura I. and Marquez G.C., Semiclassical quantization of a field theoretical model, *J.Math. Phys.* 19, 838, 1978.

3. Oficialski J. and Bialynicki-Birula I., Collisions of gaussons, *Acta*

Physica Polonica, B9, 759, 1978.

4. Newell A.C., Nonlinear Tunnelling, *J.Math. Phys.* 19, 1126, 1978.

5. Segur H. and Ablowitz J., Asymptotic solutions and conservation laws for the nonlinear Schröd. eq., *J.Math. Phys.* 17, 710 and 714, 1976.
   Pereira N.R. and Stenflo L., Nonlinear Schrödinger equations etc., *Phys. of Fluids* 20, 1733, 1977.
   Watanabe G., Solitons and Generation of tail etc. *J.Phys. Soc. Jap.* 45, 276, 1978.
   Makhankov V.G., Dynamics of classical solitons, *Phys. Rep.* 35,1,1978.

6. Zakharov J.V.E. and Shabat A.B., Exact theory of two dimensional self focusing etc. *Soviet Phys.* JETP 34, 62, 1972.

7. Goldberg A. et al., Computer-generated motion pictures of Q.M. transmission and reflection phenomena, *Am. J.Phys.* 35, 177, 1967.

8. Shimony A., Proposed neutron interferometer test for some nonlinear variants of wave mechanics, *Phys. Rev.* 20A, 394, 1979.

9. De Broglie L., *Nonlinear wave mechanics,* Elsevir, New York, 1960.

10. Andreade e Silva J., La theorie de la double solution; Fer F., Guidage de Particules, ondes singulières, *Louis de Broglie, sa conception du monde physique,* Gauthier-Villars, Paris 1973.

PART III

NON SEPARABILITY AND NON LOCALITY

Nonseparability in Quantum Mechanics

GianCarlo Ghirardi

Istituto di Fisica Teorica dell'Università and International Centre for Theoretical Physics, Trieste.

1. Introduction.

As well known, there are two main features of quantum mechanics which are quite generally considered as unsatisfactory, i.e. quantum indeterminism and quantum nonseparability. By the first expression one indicates the fact that identically prepared physical systems (even when the preparation is the most precise that the conceptual structure of the theory allows, i.e. it corresponds to the measurement of a complete set of commuting observables) can give different results when subsequently subjected to identical measurements. By the second expression one indicates the loss of individuality of a physical system as a consequence of its interaction with another one, even when the two systems are well separated in space and no longer interacting.

Hystorically, it was just the uneasiness of many physicists for the above mentioned aspects of quantum mechanics which led to the consideration of hidden-variable theories. As is well known, it has been possible to build up hidden-variable theories equivalent to quantum mechanics, but, as shown by Bell(1) and Wigner (2), such theories are necessarily non local. Therefore, even though the consideration of hidden variables allows to overcome the difficulties connected with the probabilistic character of quantum mechanics, it is not able to eliminate nonlocality, which is the direct manifestation of quantum nonseparability. Also the so-called stochastic interpretation of quantum mechanics (3) was developed with the same aims as hidden variable theories. In this case too, the difficulties connected with the undeterminism can be overcome, since the random interactions of the particles with the hypothetical underlying medium are entirely responsible for the indeterminism (so that, in a sense, it is accounted for in classical terms), but again nonlocal effects emerge (4), which are a direct consequence of quantum nonseparability. For the above reasons , these nonseparability effects are presently considered the most disturbing ones; as well known, they make quantum mechanics incompatible with the simultaneous validity of the principles of reality and locality (5,6).

These lectures will deal first of all with a discussion of the conceptual and formal aspects of quantum nonseparability and with an analysis of its consequences in various possible "classical" interpretations of quantum mechanics. The second part will be devoted to a critical discussion of a proposed (7,8,9) way to overcome these effects, i.e. the so-called hypothesis of spontaneous factorization of quantum states.

PART A.

QUANTUM NONSEPARABILITY AND ITS CONCEPTUAL CONSEQUENCES.

1.A. General Formulation of the Problem.

We shall consider ensembles of quantum systems S and we shall denote by $\mathcal{H}$ the Hilbert space associated to them. We start by distinguishing ensembles which correspond to pure states from those which are mixtures. We call pure cases the ensembles which can be described by a state vector $|\Psi\rangle$, so that the corresponding statistical operator ρ is a projection operator onto a one-dimensional linear manifold

$$\rho = |\Psi\rangle\langle\Psi| \tag{1.A.1}$$

and satisfies

$$\rho^2 = \rho. \tag{1.A.2}$$

The ensembles which cannot be associated with a state vector are called mixtures, and for them necessarily $\rho^2 \neq \rho$. For the moment we shall consider only the so-called proper mixtures or first kind mixtures, i.e. ensembles which are composed of several subensembles each of which is a pure case. They are characterized by the populations n_i and the state vectors $|\varphi_i\rangle$ corresponding to the subensembles E_i. The statistical operator is

$$\rho = \sum_i \frac{n_i}{N} |\varphi_i\rangle\langle\varphi_i|, \tag{1.A.3}$$

N being the total number of particles in the ensemble. The vectors $|\varphi_i\rangle$ are assumed to be normalized but not necessarily mutually orthogonal nor linearly independent. From the point of view of the quantum predictions, two ensembles with the same statistical operator are physically equivalent, since the expectation value of an operator $\mathcal{A}$ is given by Trace $\mathcal{A}\rho$. In particular we can then always consider in place of the ensemble described by(1.A.3) the one which has as subensembles pure ensembles associated to the eigenfunctions of ρ with weights equal to the corresponding eigenvalues.

Let us introduce now an important definition: For a physical system which is one of the members of an ensemble $\mathcal{E}$ we will say that it possesses a definite physical property a_m, whenever we can foresee with certainty the result a_m of a possible measurement of the observable $\mathcal{A}$. As well known, when we consider a pure case associated to the state vector $|\Psi\rangle$, the necessary and sufficient condition in order that the systems of the ensemble possess the property a_m is that $|\Psi\rangle$ belongs to the linear manifold spanned by the eigenvectors of $\mathcal{A}$, associated to the eigenvalue a_m

$$\mathcal{A}|\Psi\rangle = a_m|\Psi\rangle. \tag{1.A.4}$$

The indeterminacy relations limit drastically the possible sets of physical properties which can be simultaneously attributed to a system; however, provided one assumes (as we will do) that all self-adjoint operators are observable, for any pure case $|\Psi\rangle$ there is always a complete set of definite properties possessed by all systems of the ensemble, since any state vector is the unique eigenstate of a proper set of commuting observables. These definite physical properties constitute then the objective properties that quantum mechanics allows to attribute to any quantum system. To make physically more clear the above definition we can limit our considerations to the spin space of a spin $\frac{1}{2}$ particle; the previous statement then simply means that the particle "has its spin in a definite direction" since, as well known, one can always find a direction $\underline{n}$ such that $|\Psi\rangle$ is an eigenstate of $\underline{\sigma}\cdot\underline{n}$ corresponding to the eigenvalue 1.

When we pass from pure cases to first kind mixtures we can still attribute (at least conceptually) to any system of the ensemble a complete set of definite properties since it belongs to a given subensemble E_i and is then associated to a unique state vector $|\varphi_i\rangle$. Obviously in such a case we cannot say a priori to which subensemble it belongs, but there is no contradiction in stating that it possesses definite (even though un - known) physical properties.

To tackle the problem of quantum nonseparability we have now to consider an ensemble $\mathcal{E}$ of quantum systems $S=S_1+S_2$ each of which is composed of the two constituent subsystems S_1 and S_2 . As well known, the Hilbert space $\mathcal{H}^{(1,2)}$ for the description of this problem is the direct product of the Hilbert spaces $\mathcal{H}^{(1)}$ and $\mathcal{H}^{(2)}$ of the constituents

$$\mathcal{H}^{(1,2)} = \mathcal{H}^{(1)} \otimes \mathcal{H}^{(2)}. \tag{1.A.5}$$

We shall prove first of all a very simple theorem about ensembles of composite systems which are pure cases. It states that: <u>The necessary and sufficient condition in order that a complete set of commuting observables of the Hilbert space $\mathcal{H}^{(1)}$ (or $\mathcal{H}^{(2)}$) have a definite value is that the normalized state $|\Psi\rangle$ associated to the ensemble be factorized, i.e. it is the direct product of a state $|\varphi\rangle \in \mathcal{H}^{(1)}$ and a state $|\chi\rangle \in \mathcal{H}^{(2)}$</u> .

The sufficiency is obvious provided one accepts that all self -adjoint operators are observable. To prove the necessity, denoting by $|\omega^{(1)}\rangle$ the simultaneous eigenvector of the complete set in $\mathcal{H}^{(1)}$, we observe that the probability of getting the corresponding set of eigenvalues in a measurement is given by

$$P = \langle\Psi|\omega^{(1)}\rangle\langle\omega^{(1)}|\Psi\rangle. \tag{1.A.6}$$

By introducing complete orthonormal sets of states $\{|\varphi_i^{(1)}\rangle\}$ and $\{|\chi_j^{(2)}\rangle\}$ in $\mathcal{H}^{(1)}$ and $\mathcal{H}^{(2)}$, respectively, and choosing $|\varphi_1^{(1)}\rangle=|\omega^{(1)}\rangle$, we have

$$|\Psi\rangle = \sum_{ij} c_{ij}\, |\varphi_i^{(1)}\rangle \otimes |\chi_j^{(2)}\rangle, \tag{1.A.7}$$

with

$$\sum_{ij} |c_{ij}|^2 = 1. \tag{1.A.8}$$

From (1.A.6) we then get

$$P = \sum_j |c_{1j}|^2, \tag{1.A.9}$$

so that, imposing P=1 and comparing with (1.A.8), we have

$$c_{ij} = \delta_{i1}\, a_j \tag{1.A.10}$$

and finally

$$|\Psi\rangle = |\varphi_1^{(1)}\rangle \otimes \sum_j a_j |\chi_j^{(2)}\rangle. \qquad (1.A.11)$$

In ref.(10) it has been proved that, in the case of a system composed of identical constituents, the analogous requirement that one (we cannot specify which one due to the identity) of the constituent subsystems has definite physical properties implies that the state $|\Psi\rangle$ be obtained from the symmetrization or antisymmetrization (according to the statistics) of a factorized state.

We can now discuss the problem of quantum nonseparability. We consider an ensemble $\mathcal{E}$ which is a pure case associated to a nonfactorized state vector $|\Psi\rangle$, which we will write as in eq.(1.A.7). Due to the nonfactorized character of $|\Psi\rangle$, nonseparability effects arise, consisting in the fact that the individual systems S_1 (or S_2) of the ensemble cannot be thought of as having a complete set of definite properties of their own. In fact, due to the previous theorem, one could attribute definite physical properties to each subsystem S_1 (S_2) of the ensemble if and only if each composed system would be member of a subensemble which is a pure case associated to a factorized state. The whole ensemble $\mathcal{E}$ would then turn out to be described by a statistical operator

$$\bar{\rho} = \sum_{\alpha\beta} p_{\alpha\beta} |\varphi_\alpha^{(1)} \chi_\beta^{(2)}\rangle\langle\varphi_\alpha^{(1)} \chi_\beta^{(2)}|. \qquad (1.A.12)$$

Since however $\mathcal{E}$ can also be described by

$$\rho = |\Psi\rangle\langle\Psi|, \qquad (1.A.13)$$

if we require that physically the two descriptions be completely equivalent, we must impose

$$\bar{\rho} = \rho. \qquad (1.A.14)$$

However $\rho^2 = \rho$ while $\bar{\rho}^2$ cannot be equal to $\bar{\rho}$, unless only one term appears in (1.A.12). If this is the case, identifying (1.A.12) with (1.A.13) one would get that $|\Psi\rangle$ is factorized, contrary to the assumption. The analogous theorem for the case of identical constituents allows us to conclude that nonseparability effects arise also in such a case except when $|\Psi\rangle$ is obtained by symmetrization of a factorized state. We can summarize the above discussion by stating that, when an ensemble of composite systems is a pure case associated to a nonfactorized state vector, the two component subsytems S_1 and S_2 loose their individuality in the sense that one cannot think of the individual systems S_1 and S_2 as possessing a complete set of definite physical properties.

It should be obvious that this fact puts very severe limitations to the natural requirement that the description of isolated systems has an objective character. In fact, nonfactorized states emerge quite naturally in the quantum description of physical systems. Suppose that, in the remote past one has prepared two physical systems S_1 and S_2 in the states $|\varphi^{(1)}\rangle$ and $|\chi^{(2)}\rangle$, respectively. The initial state of the system is then factorized $|\psi\rangle = |\varphi^{(1)}\rangle \otimes |\chi^{(2)}\rangle$, and, when consideration is given to an ensemble of such systems, we can consider each individual system S_1 or S_2 as possessing a complete set of definite physical properties. Suppose that, due to the dynamics of the problem the two systems S_1 and S_2 interact for a certain time interval. After the interaction, quite in general, the ensemble will be described by a nonfactorized state and therefore it will be inconsistent to attribute definite physical properties to S_1 and S_2 . Obviously one can still attribute definite properties to the system $S = S_1 + S_2$ as a whole, but the constituent subsystems loose completely their individuality. This is true even if at the considered time the systems are in well separated space regions and no more interacting. This is a peculiar consequence of quantum mechanics which is very difficult to accept by itself. If one considers that all parts of the universe interact among themselves, one is led, if one accepts quantum mechanics as universally valid, to a view in which the universe constitutes a sort of "unbroken whole" whose constituent parts have no physical identity. This view is, in our opinion, hardly acceptable.

Before concluding we recall that, according to the general postulates of quantum mechanics, the possibility of considering isolated systems is regained as a consequence of a measurement process which, by reduction of the wave packet, can lead to factorized states. This is not, however, a solution for the above difficulty. Actually, it is just quantum nonseparability that gives rise to the main conceptual difficulties of the quantum theory of measurement (6).

2.A. Some Consequences of Quantum Nonseparability.

Besides the above stated features, quantum nonseparability has several other consequences which deserve to be discussed. On one side, as already stated, it has precise conceptual implications when combined with the quantum theory of measurement.

We do not want to enter here in this problem, whose essence is singnificantly summarized in the conceptual analysis leading to the E-P-R-paradox (5). For a lucid discussion of this point the reader is referred to the book by d'Espagnat (6). On the other hand, we think it is useful to stress that the "classical" counterpart of quantum nonseparability is something which is even less acceptable, i.e. nonlocality. By this we mean that whenever one tries to build up a "classical" (i.e. deterministic) interpretation of quantum mechanics, just as a consequence of the occurrence of nonfactorized states in the quantum formalism, one ends up with a theory which exhibits nonlocal effects. Interestingly enough, therefore, it is not quantum indeterminism, but it is quantum nonseparability which constitutes a deadlock for a reinterpretation of quantum mechanics in the spirit of those lines of thought indicated by d'Espagnat (6) as "realism" or "macroobjectivism" or "microobjectivism". To illustrate this point I will briefly consider hidden-variable theories and the so-called stochastic interpretation of quantum mechanics.

Hidden Variable Theories.

As well known, the basic idea of these theories is to attribute the statistical character of quantum theory to the fact that the quantum states are averages of better defined states for which the results of all possible measurements are exactly determined. The hypothetical dispersion-free states would then be specified not only by the state vector but by additional variables which are considered hidden, since we cannot prepare them. The basic formula giving the connexion with quantum theory

$$\langle \mathcal{A} \rangle = \int d\lambda \, w(\lambda) \, \mathcal{A}(\lambda), \tag{2.A.1}$$

shows that the quantum mean value of an observable $\mathcal{A}$ is obtained as a weighted average of the definite values $\mathcal{A}(\lambda)$ that the same observable would take when the hidden variables λ take the indicated value.

To discuss the consequences of quantum nonseparability in such a context, we consider a very simple case, i.e. a composed system of two spin $\frac{1}{2}$ particles in the singlet state

$$|\psi\rangle = \frac{1}{\sqrt{2}} \left\{ |s_1^+ s_2^- \rangle - |s_1^- s_2^+ \rangle \right\} \tag{2.A.2}$$

where $|s_i^\pm\rangle$ are the eigenstates of the Pauli matrices $\sigma_z^{(1)}$ and $\sigma_z^{(2)}$ for the two systems. In eq.(2.A.2) we have ignored the space variables; it is assumed, however, that the two systems are spatially well separated and non interacting. Suppose then one has two magnets M_1 and M_2 which measure $\underline{a}\cdot\underline{\sigma}^{(1)}$ and $\underline{b}\cdot\underline{\sigma}^{(2)}$, i.e. the probability that the spin of particle 1 (2) be along the direction $\underline{a}$ ($\underline{b}$). Let us denote by $P(\underline{a},\underline{b})$ the mean value of the operator $\underline{a}\cdot\underline{\sigma}^{(1)}\otimes\underline{b}\cdot\underline{\sigma}^{(2)}$, so that in the quantum formalism

$$P_Q(\underline{a},\underline{b}) = \langle\Psi|\underline{a}\cdot\underline{\sigma}^{(1)}\otimes\underline{b}\cdot\underline{\sigma}^{(2)}|\Psi\rangle. \tag{2.A.3}$$

In the hidden variable description the same quantity is

$$P_{HV}(\underline{a},\underline{b}) = \int d\lambda\, w(\lambda)\, \Sigma^{(1)}(\underline{a},\lambda)\, \Sigma^{(2)}(\underline{b},\lambda), \tag{2.A.4}$$

where for simplicity we have indicated only one hidden variable. It has to be stressed that in (2.A.4) we have explicitly assumed that the certain result $\Sigma^{(1)}(\underline{a},\lambda)$ for the variable corresponding to $\underline{a}\cdot\underline{\sigma}^{(1)}$ (and similarly for $\Sigma^{(2)}(\underline{b},\lambda)$ and $\underline{b}\cdot\underline{\sigma}^{(2)}$) in the state characterized by the value λ of the hidden variable, depend, besides λ , only from the direction $\underline{a}$ ($\underline{b}$) along which the spin component is measured and not on the setting $\underline{b}$ ($\underline{a}$) of the other apparatus. From (2.A.4) one can easily derive the fundamental Bell's inequality (1)

$$|P_{HV}(\underline{a},\underline{b}) - P_{HV}(\underline{a},\underline{c})| + |P_{HV}(\underline{a}',\underline{b}) + P_{HV}(\underline{a}',c)| \leq 2. \tag{2.A.5}$$

It is then easy to prove that, if one wants that the predictions of the hidden-variable theory reproduce all the quantum mechanical results one gets a contradiction with eq.(2.A.5). In fact , for the singlet state

$$P_Q(\underline{a},\underline{b}) = -\,\underline{a}\cdot\underline{b} = -\cos\widehat{\underline{a}\underline{b}} \tag{2.A.6}$$

Substituting P_Q in place of P_{HV} in (2.A.5) one then gets a relation which, for proper choices of $\underline{a}$, $\underline{a}'$, $\underline{b}$,and $\underline{c}$, is false.

The above incompatibility of Bell's inequality with quantum predictions is due only to the nonfactorized nature of the state $|\Psi\rangle$. In fact, if one would have only factorized states in the ensemble, so that one could write for the statistical operator

$$\rho = \int d\eta\, p(\eta)\, |\varphi^{(1)}_\eta \chi^{(2)}_\eta\rangle\langle\varphi^{(1)}_\eta \chi^{(2)}_\eta|, \tag{2.A.7}$$

one would get for $P_Q(\underline{a},\underline{b})$

$$P_Q(\underline{a},\underline{b}) = \int d\eta\, p(\eta)\, \langle\varphi^{(1)}_\eta|\underline{a}\cdot\underline{\sigma}^{(1)}|\varphi^{(1)}_\eta\rangle \langle\chi^{(2)}_\eta|\underline{b}\cdot\underline{\sigma}^{(2)}|\chi^{(2)}_\eta\rangle \tag{2.A.8}$$

which has the same formal structure as (2.A.4) and cannot give rise to contradictions. Are we then led to the conclusion that there does not exist a hidden variable theory equivalent to quantum mechanics ? This would be incorrect as has been proved by Bohm (11). Actually what we have considered above is a too restricted class of hidden variable theories, i.e. the so-called local theories. To get a complete agreement with quantum mechanics one has to allow $\Sigma^{(1)}(\underline{a},\lambda)$ $[\Sigma^{(2)}(\underline{b},\lambda)]$ to depend also from $\underline{b}$ ($\underline{a}$), i.e. in (2.A.4) we must write $\Sigma^{(1)}(\underline{a},\underline{b},\lambda)[\Sigma^{(2)}(\underline{b},\underline{a},\lambda)]$. But then we immediately see the price we have to pay for the existence of nonfactorized states (1):when forcing a classical theory to reproduce quantum mechanics we have to accept nonlocal effects. In the present case the result of the measurement of the spin of S_1 along $\underline{a}$ depends also from the direction $\underline{b}$ along which we intend to measure the spin of S_2 , no matter how far are the two measuring apparata.

Before concluding let us stress that the mere fact that a deterministic theory equivalent to quantum mechanics can be exhibited (i.e. the nonlocal version of hidden variable theories) is of great conceptual relevance and shows that quantum indeterminism is not a deadlock for a classical interpretation of quantum mechanics. On the other hand, the fact that such a theory turns out to be unavoidably nonlocal, due to quantum nonseparability effects, shows that these effects are really the crucial point for any interpretation of quantum theory in classical terms.

The Stochastic Interpretation.

Another interesting attempt at a classical interpretation of quantum mechanics comes from the existing parallelism between such a theory and the theory of stochastic processes. The conclusion of the fundamental work by Nelson (3) on this problem can be summarized by stating that for any specific quantum problem, i.e. dynamical equations and initial conditions, there exists a Markov process in configuration space which accounts for the same evolution, in the sense that the probability density $\rho(\underline{r},t)$ of finding the particle in $\underline{r}$ at time t coincides with the quantum-mechanical probability density $|\psi(\underline{r},t)|^2$.

To prove the equivalence of the Schroedinger evolution with a Markov process we start by considering a Wiener process for the position $\underline{r}(t)$, which is assumed to be a random variable

$$d\underline{r}(t) = \underline{\mu}(\underline{r},t)\,dt + d\underline{W}(t) \qquad (2.A.9)$$

Here $\underline{\mu}(\underline{r},t)$ is the forward drift and dW_i are random variables normally distributed with mean zero and variance $\sigma^2 dt$. In place of (2.A.9) one can use the backward description governed by the same equation with the backward drift $\underline{\mu}_*(\underline{r},t)$ taking the place of $\underline{\mu}(\underline{r},t)$. The forward and backward drifts are related (see eq.(2.A.11)) through the probability density $\rho(\underline{r},t)$. The key idea of the Nelson approach (3) is that of characterizing the stochastic process (2.A.9) not by assigning, as usual, the drift $\underline{\mu}(\underline{r},t)$, but by a time symmetric dynamical condition which takes the form of Newton's equation of motion

$$\frac{1}{2}\left(DD^* + D^*D\right)\underline{r}(t) = \underline{F}/m \tag{2.A.10}$$

where D and D^* are the forward and backward stochastic derivatives. Let us use, in place of $\underline{\mu}$ and $\underline{\mu}_*$, the current velocity $\underline{v}$ and the stochastic velocity $\underline{u}$ defined as

$$\underline{v} = \frac{1}{2}\left(\underline{\mu}+\underline{\mu}_*\right), \quad \underline{u} = \frac{1}{2}\left(\underline{\mu}-\underline{\mu}_*\right) \equiv \frac{\sigma^2}{2}\,\mathrm{grad}\ln\rho(\underline{r},t). \tag{2.A.11}$$

The forward derivative can be written as

$$D f(\underline{r}(t),t) = \left[\frac{\partial}{\partial t} + \underline{\mu}\cdot\mathrm{grad} + \frac{\sigma^2}{2}\Delta\right] f(\underline{r}(t),t), \tag{2.A.12}$$

and the backward one D^* has the same expression with $\underline{\mu}\rightarrow\underline{\mu}_*$ and $\sigma^2\rightarrow-\sigma^2$. The dynamical equation (2.A.10) becomes then

$$\frac{\partial\underline{v}}{\partial t} = \frac{\underline{F}}{m} + (\underline{u}\cdot\mathrm{grad})\underline{u} - (\underline{v}\cdot\mathrm{grad})\underline{v} + \frac{\sigma^2}{2}\Delta\underline{u}, \tag{2.A.13}$$

which, together with the continuity equation

$$\frac{\partial\rho}{\partial t} = -\mathrm{div}(\rho\underline{v}), \tag{2.A.14}$$

(which follows from the Fokker-Planck equations) are the basic equations which characterize completely the Markov process. In fact, assigning $\rho(\underline{r},t_0)$ and $\underline{v}(\underline{r},t_0)$ one can determine $\underline{u}(\underline{r},t_0)$ by (2.A.11) and then solve the set of coupled equations for $\rho(\underline{r},t)$ and $\underline{v}(\underline{r},t)$ which characterize completely the Markov process.

To prove the equivalence with the Schroedinger equation we put

$$R(\underline{r},t) = \frac{1}{2}\ln\rho(\underline{r},t), \tag{2.A.15}$$

so that

$$\underline{u}(\underline{r},t) = \sigma^2\,\mathrm{grad}\,R(\underline{r},t), \tag{2.A.16}$$

and we assume that the current velocity field is also a gradient

$$\underline{v}(\underline{r},t) = \sigma^2\,\mathrm{grad}\,S(\underline{r},t). \tag{2.A.17}$$

Eq. (2.A.13, 14), when $\underline{F} = -\mathrm{grad}\,V$, become

$$\begin{aligned}\frac{\partial R}{\partial t} &= -\frac{\sigma^2}{2}\Delta S - \sigma^2\,\mathrm{grad}\,R\cdot\mathrm{grad}\,S\\ \frac{\partial S}{\partial t} &= -\frac{1}{\sigma^2 m}V - \frac{\sigma^2}{2}\left[(\mathrm{grad}\,S)^2 - (\mathrm{grad}\,R)^2\right] + \frac{\sigma^2}{2}\Delta R.\end{aligned} \tag{2.A.18}$$

Taking the sum of the first equation with the second one multiplied by the imaginary unit one gets

$$\frac{\partial e^{R+iS}}{\partial t} = \frac{i}{2}\sigma^2\Delta e^{R+iS} - \frac{i}{\sigma^2 m}V e^{R+iS} \tag{2.A.19}$$

which coincides with the Schroedinger equation if one puts

$$\sigma^2 = \frac{\hbar}{m} , \quad \psi = e^{R+iS} . \qquad (2.A.20)$$

Some comments about the above derivation are necessary. Due to (2.A.20), the assignement of $\psi(\underline{r},0)$ determines $R(\underline{r},0)$ and $S(\underline{r},0)$ and also $\rho(\underline{r},0)=\exp[2R(\underline{r},0)]$ and $\underline{v}(\underline{r},0)=\frac{\hbar}{m}\,\mathrm{grad}\,S(\underline{r},0)$. Going to the considered Markov process and solving the dynamical equations, we get a probability density and current velocity which are related at any time to the solution of the Schroedinger equation by the previous relations. We can then identify the Schrödinger evolution with the unfolding of the so determined Markov process. However it has to be stressed that in this way the drift is not preassigned as in the classical diffusion processes but it depends on the initial wave function $\psi(\underline{r},0)$.

As exhaustively discussed in ref.(4) it is just this dependence of the drift from the initial conditions which makes problematic also the stochastic interpretation. In fact, when one is dealing with noncomposite systems, the stochastic interpretation can be considered as consistent provided one accepts that the preparation and measurement processes affect strongly the underlying medium in which the particles propagate. In such a case one can also give a precise conceptual meaning to particle trajectories, even though they turn out to be non observable. One can also prove that in such an interpretation all quantum indeterminism is due to the random interactions with the medium. However, when composite systems are considered, again difficulties arise from quantum nonseparability. In fact (4) if one considers the system of two freely propagating particles which at time t=0 are described by a nonfactorized wave function corresponding to mean value 0 and variance $\hbar^2/2\Delta^2$ for the total momentum and $\bar{p}$ and $\hbar^2/2\delta^2$ for the relative momentum, one discovers that, in the evolution, there is a strict correlation between the positions of the particles for subsequent times. In particular choosing $\delta^2/\Delta \gg \hbar/\bar{p}$ there is a time after which the two particles are in well separated space regions and it is impossible to explain the correlations by ascribing definite momenta to the particles at the initial time. One is then compelled to conclude that the diffusion processes of the pairs of the ensemble are strongly correlated even when they are far apart. This is a clear manifestation of nonlocal effects. Another way of evidenciating the nonlocal nature of the process comes from the observation that, for the considered problem, the equivalent stochastic processes can be written, in terms of the two random variables $\underline{r}_1$ and $\underline{r}_2$ (the position variables) as

$$d\underline{r}_1(t) = \underline{\mu}_1(\underline{r}_1,\underline{r}_2,t)\,dt + d\underline{W}_1$$
$$d\underline{r}_2(t) = \underline{\mu}_2(\underline{r}_1,\underline{r}_2,t)\,dt + d\underline{W}_2 \tag{2.A.21}$$

The two Wiener processes $d\underline{W}_1$ and $d\underline{W}_2$ are uncorrelated, but, the dependence of $\underline{\mu}_i(\underline{r}_1,\underline{r}_2,t)$ from both variables, which is a consequence of the nonfactorized nature of the initial wave function, couples the two equations even for large values of the distance $|\underline{r}_1-\underline{r}_2|$. The same fact can be evidenciated by observing that the considered stochastic process can also be described through two uncoupled stochastic equations for the independent random variables $\underline{R}$, the centre-of-mass position, and $\underline{r}_2-\underline{r}_1$, the relative coordinate. One can then state that everything goes as if the medium would affect independently the centre of mass of the system and the relative particle. The system then behaves like a unique whole even when the two particles are spatially well separated and non interacting. The stochastic interpretation meets therefore the same difficulties as hidden variable theories when confronted with quantum nonseparability.

PART B.

THE HYPOTHESIS OF SPONTANEOUS FACTORIZATION

1.B. Formulation of the Problem.

The second part of this work is devoted to a discussion of the so-called hypothesis of spontaneous factorization of quantum states. The many conceptual difficulties of quantum mechanics arising essentially from quantum nonseparability have led several physicists to the conclusion that, since the peculiarities or paradoxical implications of the formalism cannot be cured by considering equivalent (from the predictive point of view) but different (from the point of view of the interpretation) descriptions of quantum phenomena, one should face directly the problem of considering alternative theories which are in disagreement with quantum mechanics, and in which the unpleasant phenomenon of quantum nonseparability is absent. Obviously the disagreement should be kept as small as possible owing to the very high accuracy of experimental verification of quantum predictions in an incredibly large variety of physical phenomena. A very natural alternative would be to consider local hidden variable theories, which, as stated above, can be made equivalent to quantum mechanics for non composite systems, but differ from it in the case of composite systems, due to the validity of Bell's inequality. It has to be remarked that the discrepancies from quantum mechanics of such theories for composite systems, emerge only with respect to correlation experiments between the observables of the constituents. For this reasons there has been a lot of interesting experimental work (12-17) aimed to test Bell's inequality. The results of these experiments cannot yet be considered as definitive, however they point, with few exceptions, at a confirmation of quantum predictions and at an exclusion of local hidden variables.

We stress that, due to the previously remarked strict formal similarity between the expectation values of correlation operators in hidden variable theories and in quantum theory when only factorized states appear, the formal analogue of these theories would be a quantum theory in which only factorized states occur. Actually, such an assumption has been made by Jauch (9). However, since this hypothesis alters completely the formal structure of quantum mechanics, it seems to be too restrictive and would yield serious difficulties when confronted with the higly accurate predictions of quantum mechanics for example for bound states of many particle systems. One has to keep in mind that, after all, the consequences of quantum nonseparability become unpalatable when the constituent subsystems are far apart from each other and non interacting. It is under these conditions that one would like to speak of an isolated system and of its physical properties, which is forbidden by the quantum formalism if the system has been in interaction with other systems even in the remote past.

For the above reasons some attention has been paid to an interesting hypothesis put forward by Furry (7). It consists in assuming that quantum theory is valid for small separations of the constituents subsytems of a composite system, but that, when they are far apart, the nonfactorized state vector associated to the ensemble of composite systems changes into a statistical mixture of factorized states. In this way each individual subsystem regains definite physical properties. Obviously, to keep the violation of quantum mechanics minimal, one assumes that the results of all possible measurements of observables of each one of the constituent subsystems on the ensemble should coincide with those given by the nonfactorized state. If this would not be the case such a violation of quantum predictions would have already emerged experimentally. On the contrary, with the above restriction the discrepancies from quantum predictions will appear only in correlation experiments, which are not very common and quite difficult to perform.

Let us remark that the spontaneous factorization hypothesis for widely separated subsytems yields expressions for the correlation functions such that Bell's type inequalities hold. Tests of local hidden variable theories are then also tests for the spontaneous factorization. In spite of the formal resemblance of the formulae for hidden variables and for the suggested modified quantum theory it must be born in mind that the interpretation of the formulae are quite different in the two cases. Instead of attempting to reduce the observed results for statistical measurements on quantal systems to an underlying deterministic classical structure (hidden variables) the spontaneous factorization hypothesis has a purely quantal meaning and origin. The basic laws for microphysics are considered quantal and quantum mechanics is modified only for large separations.

2.B. Implications of the Assumption of Spontaneous Factorization.

Due to the observed analogy of the predictions of local hidden variable theories with those following from spontaneous localization, the attention of physicists has been more concentrated on the experimental consequences of factorization than on its conceptual implications or on the problem of constructing explicit dynamical models leading to it. We want to prove here, following ref.(18), than when this hypothesis is accepted, relevant changes must be made in the quantum formalism to account for it.

To this purpose we consider an ensemble of composed quantum systems $S = S_1 + S_2$ which is, at the initial time t=0, a pure case associated to the state vector $|\psi(1,2)\rangle$. Conventional quantum mechanics accounts for the evolution of the ensemble by the mapping

$$|\psi(1,2)\rangle \longrightarrow |\psi(1,2,t)\rangle = e^{-iHt}\,|\psi(1,2)\rangle. \qquad (2.B.1)$$

According to the spontaneous factorization hypothesis the evolution can be represented completely in general by the mapping

$$|\psi(1,2)\rangle \longrightarrow E_\psi(t), \qquad (2.B.2)$$

from state vectors to statistical ensembles. We make now the necessary assumptions on the mapping (2.B.2). As discussed we want that

1. Whenever the quantum mechanical wave function $|\psi(1,2,t)\rangle$ of (2.B.1) describes two spatially well separated subsytems, the corresponding $E_\psi(t)$ of (2.B.2) be a mixture of factorized states.

2. At all times the predictions regarding all possible measurements of observables of one of the two subsystems which can be deduced from $E_\psi(t)$ coincide with those given by $|\Psi(1,2,t)\rangle$
3. When the original ensemble, instead of being a pure case, is a statistical mixture, the corresponding ensemble at time t be the union of the ensembles corresponding according to (2.B.2) to the pure states which are present in the original ensemble.

We stress that, apart from the first requirement which embodies the hypothesis under discussion, the other two correspond to basic physical requirements. We also stress that, since at least some pure states must actually be mapped into statistical mixtures, the mechanism must be described at least at the level of statistical operators.

The problem which we want to discuss is now the following: is it possible to account for the spontaneous factorization of quantum states within the formalism of statistical operators? In fact, since there are physically different ensembles described by the same statistical operator, it could happen that the considered mapping requires to distinguish between physically non identical ensembles corresponding to the same statistical operator. As we shall see, this is actually the case. Since the correspondence (2.B.1) is one-to-one, the mapping (2.B.2) defines, for any given t, a mapping

$$|\Psi(1,2,t)\rangle \longrightarrow E_\psi(t) \tag{2.B.3}$$

between the evolved of pure states at time t and statistical ensembles. Let us denote by ρ and $\rho^\#$ the statistical operators corresponding to the left and right-hand sides of (2.B.3), respectively. The problem can then be reformulated as follows: does it exist a mapping

$$\rho \longrightarrow \rho^\# \tag{2.B.4}$$

of any ρ into $\rho^\#$ such that

$$\mathrm{Trace}^{(\alpha)}\rho = \mathrm{Trace}^{(\alpha)}\rho^\#, \ \alpha = 1,2, \tag{2.B.5}$$

where trace$^{(\alpha)}$ denotes the trace on Hilbert space $\mathcal{H}^{(\alpha)}$, and such

$$(a\rho + b\rho')^\# = a\rho^\# + b\rho'^\#, \ \forall a,b > 0,\ a+b=1. \tag{2.B.6}$$

Equations (2.B.5) and (2.B.6) embody the requirements 2 and 3.

We use now a theorem (12,13) on trace preserving completely positive linear mappings among statistical operators. It sta tes that the most general mapping of this kind has the form

$$\rho^{\#} = \sum_i A_i \rho A_i^{\dagger} \;, \quad \sum_i A_i^{\dagger} A_i = \mathbb{1} \,, \tag{2.B.7}$$

where the A_i are bounded operators. Let us now introduce two complete orthonormal sets $\{|\varphi_i^{(1)}\rangle\}$ and $\{|\psi_j^{(2)}\rangle\}$ in the Hilbert spaces of subsystems S_1 and S_2, respectively. Let us choose for $|\Psi(1,2,t)\rangle$ the factorized state

$$|\psi(1,2,t)\rangle = |\varphi_\ell^{(1)}\rangle \otimes |\psi_m^{(2)}\rangle . \tag{2.B.8}$$

The corresponding $\rho^{\#}$ can then be written

$$\rho^{\#} = \sum_i A_i \, |\varphi_\ell^{(1)} \psi_m^{(2)}\rangle \langle \psi_m^{(2)} \varphi_\ell^{(1)}| \, A_i^{\dagger} . \tag{2.B.9}$$

Imposing condition (2.B.5) one gets

$$\sum_i \sum_k \langle \psi_k^{(2)}| A_i |\varphi_\ell^{(1)}\rangle |\psi_m^{(2)}\rangle \langle \psi_m^{(2)}| \langle \varphi_\ell^{(1)}| A_i^{\dagger} |\psi_k^{(2)}\rangle = |\varphi_\ell^{(1)}\rangle\langle \varphi_\ell^{(1)}| \tag{2.B.10}$$

Let us take the expectation value of (2.B.10) on $|\varphi_n^{(1)}\rangle$, with $n \neq \ell$:

$$\sum_i \sum_k |\langle \varphi_n^{(1)}| \langle \psi_k^{(2)}| A_i |\psi_m^{(2)}\rangle |\varphi_\ell^{(1)}\rangle|^2 = 0 \tag{2.B.11}$$

so that

$$\langle \varphi_n^{(1)}| \langle \psi_k^{(2)}| A_i |\psi_m^{(2)}\rangle |\varphi_\ell^{(1)}\rangle = 0, \; n \neq \ell . \tag{2.B.12}$$

Analogously we get

$$\langle \varphi_n^{(1)}| \langle \psi_k^{(2)}| A_i |\psi_n^{(2)}\rangle |\varphi_\ell^{(1)}\rangle = 0, \; k \neq m , \tag{2.B.13}$$

so that A_i is diagonal in the considered basis, which, being an arbitrary factorized basis, implies that A_i is a constant multiple of the identity for any i. Since the trace is preserved, we get

$$\rho^{\#} = \rho \tag{2.B.14}$$

so that the only mapping satisfying our conditions is the identical mapping which cannot lead to the factorization of a non-factorized state.

The conclusion is that, to account for the spontaneous factorization hypothesis one has to go beyond the statistical operator formalism and to take explicitly into account the detailed composition of statistical ensembles. The previous theorem amounts in fact to state that necessarily at least some physically different statistical ensembles corresponding to the same statistical operator must evolve into ensembles corresponding to different statistical operators. To avoid the possibility that physically indistinguishable initial conditions lead to physically distinguishable final situations, the theoretical framework accounting for the spontaneous factorization has to deal differently with physically different ensembles corresponding to the same statistical operator. The hypothesis under discussion requires therefore, not only a violation of quantum mechanics at large separations but a reformulation of quantum theory which goes beyond the statistical operator formalism. To show how this can be done is the subject of the next section.

3.B. A Reformulation and a Generalization of Quantum Mechanics.

According to the discussion of the previous section we are led to the problem of overcoming the statistical operator formalism in such a way that physically different ensembles associated to the same statistical operator are described differently in the mathematical formalism. We can observe that an ensemble is characterized by the states which appear in it and by the corresponding weights. The simplest way to describe the ensemble is then that of assigning, for any vector (or better for any ray) of the Hilbert space of the system a real positive number which represents the weight (possibly equal to zero) with which such a state enters into the ensemble. One then has to rewrite the quantum evolution equations in terms of these new quantities (which we shall call ray functionals or density functionals and denote by $p(\psi)$) and, if possible, to generalize the equations to account for the spontaneous factorization of quantum states.

Let us therefore make the following assumption: The statistical ensembles, which are the basic physical objects, are associated to real nonnegative functionals defined on Hilbert space rays, or equivalently on vectors ψ , and satisfying

$$p(c\psi) = p(\psi), \quad \forall \text{ complex } c \neq 0. \tag{3.B.1}$$

Particular cases are: δ -function like functionals corresponding to ensembles which are pure cases; functionals different from zero only over a finite set of states, corresponding to the usual mixtures; and functionals different from zero on a continuous set of states corresponding to continuous mixtures, a typical example of this kind being the isotropic mixture, already considered in refs.(8) and (9), in which all possible spin orientations appear with equal weights.

Obviously $p(\psi)$ must be normalized in some way in order to allow a comparison between the weights associated to different vectors. This leads to the problem of defining an integration of $p(\psi)$ on $\mathcal{H}$. We shall confine our considerations, for simplicity, to a finite dimensional Hilbert space (as for instance the space referring only to spin degrees of freedom). The generalization to the infinite dimensional case is possible and requires the consideration of measures on $\mathcal{H}$.

For an N-dimensional Hilbert space we consider the volume element

$$d\mathcal{H}^{(N)} = dx_1 dx_2 \dots dx_N \, dy_1 dy_2 \dots dy_N \tag{3.B.2}$$

where

$$x_i + i y_i = z_i = \langle e_i | \psi \rangle, \tag{3.B.3}$$

being $(|e_i\rangle, i=1,2,\dots,N)$ a complete orthonormal set in $\mathcal{H}^{(N)}$. Note that $d\mathcal{H}^{(N)}$ is invariant under unitary transformations (changes of basis) since any unitary transformation on z_i induces an orthogonal transformation in the Euclidean 2N-dimensional space (x_i, y_i). In polar coordinates we write

$$d\mathcal{H}^{(N)} = \rho_1 d\rho_1 \, d\theta_1 \, \rho_2 d\rho_2 \, d\theta_2 \dots \rho_N d\rho_N \, d\theta_N, \tag{3.B.4}$$

where

$$\rho_i e^{i\theta_i} = z_i . \tag{3.B.5}$$

Being $p(\psi)$ a ray function, when expressed in terms of the variables ρ_i and θ_i it is independent of $\sum_K \rho_K^2$ and $\sum_i \theta_i$ (in fact $\psi \to e^{i\alpha}\psi$ amounts to $\sum\theta_i \to \sum_i \theta_i + N\alpha$). Let us consider then the hyperplane $\sum_i \theta_i = 0$. On the lines orthogonal to it $p(\psi)$ takes a constant value. Moreover when only one of the variables ϑ_i is increased by 2π the Hilbert space ray does not change and so $p(\psi)$ keeps the same value. Let us then consider the hypercube whose sides have a lenth 2π and are parallel to the axes ϑ_i . All rays cross this hypercube and the length of the lines corresponding to a ray, lying within the hypercube is constantly equal to $2\pi\sqrt{N}$. In the figure we have illustrated the situation for N=2. The bold face line represents a single ray.

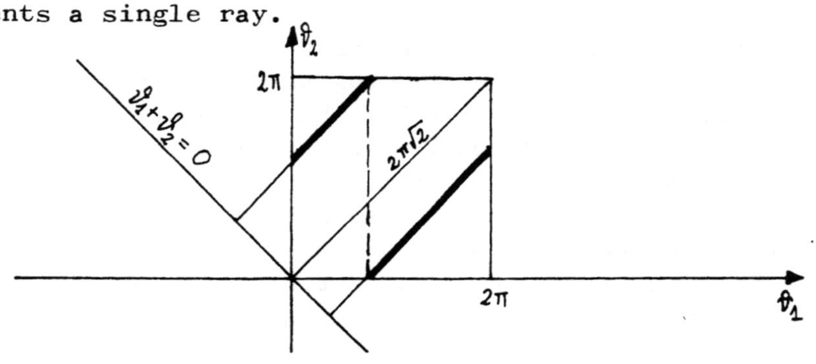

We can then perform the integration using the volume element

$$d\psi = \delta\left(\sum_K \rho_K^2 - 1\right) d\mathcal{H} \tag{3.B.6}$$

and integrate over any hypercube of side 2π. For any real valued function $f(\psi)$ which is a ray functional we will then write

$$\int f(\psi)\, d\psi = \int_0^\infty \rho_1 d\rho_1 \cdots \int_0^\infty \rho_N d\rho_N \int_0^{2\pi} d\theta_1 \cdots \int_0^{2\pi} d\theta_N\, \delta\left(\sum_K \rho_K^2 - 1\right) f(\rho_1, \ldots \rho_N, \theta_1, \ldots \theta_N). \tag{3.B.7}$$

The δ-function restricts the integration to normalized vectors. The invariance of the definition (3.B.7) for changes of basis in $\mathcal{H}^{(N)}$ is a consequence of the invariance of $d\mathcal{H}^{(N)}$ and $\delta(\sum_K \rho_K^2 - 1)$ and of the fact that if we displace the hypercube keeping its sides parallel to the θ_i axes, any ray crosses it, the length of the lines corresponding to a ray lying within the hypercube being again equal to $2\pi\sqrt{N}$. By using the above integration we will normalize our density functional to one:

$$\int p(\psi)\, d\psi = 1. \tag{3.B.8}$$

We rewrite now the conventional quantum formulae in the new language. For the quantum expectation value of an observable we now have

$$\langle A \rangle = \int d\psi\, p(\psi) \langle \psi | A | \psi \rangle, \tag{3.B.9}$$

which is easily seen to coincide for pure states or finite mixtures with the standard formula. Note that $\langle \psi | A | \psi \rangle$ is a ray functional. The statistical operator can be written as

$$\rho = \int d\psi\, p(\psi)\, |\psi\rangle\langle\psi|. \tag{3.B.10}$$

Comparing with (3.B.9) one gets

$$\langle A \rangle = \mathrm{Trace}\, A\rho, \tag{3.B.11}$$

as in the standard case. The basic point is now to rewrite the Schroedinger equation in the language of density functionals. To this end we observe that Schroedinger evolution is characterized by the requirement that $p(\psi)$ takes at time t the same value on $e^{-iHt}\psi$ that it takes at time 0 on ψ. We then have

$$p(e^{-iHt}\psi, t) = p(\psi, 0) \tag{3.B.12}$$

which implies

$$\frac{dp(e^{-iHt}\psi, t)}{dt} = 0 \tag{3.B.13}$$

Let us now use the complete orthonormal set $\{|e_K\rangle\}$ writing

$$\langle e_K | e^{-iHt} | \psi \rangle \equiv \langle e_K | \psi(t) \rangle = Z_K(t) = X_K(t) + i\, Y_K(t). \tag{3.B.14}$$

From (3.B.13) we then get

$$\frac{\partial p}{\partial t} + \sum_K \left(\frac{\partial p}{\partial x_K} \dot{x}_K + \frac{\partial p}{\partial y_K} \dot{y}_K \right) = 0 \qquad (3.B.15)$$

the dot denoting as usual differentiation with respect to t. Let us consider now the quantity $\mathrm{Grad}\, p(\psi)$ defined by

$$\langle e_K | \mathrm{Grad}\, p(\psi) \rangle = \frac{1}{2} \left(\frac{\partial p}{\partial x_K} + i \frac{\partial p}{\partial y_K} \right), \qquad (3.B.16)$$

which is easily seen to behave like an Hilbert space vector under changes of basis. Eq.(3.B.15) becomes then

$$\frac{\partial p}{\partial t} + \langle \mathrm{Grad}\, p(\psi) | \dot{\psi} \rangle + \langle \dot{\psi} | \mathrm{Grad}\, p(\psi) \rangle = 0 \qquad (3.B.17)$$

Use of the equation $\dot{\psi} = -i H \psi$ in (3.B.17) finally gives

$$\frac{\partial p(\psi,t)}{\partial t} = 2\, \mathrm{Im} \langle \psi | H | \mathrm{Grad}\, p(\psi,t) \rangle, \qquad (3.B.18)$$

which is the Schroedinger equation in the new language. If one writes (3.B.18) formally as

$$\frac{\partial p(\psi,t)}{\partial t} = \mathcal{S}\, p(\psi,t), \qquad (3.B.19)$$

the mapping $\mathcal{S}$ among density functionals turns out to be linear with respect to real numbers. We note that (3.B.18) has obviously the property of transforming pure states into pure states, i.e. it maps δ-like density functionals onto themselves. The forms (3.B.18-19) of the Schroedinger equations are particularly apt as starting points for generalizations. How to do this has been discussed in great detail in ref.(21). The basic point is the construction of a mapping $\mathcal{B}$ among density functionals defined on the Hilbert space $\mathcal{H}^{(1,2)}$ of a composite system

$$p^{\#}(\psi) = \mathcal{B}\, p(\psi), \qquad (3.B.20)$$

with the following properties:

i.-It is linear,

ii.-If one denotes by ρ and $\rho^{\#}$ the statistical operators associated to $p(\psi)$ and to its image $p^{\#}(\psi)$, respectively, $\mathcal{B}$ is such that

$$\mathrm{Trace}^{(\alpha)} \rho = \mathrm{Trace}^{(\alpha)} \rho^{\#}, \qquad (3.B.21)$$

iii.-For any $p(\psi)$, its image is a factorized mixture, i.e. takes nonzero values only on factorized states of $\mathcal{H}^{(1,2)}$. When acting on a factorized functional, $\mathcal{B}$ is the identical mapping.

In terms of such a mapping it is then easy to write the dynamical equation yielding the spontaneous factorization process. Let us write

$$p(\psi,t+dt)=(1-\lambda\, dt)[p(\psi,t)+S\,p(\psi,t)\,dt]+\lambda dt\; B\,p(\psi,t), \tag{3.B.22}$$

where S corresponds to the pure Schroedinger evolution(see (3.B.19)). The meaning of (3.B.22) is very simple; we have a system subjected to the pure Schroedinger evolution which suffers at random times, with mean frequency λ an action which transforms $p(\psi)$ into $B\,p(\psi)$. Eq.(3.B.22) is equivalent to

$$\frac{\partial p(\psi,t)}{\partial t}=S\,p(\psi,t)-\lambda(1-B)p(\psi,t), \tag{3.B.23}$$

which is then an actual generalization of the Schroedinger equation (3.B.19). Note that, as a consequence of property ii), the statistical operator associated to $p(\psi,t)$ has partial traces, at all times, which coincide with those of the statistical operator generated by the pure quantum evolution. This means that the differences between the dynamics implied by (3.B.23) and Schroedinger evolution can emerge only in correlation measurements between the two systems. If one puts

$$p(\psi,t)=e^{St}\,\bar{p}(\psi,t) \tag{3.B.24}$$

one gets

$$\frac{\partial \bar{p}(\psi,t)}{\partial t}=-\lambda\bar{p}(\psi,t)+\lambda e^{-St}B\,e^{St}\bar{p}(\psi,t) \tag{3.B.25}$$

When the two subsystems are not interacting one has

$$B\,e^{St}=e^{St}\,B \tag{3.B.26}$$

so that

$$\frac{\partial \bar{p}(\psi,t)}{\partial t}=-\lambda(1-B)\bar{p}(\psi,t) \tag{3.B.27}$$

with the initial condition $\bar{p}(\psi,0)=p(\psi,0)$. Let us put now

$$p(\psi,0)\equiv\bar{p}(\psi,0)=p_f(\psi)+p_{nf}(\psi), \tag{3.B.28}$$

where $p_f(\psi)$ is different from zero only on the factorized states and $p_{nf}(\psi)$ only on nonfactorized ones. Note that $B\,p_f=p_f$. If we look for a solution of the type

$$\bar{p}(\psi,t)=p_f(\psi)+\alpha(t)p_{nf}(t)+[1-\alpha(t)]B\,p_{nf}(\psi), \tag{3.B.29}$$

since $B^2=\mathbb{1}$ one gets

$$\frac{d\alpha(t)}{dt}[p_{nf}-B\,p_{nf}]=-\lambda\alpha(t)[p_{nf}-B\,p_{nf}], \tag{3.B.30}$$

and being $[p_{nf}-B\,p_{nf}]\neq 0$ one gets

$$\frac{d\alpha(t)}{dt}=-\lambda\alpha(t),\qquad \alpha(t)=e^{-\lambda t}$$

The solution of eq.(3.B.27) is then

$$p(\Psi,t)=e^{St}p_f(\Psi)+e^{-\lambda t}e^{St}p_{nf}(\Psi)+(1-e^{-\lambda t})\,B\,e^{St}p_{nf}(\Psi), \qquad (3.B.31)$$

and the dynamics induces gradually a transition to a factorized $p(\Psi,t)$. As presently formulated the dynamical equation induces the spontaneous factorization even for states for which the subsystems are not spatially separated. We can overcome this limitation by replacing in eq.(3.B.23) the mapping B with a new mapping $B_>$. To this purpose we consider the operator $P_>$ which projects outside a certain relative distance a of the component subsystems

$$P_>\,\psi(\underline{r}_1,\underline{r}_2)=\begin{cases}0, & |\underline{r}_1-\underline{r}_2|<a\\ \psi(\underline{r}_1,\underline{r}_2), & |\underline{r}_1-\underline{r}_2|>a\end{cases}$$

and define the mapping $B_>$ by

$$p^*(\psi)=B_>\,p(\psi)=\int\left[\langle\bar\Psi|P_>|\bar\Psi\rangle\,p^{\#}_{\bar\Psi}(\psi)+\langle\bar\Psi|(1-P_>)|\bar\Psi\rangle\,p_{\bar\Psi}(\psi)\right]p(\bar\Psi)\,d\bar\Psi. \qquad (3.B.32)$$

In this equation we have denoted by $p_{\bar\Psi}(\psi)$ the density functional corresponding to the pure state $\bar\Psi$, and by $p^{\#}_{\bar\Psi}(\psi)$ its image under B. The mapping $B_>$ has then the following property

$$B_>\,p(\psi)=\begin{cases}B\,p(\psi) & \text{for } \langle\psi|(1-P_>)|\psi\rangle\,p(\psi)=0\\ p(\psi) & \text{for } \langle\psi|P_>|\psi\rangle\,p(\psi)=0\end{cases} \qquad (3.B.33)$$

so that when $p(\psi)$ is different from zero only on states which are localized within $|\underline{r}_1-\underline{r}_2|<a$ the additional term in the evolution equation is not effective, while when the states develope components on $|\underline{r}_1-\underline{r}_2|>a$ the mechanism leading to the mixture of factorized states begins to act.

This completes our program of building up a dynamical model which describes the spontaneous factorization process. The model contains two parameters a and λ, which are related to the distance after which the factorization starts to take place and to the time which is necessary in order that the factorization be practically completed. In actual cases, as for instance in the two photon correlation experiments, a could be chosen to be the coherence length of the photon, while λ plays the role of a relaxation time for the process.

References.

1. J.S.Bell, Physics, 1, 195 (1964).
2. E.P.Wigner, Amer. Journ. Phys., 38, 1005 (1970).
3. E.Nelson, Phys. Rev., 150, 1079 (1966).
4. G.C.Ghirardi, C.Omero, A.Rimini, T.Weber, Rivista del Nuovo

Cimento, vol.1,N°3 (1978).
5. A. Einstein,B.Podolsky,N.Rosen,Phys.Rev.,47, 777 (1935).
6. B. d'Espagnat, Conceptual Foundations of Quantum Mechanics, (Menlo Park, Cal.,1971).
7. W.H.Furry, Phys.Rev.,49,393,476 (1936).
8. D.Bohm,Y.Aharonov, Phys.Rev., 108, 1070 (1957).
9. J.M.Jauch, Rendiconti S.I.F.,Course IL (1971).
10.G.C.Ghirardi,A.Rimini,T.Weber,C.Omero, Nuovo Cimento, 39B, 130 (1977).
11.D.Bohm, Phys.Rev., 85, 160 (1952).
12.H.Langhoff, Zeits.Phys., 160, 186 (1960).
13.C.A.Kocher, E.D.Commins, Phys.Rev.Letters, 18, 575 (1967).
14.S.J.Friedman, J.F.Clauser,Phys.Rev.Letters, 28, 938 (1972).
15.L.Kasday,Rendiconti S.I.F., Course IL (1971).
16.G.Faraci,D.Gutkowsky,S.Notarrigo,A.R.Pennisi, Lett.Nuovo Cimento, 9, 607 (1974).
17.L.R.Kasday,J.D.Ullmann,C.S.Wu,Nuovo Cimento, 25B, 633(1975).
18.G.C.Ghirardi,A.Rimini,T.Weber,Nuovo Cimento, 31B, 177(1976).
19.K.Kraus,Ann.of Phys. N.Y., 64, 311 (1971).
20.G.Lindblad,Comm.Math.Phys., 48, 119 (1976).
21.G.C.Ghirardi,A.Rimini,T.Weber,Nuovo Cimento, 36B, 97 (1976).

THE 1927 EINSTEIN AND 1935 E.P.R. PARADOX

O. Costa de Beauregard
Institut Henri Poincaré
11 rue Pierre et Marie Curie,75231 Paris Cedex 05.

I. INTRODUCTION

Eberhard[1] concludes a recent paper by stating that the 1927 Einstein[2] and 1935 E.P.R.[3] paradox, now very well substantiated experimentally, admits only four possible answers:

-I.- Just compute, do not think, and thus avoid headaches. This is playing the ostrich, and the majority's choice.

-II.- It may be that quantum mechanics turns out wrong, and that the distant correlation dies out with increasing distance[4] or under sophisticated conditions[5].

-III.- It may be that relativity theory is wrong in that the wave collapse is an instantaneous, non covariant process[6].

-IV.- The accepted causality concept must be replaced by a new one.

I reject issues I as eluding the problem, II and III as respectively denying two theories the schemes of which have never been found faulty. I intend to show that relativistic quantum mechanics, both in the form of the Schwinger-Feynman-Dyson S-matrix formalism, and in the form of covariant first quantization, does provide the full mathematical solution to the problem, and, then, to outline the overall interpretation of that solution.

II. S-MATRIX FORMALIZATION OF THE EINSTEIN CORRELATION

Consider[7] an expansion of the form (summation sign upon repeated indexes omitted)

$$|\Phi\rangle = c^{ij\ldots} \, |\phi_i\rangle \, |\psi_j\rangle \ldots \qquad (1)$$

where the $|\phi\rangle$'s span disjoint Hilbert spaces, and set

$$M = p \otimes q \otimes \ldots \qquad (2)$$

where the p's... denote Hermitean operators operating respectively in the spaces $|\phi\rangle$... The correlated mean value

$$\langle\Phi|M|\Phi\rangle = c^{*i'j'\ldots} \, c^{ij\ldots} \, \langle\phi_{i'}|p|\phi_i\rangle\langle\psi_{j'}|q|\psi_j\rangle \ldots \qquad (3)$$

comprizes a diagonal contribution

$$\langle\Phi|M|\Phi\rangle_o = \omega^{ij\ldots} \, \langle p_i\rangle\langle q_j\rangle \ldots \qquad (4)$$

with (no summation here on repeated indexes!)

$$\omega^{ij\ldots} \equiv c^{*ij\ldots} \, c^{ij\ldots} \qquad (5)$$

$$<p_i> \equiv <\phi_i|p|\phi_i>, \quad <q_i> \equiv <\psi_j|q|\psi_j>, \ldots \qquad (6)$$

plus an off-diagonal contribution

$$\Delta<\Phi|M|\Phi> = <\Phi|M|\Phi> - <\Phi|M|\Phi>_o \qquad (7)$$

The latter is rendered zero if, and only if, the representation diagonalizes all the operators p..., in which case the quantal mean value (3) assumes the form (4) corresponding to a classical mixture. But then it is merely the semblance of a mixture, being relative to the reference frame (the basis). The difference between the "wavelike probability calculus" defined in 1926 by Born (and implicitely by Einstein in his 1905 theory of the photon), as the statistical interpretation of quantum mechanics, is here expressed as the difference between the addition (1) of partial amplitudes and that (4) of partial probabilities, entailing the existence of the interference terms (7).

Now we go from the abstract quantal formalism to the Schwinger-Feynman-Dyson S-matrix formalism. The point is that the Feynman transition probability between a set of incoming and a set of outgoing substates , when expressed either for prediction in terms of the outgoing, or for retrodiction in terms of the incoming, substates, is precisely of the form (3). The Einstein paradox consists in the interference (7) existing between either the final measurements (outgoing states) or the initial preparations (incoming states), although they are spatially separated from each other (and described as such in the interaction picture). This interference, or Einstein correlation, is clearly tied by the Feynman zigzags, and thus consists of a non-separability of either distant measurements tied by their common past (Einstein correlation proper) or distant preparations tied by their common future (time reversed Einstein correlation).

Both forms of the correlation are very well substantiated experimentally, the first one as a correlation between polarization states[8], the second one as a correlation between occupation numbers at emission[9]. Of course, both types of experiments could be time reversed.

This formalism shows, and experiments verify[10], that the correlation is invariant with respect to arbitrary displacements of the preparing or measuring devices along the corresponding beams. Even, the separation between two such devices, which is necessarily spacelike if the particles used have zero rest mass (e.g. photons), could be made timelike in the case of heavy particles (or even, in the case of photons, by folding one beam by means of a mirror). In this consists an extremely strong objection against those proposals falling in either of the categories II[4 5] or III[6] stated in the Introduction.

Concluding this Section:

-I.- The Einstein paradox is born from the union of two previous paradoxes: Intrinsic time symmetry, as existing in statistical mechanics (Loschmidt, Zermelo), and even in the calculus of probabilities itself; Born's rule of adding partial amplitudes rather than probabilities. One need not say that both traits are inherent in the Feynman formalism, so that the Feynman zigzag is the agent of the Einstein correlation and the wizzard of the Einstein paradox.

-II.- Full relativistic and full quantal covariance are traits of the Einstein correlation. This is not surprising, as both relativity theory and quantum theory are daughters of the physics of waves.

-III.- The presence of the off diagonal terms (7) entails that (so to speak) "the dice are not cast when shaken in the cup" (the space time region where the beams meet) but where and when they stop rolling on the table (where and when the measuring devices work). Nevetheless they are correlated, and this is the Einstein[2] or EPR[3] paradox! Two consequences then follow: a) time-symmetry of the causality concept at the elementary level (see point IV in the Introduction); b) impossibility to think of the micro-entities as of objects separately endowed with properties[11].

III. COVARIANT FIRST QUANTIZATION

I have shown elsewhere[11, 12] how the reciprocal Fourier transforms and the Parseval equality (Hermitean scalar product) for solutions of Klein-Gordon, Dirac, Petiau-Duffin-Kemmer, etc... equations, written in the Dirac notation as

$$\langle x|a\rangle = \langle x|k\rangle \langle k|a\rangle , \qquad \langle k|a\rangle = \langle k|x\rangle \langle x|a\rangle, \tag{8}$$

$$\langle a|b\rangle = \langle a|x\rangle \langle x|b\rangle = \langle a|k\rangle \langle k|b\rangle, \tag{9}$$

can receive a covariant form; $|x\rangle \langle x|$ is a fluxlike integral (implying the conserved current operator) over an arbitrary spacelike surface σ in space-time, and $|k\rangle\langle k|$ an integral over both sheets of the mass shell in k-space. By substitution one gets

$$\langle x|a\rangle = \langle x|x'\rangle\ \langle x'|a\rangle \tag{10}$$

where $\langle x|x'\rangle$ is nonelse than the Jordan-Pauli propagator , such that

$$\langle x'|x''\rangle \equiv \langle x'|k\rangle\langle k|x''\rangle = \langle x'|x\rangle\ \langle x|x''\rangle, \tag{11}$$

which is zero outside the light cone, and non-zero inside both past and future. Formula (10) solves the Cauchy problem, and also is the expansion of the wave function at any space-time point upon the complete set of orthogonal Jordan-Pauli propagators with apexes on σ', the coefficients of the expansion being the values of the wave function on σ'.

Formula (10) is the covariant extension of the non-relativistic formula

$$\psi(\vec{x}) = \iiint \delta(\overrightarrow{x - x'})\ \psi(\vec{x}')\ d^3x', \tag{12}$$

so that $\langle x|x'\rangle$ is the eigenfunction corresponding to the question "Is the particle found crossing a given element of σ"; the corresponding position operator is x^λ $(\lambda = 1,2,3,4;\ x^4 = ict)$ provided that x^λ ends on σ (that is, 3, not 4, degrees of freedom)* . An answer yes implies that the particle has come inside the past, and will go inside the future, half light cone, which is trivial. What is not trivial, however, is that the collapse of $\langle x|a\rangle$ into $\langle x|x'\rangle$ affects symmetrically both past and future, which again entails the time symmetric causality concept proposed[11 13] for solving the Einstein paradox. Clearly, this concept is imposed by the very mathematical formalism.

* This statement is not contradictory to Newton and Wigmer's definition of a position operator, because I accept negative frequencies.

IV. CONCLUSIONS

In almost all dictionaries the fundamental meaning of the word paradox is defined as "a surprising but perhaps true statement". Copernicus' heliocentrism has been such a paradox; Planck's quantum and Einstein's relativity also. A true paradox of this sort is not "reducible". It must be adequately formalized, both mathematically and conceptually, thus producing , in Kuhn's[14] words, a new paradigm[15]. Incidentally, Kuhn's doctrine of "scientific revolutions" is essentially the same as Duhem's[16].

With the Einstein or E.P.R. paradox we are certainly facing a "scientific revolution". The experimental facts are there, together with a quite appropriate mathematical formalism. Not two theorists agree, however, on its acceptable interpretation, although all agree that a change in paradigm is inevitable.

The interpretation I advocate is a straight-forward lecture of relativistic quantum mechanics. I deam that the basic relativistic and quantum schemes are both essential in this problem.

The one point, however, I am changing in the usual discourse is, on the relativistic side, Einstein's prohibition to "telegraph into the past" and, on the quantal side, the Copenhagen conception that the wave collapse affects the future and not the past. In other words, I conceive of the individual quantal transition (or measurement, or preparation) as intrinsically time symmetric[11] - as is the individual collision in classical statistical mechanics and, indeed, the individual stochastic evént in the classical calculus of probabilities. However, the blending of intrinsic time symmetry with Born's rule of adding partial amplitudes rather than probabilities renders the sting of the Einstein paradox much more painful than that of the Loschmidt or Zermelo ones.

I am thus advocating a new-time symmetric basic causality concept, and arguing that the time asymmetry of macroscopic causality, as we know it ,is essentially the same as the one occurring in the probability calculus and in the physics of waves (combined together by Born). In Mehlberg's[17] words, it is "factlike and not lawlike", and macroscopically expressed as a (time asymmetric) boundary condition.

Such a situation is extremely reminiscent of the one encountered in elementary particle theory, where we have intrinsic symmetry and factlike asymmetry between particles and antiparticles. Moreover, there is a partial binding between the two problems, as the Jordan-Pauli propagator admits of the two expressions $D_{ret}+D_{adv}$ and $D_{+}-D_{-}$.

I am therefore submitting that anti-Carnot and advanced waves processes are, macroscopically speaking, not strictly forbidden, and could perhaps be observed, or produced, in appropriate contexts. An investigation of the intrinsic symmetry somewhat hidden in the field of particles has been very rewarding; why not so in the field of causality, that is, in the field of information tied with the concept of probability? This I have discussed elsewhere[18].

REFERENCES

1. Eberhard, P., Nuovo Cim.,46B, 392, 1978.
2. Einstein, A., in Rapports du 5e Conseil Solvay, 1928, 253-256.
3. Einstein, A., Podolsky, B. and Rosen, N., Phys. Rev.,47, 777, 1935.
4. Schrödinger, E., Proc. Camb. Phil. Soc.,31, 555, 1935;
   Furry, W., Phys. Rev.,49, 393, 1936.
5. Selleri, F., Found. Phys.,8, 103, 1978.
6. D'Espagnat, B., Conceptual Foundations of Quantum Mechanics,2nd ed. Benjamin, New York, 1976, p.90, 238, 265 and 281.
7. Costa de Beauregard, O., Phys. Lett.,67A, 171, 1978; Nuovo Cim. 1979.
8. Freedman, S.J. and Clauser, J.F., Phys. Rev. Lett.,28, 938, 1972;
   Clauser, J.F., Phys. Rev. Lett.,36, 1223, 1976;
   Fry, E.S. and Thompson, R.C., Phys. Rev. Lett.,37, 465, 1976.
9. Pflegor, R.L. and Mandel, L., Phys. Rev.,159, 1084, 1967.
   J. Opt. Soc. Amer.,58, 946, 1968.
10. Wilson, A.R., Lowe, J. and Butt, D.K., J. Phys. G2,2613, 1976;
    D'Agostino, M. and Maroni, C., Nuovo Cim.,40B, 143, 1977.
11. Costa de Beauregard, O., Nuovo Cim.,42B, 41, 1977.
12. Costa de Beauregard, O., Précis de mécanique quantique relativiste, Dunod, Paris, 1967.
13. Stapp, H.P., Nuovo Cim., 29B, 270, 1975.
    Davidon, W.C., Nuovo Cim., 36B, 34, 1976.
14. Kuhn, T.S., The structure of scientific revolutions, 2nd ed., University Press, Chicago, 1970.

15. Paradox and Paradigm, Colodny R.G. ed., University Press, Pittsburgh, 1973.

16. Duhem, P., The aim and structure of physical theory, University Press, Princeton, 1954, Part. II, Chap. lV and VI.

17. Mehlberg, H., Physical laws and time's arrow, in Current issues in the philosophy of science, Feigl, H. and Maxwell, G., eds. Holt, Rinehart and Winston, New York, 1961.

18. Costa de Beauregard, O., Found. Phys., 6, 539, 1976; Synthese, 35, 129, 1977.

EINSTEIN LOCALITY AND THE QUANTUM MECHANICAL LONG-DISTANCE EFFECTS

F. Selleri
Istituto di Fisica-Università di Bari
INFN - Sezione di Bari

1 - INTRODUCTION

Some of the great physicists who contributed to the development of quantum theory were not satisfied with the final formulation of quantum mechanics give by the Copenhagen and Göttingen schools.

The best known cases are those of Planck, Einstein, Schrödinger and de Broglie.

The first criticism of the theory, on the ground of physics, was the famous 1935 article[1] by Einstein, Podolsky and Rosen (EPR), whose far-reaching implications are only now beginning to be understood.

The EPR paper called attention, for the first time, to the quantum mechanical treatment of widely separated events and stressed the necessity that any reasonable physical theory treats them as independent: if α and β are two systems that have interacted in the past, but are now arbitrarily distant, Einstein stressed that "the real, factual situation of system α does not depend on what is done with system β which is spatially separated from the former[2]".

In the following we review briefly the quantum mechanical treatment of two distant correlated atomic systems α and β :suppose that α is in state $|\psi_\alpha>$ and β in state $|\psi_\beta>$ .Then the global system $\alpha + \beta$ has as a state-vector

$$|\psi_\alpha> \quad |\psi_\beta> \tag{1}$$

Considering now a system Σ with spin zero disintegrating spontaneously into two spin - $\frac{1}{2}$ systems α and β , let us say to the $l = 0$ state of α and β , the spins of α and β must be in the singlet state, which means that the final state-vector is

$$|\psi_S> = \frac{1}{\sqrt{2}} \{|u^+> \ |v^-> - |u^-> \ |v^+>\} \tag{2}$$

where $|u^+>$ is the state for α with z-component of the spin equal to $+\frac{1}{2}$ and so on.

Notice that if $\vec{S}_\alpha$ and $\vec{S}_\beta$ are the spin operators of particles α and β , respectively,and if $\vec{S} = \vec{S}_\alpha + \vec{S}_\beta$:

$$\begin{aligned} \vec{S}^2 \ |\psi_S> &= 0 \\ S_3 \ |\psi_S> &= 0 \end{aligned} \tag{3}$$

In order that S_3 applied to $|\psi_\alpha> \ |\psi_\beta>$ gives zero,one would have to ensure that no terms of the type $|u^+> \ |v^+>$ (having $S_3=+1$) and no terms $|u^-> \ |v^->$ (having $S_3= -1$)appear.Therefore $|\psi_\alpha>$ must reduce to just $|u^+>$ and $|\psi_\beta>$ to $|v^->$ or,alternatively, $|\psi_\alpha>$ must reduce to $|u^->$and $|\psi_\beta>$to $|v^+>$ Thus the only states of the $|\psi_\alpha>|\psi_\beta>$ type giving $S_3= 0$ as a result of measurement are

$$|u^+>|v^-> \quad \text{and} \quad |u^->|v^+> \tag{4}$$

But if one introduces the S = 0 triplet state,given by

$$|\psi> = \frac{1}{\sqrt{2}} \{|u^+>|v^->+|u^->|v^+>\} \tag{5}$$

one sees that one can write

$$|u^+>|v^->=\frac{i}{\sqrt{2}}\{|\psi_t> + |\psi_S>\}$$
$$|u^->|v^+>=\frac{1}{\sqrt{2}}\{|\psi_t> - |\psi_S>\} \qquad (6)$$

Therefore the states (4) are superpositions of $S^2 = 0$ and $S^2 = 1(1+1)\hbar^2$ states and a measurement of S^2 on them can give a result different from 0, which is instead what one always obtains with $|\psi_S>$. The conclusion is that the states(4) are observably different from $|\psi_S>$.

2 - THE E.P.R. PARADOX

Let a molecule M be given with spin 0, capable of decaying to the singlet state of two spin - ½ atoms α and β.

Consider a very large number (N) of such decays $M \rightarrow \alpha+\beta$ and repeat on each pair the following reasoning[3].

1 - At time t_o a measurement of the third component of the spin is performed on α. Suppose + ½ is obtained.

2 - We are then sure that a future ($t > t_o$) measurements of the third component of the spin of β will give - ½, because this is predicted to be so from quantum mechanics which we assume to be correct.

3 - But, at time $t = t_o$ when α interacts with an instrument nothing can happen to particle β which can be as faraway as one wishes from α.

4 - Certainty of obtaining $S_{\beta 3} = -\frac{1}{2}$ results is a state vector $|v^->$ for the atom β. Because of the previous point this must be true before and after time t_o.

5 - But quantum mechanics predicts that the third component of the total spin of the two atoms α and β must be zero before t_o and that it remains so even after the measurement of $S_{\alpha 3}$ an α at $t = t_o$.

6 - The only state-vector for α + β which describes β as $|v^->$ and which gives zero for the third component of the two atoms α and β is $|u^+>|v^->$

7 - Repeating the above reasoning for every one the N pairs of atoms α and β we conclude that the statistical ensemble which they form is

described as a mixture, with equal statistical weights (= 1/2) of

$$|u^+>|v^-> \text{ and } |u^->|v^+>$$

even before any measurement is performed.

8 - The latter conclusion contradicts however in an observable manner the description given by $|\psi_S>$.We arrive thus at a paradox.

This famous paradox does not arise if all the reasoning is carried on strictly within quantum mechanics as Bohr showed in his 1935 reply[4] This means that in the previous points there are some which contain elements foreign to and incompatible with quantum mechanics.A quick look will convince the reader that points 1,2,5 above are strict consequences of quantum mechanics and 7 and 8 are conclusive points completely deducibile from the first six points.

The foreign elements must therefore have been introduced in points 3,4 and 6.But 4 and 6 are consequences of 3 and of quantum mechanics. Therefore the statement incompatible with quantum mechanics is 3.

This statement consists of three parts

3a) At $t=t_o$ α interacts with an instrument.

3b) At $t=t_o$ nothing can happen to β which is very far away from α.

3c) What is true for α at times $t>t_o$ must therefore be true before t_o.

Nothing can obviously be wrong with 3a) which is simply a description of the time at which a measurementes is performed on α.Furthermore 3c) is simply a rephrasing of 3b),it defines what is meant by the words "nothing can happen".

The conclusion is therefore that the statement 3b) is incompatible with quantum mechanics.

There are essentially two ways of denying 3b). The first consists of the statement that at time $t = t_o$ system β is not observed and that 3b),like every assumption about the unobserved "objective reality" should be avoided.This is the standard positivistic view point.The second way to deny 3b is to assume that α acts at a distance on β and therefore that something happens to β because of the measurement performed on α.The action must be istantaneous and its efficiency must be independent on distance.

3 - BELL'S INEQUALITY

In the previous paragraph we saw that a contradiction exists between the quantum mechanical formalism and the statement "...at time $t=t_o$ when α interacts with an instrument nothing can happen to particle β which can be as far away as one wishes from α".

This statement together with a very general idea of causality embodies the concept of Einstein locality[5] and can be used in order to deduce experimental predictions, such as Bell's inequality[6], which contradict quantum mechanics.

Such predictions are based on correlation functions which are defined as follows: if an observer O_A measures the dicotomic observable A on particle α and another observer O_B measures the similar observable B on the particle β the correlation function P(AB) is, by definition, the average product of A and B:

$$P(AB) = \langle AB \rangle$$

Bell's inequality reads then

$$|P(AB) - P(AB')| + |P(A'B)+P(A'B')| \leq 2 \qquad (7)$$

At it is well known quantum mechanics predicts for the l.h.s. of (7) values as high as $2\sqrt{2}$, with a ~ 41% violation of the inequality.

In the following we will discuss the proof of Bell's inequality following Bell's "Theory of local beables"[7] which seems of great generality since it is based only on the assumption of relativistic causality, i.e. on the idea that an event in space-time is determined exclusively by the events of its backward light-cone.

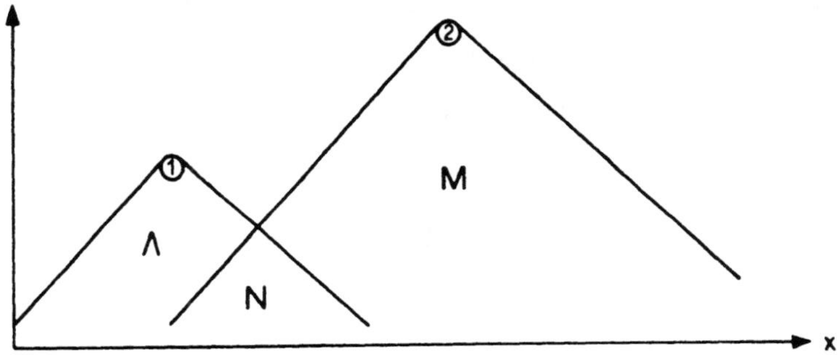

FIG.1

Let A and B be two events taking place in two spatially separated regions 1 and 2 (fig.1). Let Λ and N (M and N) provide a complete specification of all events and processes having taken place in the backward light-cone of A(B). According to special relativity only Λ and N(M and N) can influence the event A(B). In a deterministic approach one could write

$$\begin{aligned} A &= A(\Lambda,N) \\ B &= B(M,N) \end{aligned} \tag{8}$$

Notice that events of type N have only events of the same type in their backward light-cone. Events of type Λ have instead both Λ and N in their backward light-cone. Events of type M are similarly determined by M + N.

Therefore, if we refer to a statistical ensemble of situations like the one shown in fig.1 and can have in every single situation different values for the "beables" Λ,M,N, it is possible to assume for the overall probability density $\rho(\Lambda,M,N)$ a factorization of the following type:

$$\rho(\Lambda,M,N) = \rho_1(\Lambda,N) \quad \rho_2(M,N)\ \rho_o(N) \tag{9}$$

If the "beables" A and B (for instance, results of measurements) have different possible values (let us consider $\pm$ 1 as a dicotomic case) depending, according to (8), on Λ,M and N, the correlation function is given by

$$P(A,B) = \int d\Lambda dM dN \rho_1(\Lambda N)\rho_2(MN)\rho_o(N)A(\Lambda N)B(MN) \tag{10}$$

Notice that the deterministic formula(10) can be used to deduce a " probabilistic " formulation of the correlation function[8] . Suppose, in fact, that $\Lambda = \{a, \lambda\}$ where a is fixed while λ varies over the statistical ensemble. Suppose, furthermore, that $M = \{b,\mu\}$ with fixed b and variable μ, and suppose finally that $N = \upsilon$. One obtains from (10)

$$\begin{aligned} P(ab) &\equiv P\{A(a), B(b)\} \\ &= \int d\lambda d\mu d\upsilon\ \rho_o(\upsilon)\ \rho_1(a\lambda\upsilon)\rho_2(b\mu\upsilon)A(a\lambda\upsilon)B(b\mu\upsilon) \end{aligned}$$

whence

$$P(ab) = \int d\upsilon\ \rho_o(\upsilon)p(a\upsilon)q(b\upsilon) \tag{11}$$

where

$$p(a\upsilon) = \int d\lambda\ \rho_1(a\lambda\upsilon)A(a\lambda\upsilon)$$
$$q(b\upsilon) = \int d\mu\ \rho_2(b\mu\upsilon)B(b\mu\upsilon) \tag{12}$$

obviously

$$-1 \le p(a\upsilon) \le +1$$
$$-1 \le q(b\upsilon) \le +1 \tag{13}$$

Bell's inequality can be deduced directly from (11) and (13),as it will be shown next.

Consider an inequality of the type[9]

$$\sum_{ij} c_{ij}\ P(a_i b_j) \le M. \tag{14}$$

Notice that (14) can be true,given (11) if and only if

$$L \equiv \sum_{ij} c_{ij}\ p(a_i\lambda)\ q(b_j\lambda) \le M \tag{15}$$

In fact, if (15) is true it is enough to multiply it by $\rho(\lambda)$and integrate to get (14).Conversely, if (14) has to be true for any conceivable distribution of hidden variables, it is enough to choose $\rho(\lambda) = \delta(\lambda-\lambda_0)$ to deduce (15) from (14).This important theorem allows us to obtain M : in fact the most stringent inequality is found when M is taken equal to the maximum value of the l.h.s. of (15).Therefore

$$M = \text{Max}\left\{\sum_{ij} c_{ij}\ p(a_i\lambda)q(b_j\lambda)\right\} \tag{16}$$

where the maximum has to be taken over all the conceivable dependences of p and q on λ.

The largest value of L is therefore in all cases in one of the vertices of a cube in the space having all the $p(a_i)$ and all the $q(b_j)$ as coordinates where

$$p(a_i\lambda) = \xi_i = \pm 1$$
$$q(b_j\lambda) = \eta_j = \pm 1 \tag{17}$$

so that

$$M = \underset{\xi,\eta}{\text{Max}}\left\{\sum_{ij} c_{ij}\ \xi_i\eta_j\right\} \tag{18}$$

This is our main result.

It can be shown that only trivial or reducible inequalities exist for linear combinations of 1,2,3,5 correlation functions. In the case of four correlation functions with coefficients having nonfactorable signs one deduces

$$c_{11}P_{11} + c_{12}P_{12} + c_{21}P_{21} + c_{22}P_{22} < \sum_{ij=1}^{2} |c_{ij}| - 2 \underset{l,m}{\mathrm{Min}} |c_{lm}| \qquad (19)$$

where $P_{ij} = P(a_i b_j)$. The previous inequality reduces to Bell's inequality if $c_{11} = c_{12} = c_{21} = +1$ and $c_{22} = -1$.

Similar inequalities can be deduced for linear combinations of six or more correlation functions.

It should be noticed that the previous technique allows one to obtain, in principle, all the possible (infinite) inequalities deducible from Einstein locality. In practice it can be interesting to investigate numerically the simplest ones and to try to understand if a different more synthetic way existes to express the physical content of Einstein locality.

Three types of experiments have been performed in order to study the contrast between Einstein locality and quantum mechanics expressed by Bell's inequality.

A) CASCADE PHOTON EXPERIMENTS

Following the original suggestion by Clauser, Horne, Shimony and Holt[10] experiments on the transmission/absorption of two correlated photons emitted in the same atomic cascade (for instance in a Ca J=0 to J=0 two photon trasition) have been performed by Freedman and Clauser[11], Holt and Pipkin[12], Clauser[13] and Fry and Thompson[14].

With the exception of the Holt-Pipkin experiment, agreement has always been found with the quantum mechanical predictions.

B) POSITRONIUM ANNIHILATION EXPERIMENTS

A test of Bell's inequality using the high-energy photons produced by positronium annihilation ($e^+e^- \to \gamma\gamma$) is possible if one studies through correlated Compton scatterings the polarization correlations of the two γ rays. The first experiments of this type were performed by Kasdy, Ullman and Wu[15] followed by Faraci, Gutkowski, Notarrigo and Pennisi[16], by

Wilson,Lowe and Butt[17] and by Bruno,d'Agostino and Maroni[18].

With the exception of the Catania experiment good agreement has been found with the predictions of quantum mechanics.

C) PROTON - PROTON SCATTERING EXPERIMENTS

Following the suggestion originally made by Fox[19] an experiment designed to test the validity of Bell's inequality for spin correlations of two protons has been carried out by Lamehi-Rachti and Mitting[20].Once more, agreement with quantum mechanics was obtained within the limited statistics of this experiment.

We know of three further experiments being run or prepared now[21]:

Two comments should be added to the published experiments.Firstly the majority of them agrees with quantum mechanics,but no satisfactory explanation has been found for the reported disagreements , particularly of the experiment by Holt and Pipkin[12].

Considering the foundamental nature of the information that one tries to obtain from these experiments,one is certainly not happy to decide on a majority basis.Secondly,all the performed experiments have not been direct experimental controls of the contrast between Einstein locality and quantum mechanics, but have always needed additional assumptions in order to compare theory and experiment.

For instance the CHSH[10] assumption is,given that a pair of photons emerges from two polarizers,that the probability of their joint detection from two photomultipliers is independent of the polarizer orientations.

As Clauser and Shimony[22] noted, it is noteworthy that there exists an important hidden-variables theory - the semiclassical radiation theory - which correctly predicts a large body of atomic physics data,but which denies the CHSH assumption.We can add that such an assumption is contrary to the spirit of all hidden-variables theories: emergence from two polarizers should in all cases imply a selection of the two-photon hidden variables,but these variables could well be those that determine the photomultiplier discharge. In this way the latter effect could become dependent on the polarizers' orientation which selects the hidden variables.

An alternative assumption has been formulated by CH[8] and consists of the idea that for every pair of particles the probability of a count with the polarizer in place is less than or equal to the corresponding probability with the polarizer removed. Again it can be objected that the polarizer in place implies a selection of the hidden variables and that the probability of a count may be larger with the selected rather with the "normally distributed" hidden variables.

In view of these considerations it is in our opinion urgent that more theoretical and experimental research on the verificability of Einstein locality be carried out. One step in this direction has been taken by Livi[23] who proposes to use molecular predissociation of the NO molecule to test Bell's inequality.

4 - SUPERLUMINAL SIGNALS

A development of the EPR-paradox which allows one at least in principle to send arbitrarily complicated signals with velocity larger than that of light over arbitrarily large distances has recently been found by A. Garuccio and the present author[24].

The physical situation is the same as in the previous sections: two distinguishable spin - $\frac{1}{2}$ particles α and β are produced at time to in a certain region R_o of space from the

$$M \to \alpha + \beta$$

decay of a system M. The spin/parity properties of M and of the hamiltonian generating the previous decay are such that the (α,β)- pair is produced in a state with zero orbital angular momentum and with total spin zero [singlet-state $|\psi_s>$ of eq.(2)]

As one sees from(3)the total squared spin in the initial state is zero. This is so also for

$$J^2 = (\vec{L}+\vec{S})^2$$

(where $\vec{L}$ is the orbital angular momentum of the (α,β)-pair), as a simple calculation shows, as a consequence of the fact that the initial state is an $l=0$ state.

Two observers O_A and O_B control two instruments A and B put on the paths of the particles α and β, respectively, in such a way that A is going to perform its measurement on α <u>before</u> B performs its measurement on β.

A measures $S_{\alpha 3}$ (third component of the spin of α) at time t_A.
B measures $S_{\beta 3}$ (third component of the spin of β) at time t_B.
The geometry of the experiment is such that $t_A < t_B$:

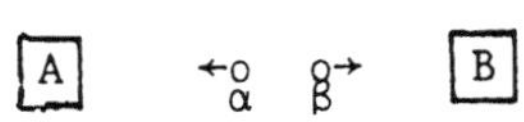

Quantum mechanics implies that the A measurement of $S_{\alpha 3}$ will give $+\frac{\hbar}{2}$

and $-\frac{\hbar}{2}$ with equal (50%) probability, as it follows from eq(2)Furthermore the reduction of the state vector,which is a well known consequence of the theory,implies that the following change of state will take place

Result	Probability	Final State
$+\hbar/2$	50%	$\lvert u^{+}>\lvert v^{-}>$
$-\hbar/2$	50%	$\lvert u^{-}>\lvert v^{+}>$

As noticed in the first section,however, the previous states are superpositions of "singlet" ($J^2 = 0$) and "triplet" ($J^2 = 2\hbar^2$) states so that in the previous mixture on has

$$<J^2> = \hbar^2$$

Notice that the $S_{\alpha 3}$ -measurement can be made without perturbing in any way the orbital angular momentum $\vec{L}$ of the (α,β) -pair since the latter operators commute with $S_{\alpha 3}$. Therefore, also after the first measurement we have the right to assume that we are dealing with an $l = 0$ state. Therefore the change of spin implies an identical change of the <u>total</u> angular momentum of the (α,β) -pair.

Since all the known interactions conserve the total angular momentum a corresponding change is going to take place in the apparatus.Pictorially we are thus led to the conclusion that the act of measurement sets A in rotation.Mathematically this is a consequence of the reduction of the state-vector from $\lvert\psi_s>$ to one of the states $\lvert u^{+}> \lvert v^{-}>$ and $\lvert u^{-}>\lvert v^{+}>$. Notice that the net effect of the measurements performed by the instrument A on N independent (α,β) -pairs is the exchange of N $\hbar^2$squared total angular momentum.This fact, which is proved mathematically elsewhere, is of intuitive validity,since J^2 is a positive definite scalar quantity.

Therefore the physical effect on the apparatus can be as large as one whishes, if one performs similar measurements on a large number of (α,β) -pairs all described by the "singlet" state-vector.

It is precisely this fact that makes the effect at least in principle observable, since the exchange of a single (or of a few) quanta between atomic systems and a macroscopic body results in unobservable changes in the latter which is in practice always in interaction with the rest of the world through different mechanisms such as the exchanges of heat , infra-red waves, cosmic rays, and so on.

This is the essential difference between our proposal and similar ones[25] advanced recently: in pictorial terms we may say that with our mechanism the apparatus "rotates as fast as one whishes".

The second measurement, performed by B on β in order to find $S_{\beta 3}$ is not going to generate any reduction of the state-vector. In fact the states $|u^+\rangle\ |v^-\rangle$ or $|u^-\rangle\ |v^+\rangle$, one of which describes the (α,β) -pair after the first measurement has been performed, are eigenstates of $S_{\beta 3}$

$$S_{\beta 3}\left\{ |u^+\rangle \ \ |v^-\rangle \right\} = -\frac{\hbar}{2}\,|u^+\rangle\,|v^-\rangle$$

$$S_{\beta 3}\left\{ |u^-\rangle \ \ |v^+\rangle \right\} = \frac{\hbar}{2}\,|u^-\rangle\,|v^+\rangle$$

According to quantum Mechanics they are therefore left totally unchanged by the act of measurement made by B. No variation in angular momentum of the (α,β) -pair is therefore going to take place. Correspondingly, no rotation of B is being generated. The only difference between A and B is that A performs the first measurement, as a consequence of which it reduces the singlet-state, exchanges angular momentum, rotates. If B had performed the first measurement it would have been set in rotation instead of A. Therefore O_A can decide to set in rotation B by shifting the apparatus A away from the path of α just before time t_A.

At time t_B observer O_B will see the apparatus B is set in rotation. The time difference $t_B - t_A$ can be made as small as one whishes, while the AB distance can be arbitrarily large. Obviously, one is here dealing with a signal from O_A to O_B which propagates instantaneously.

As was said above, N (α,β) -pairs can be used to transmit one bit of information between O_A and O_B. For instance, the two observers could have agreed on the following meaning

ROTATION OF B	→	DOT
LACK OF ROTATION OF B	→	DASH

If n groups of N (α,β) -pairs each are sent from R_o toward the instruments

A and B, Morse alphabet can obviously be used in order to send instantaneously any messages consisting of n bits from O_A to O_B.

The previous reasoning can be carried out by using photons with correlated polarization states, instead of spin - $\frac{1}{2}$ particles. For instance it is well known that the (J=0) → (J=0) Ca deexitation generates two photons γ and γ' with over-all polarization state:

$$|\eta\rangle = \frac{1}{\sqrt{2}} \left\{ |x\rangle\, |y'\rangle - |y\rangle\, |x'\rangle \right\}$$

where $|x\rangle$ and $|y\rangle$ are linear polarization states for γ along the x- and y- axis, respectively, and so on. Also in this case one concludes that messages can be sent from O_A to O_B with infinite velocity.

One is here dealing with quantum electrodynamics and therefore with a part of relativistic quantum theory.

5 - NONLOCALITY AND RELATIVITY

The question whether or not the eventual violations of Bell's inequality can carry a signal has discussed by Bohm and Hiley[26] who commented: "If it can, we will be led to a violation of the principles of Einstein's theory of relativity, because the instantaneous interaction implied by the quantum potential will lead to the possibility of a signal that is faster than light".

We saw that signals are indeed exchanged istantaneously in the fourth section. Therefore,there seem to be problems in reconciling with special relativity a full validity of the quantum mechanical treatment of distant systems.

Bohm and Hiley showed that the two-body Schrödinger equation for the wave fuction $\psi(\vec{x}_1\ \vec{x}_2 t) = R \exp\{i\ S/\hbar\}$ can be transformed into two equations,one of which is a Hamilton - Jacobi equation for the system of two particles,acted on not only by the classical potential V but also by the quantum potential $Q\ (\vec{x}_1 \vec{x}_2 t)$.

The latter has strange nonlocal properties since

(a) It does not in general produce a vanishing interaction between the two particles when $|\vec{x}_1 - \vec{x}_2| \to \infty$.

(b) It cannot be expressed as a universal function of the coordinates as it can be done with usual potentials.

(c) It depends on $\psi(\vec{x}_1\vec{x}_2 t)$ and therefore on the quantum system as a whole. Since $R=R(\vec{x}_1\vec{x}_2 t)$ the force acting on particle 1 depends on the simultaneous position of particle 2 and viceversa.

Bohm and Hiley basically accept the nonlocal effects which consider the essential new quality implied by quantum theory and try to develop a physical picture of the world based on the notion of "unbroken wholeness" which they attribute to correlated quantum systems.

Nevertheless,the problem of reconciling nonlocal effects with relativity remains unsolved.In a recent paper Hiley[27]quotes the following opinion, expressed in 1972 by Dirac: "It (nonlocality) is against the spirit of relativity but is the best we can do at the present time ....

and of course one is not satisfied with such a theory. I think one ought to say that the problem of reconciling quantum theory and relativity is not solved".

In recent papers Vigier[28] has noted that the validity of the quantum mechanical predictions for correlated particles implies "a destruction of the Einsteinian concept of material causality in the evolution of Nature".

Vigier's theory contains three fundamental elements:

(a) extended "rigid" particles which move always with subluminal velocity, but which can propagate within their interiors signals with superluminal velocity.

(b) a physical vacuum viewed as a thermostat of such rigid particles, which provides the basis, in the spirit of the older Bohm-Vigier proposal, to the probabilistic properties of quantum phenomena.

(c) waves, which propagate as real physical collective excitations (i.e. as density waves) on the top of the previous thermostat.

In this way information starting on the ψ wave's boundary (such as the opening or closing a slit in the double slit Young hole interference experiment) reacts with superluminal velocity (via the quantum potential) on the particles motions which move with subluminal group velocities along the lines of flow of the quantum mechanical ψ waves.

This theory, if really consistent with special relativity, could generate instantaneous interactions between distant particles as implied by the quantum mechanical correlations.

Several authors have advanced solutions of the EPR-paradox which try to incorporate in the theory a fully time-symmetric formalism.

A very interesting proposal is the one by Rayski[29] who considered the oldfashioned concept of state to be "inadequate and misleading". He proposed to consider a measurement at a given time as serving two purposes: yield information about the system comparable with any preceding information and, simultaneously, prepare the initial state for the future. If a measurement of A at time t, has given the eigenvalue A_1 relative to the eigenstate $|A_1>$ and if a measurement of B at time $t_2 (t_2 > t_1)$ has given

B_m relative to the eigenstate $|B_m\rangle$,Rayski proposes to use both vectors $|A_1\rangle$ and $|B_m\rangle$ in the time interval (t_1,t_2).In this way a full invariance of the theory under time-reversal is built in from the outset.These formal assumptions are based on Rayski's physical idea that "the measurement yields information about some preexisting values of the measured quantities and,at the same time,causes a perturbation and produces some new, unknown but nonetheless existing,values of other observables". Quantum mechanics and realism are thought in this way to have been reconciled, because this new interpretation is thought not to contradict the formalism of quantum mechanics,nor to be in conflict with any of its experimentally verifiable consequences.We found Rayski's proposal very interesting for opposite reasons,because, as was shown elsewhere this reconciliation with realism is obtained at the price of building a theory which can never violate Bell's inequality[30].

Other proposals of time-symmetric theories imply,in a way or another,transmission of signals toward the past. A very nice paper along such lines has been written by Rietdijk[25]who argued that a full acceptance of a realistic description of atomic objects and of quantum mechanics leads one to the conclusion that the human choice of the observables to be measured on a beam of particles acts retroactively in time on the production events forcing them to generate particles in eigenstates of the observables to be measured.A similar proposal was advanced by Stapp[31]. According to him "Bell's theorem shows that no theory of reality compatible with quantum theory can allow the spatially separated parts of reality to be independent: these parts must be related some way that goes beyond the familiar idea that causal connections propagate only into the forward ligt-cone".

Fundamental in Stapp's theory is the idea that information flows from an event both forward in time to its potential successors and backard to its antecedents; it is however not clear how this information is propagated and in this respect Stapp's theory is less complete than Costa de Beauregard's[32]where propagation of signals toward the past takes place physically through the propagation of waves and particles.

6 - CONCLUSIONS

The contrast between Einstein locality and quantum mechanics is now becoming increasingly clear.For instance Wightman [33]stated: "The ERP Paradox arises from .. the assertion: the state of one fragment depends on what experiment is chosen to be done on the other, even though it may happen that there is no time for a light signal to travel from one fragment to the other to communicate the choice."

If Einstein locality will be found to be violated in nature, as the preliminary experimental evidence seems to imply, istantaneous transmission of signals between points with arbitrarily large distance shall have to be admitted.

The least that it can be said is that such a possibility is against the spirit of special relativity. Theoretical investigations of "tachyonic" effects have in recent years shown that their existence is compatible with the formalism of the theory of relativity: one has however a reversal of causes and effects and very funny descriptions of physical reality should be accepted.

If Einstein locality survives as a true property of nature then quantum mechanics shall have to be modified. Such an idea does not seem terribly shocking to Dirac who wrote in 1975[34]"... I think it might turn out that Einstein will prove to be right,because the present form of quantum mechanics should not be considered as the final form. There are great difficulties... in connection with the present quantum mechanics. It is the best that one can do up till now.But,one should not suppose that it will survive indefinitely into the future.And I think that it is quite likely that at some future time we may get an improved quantum mechanics in which there will be a return to determinism and which will, therefore, justify the Einstein point of view."

REFERENCES

1) A.Einstein,B.Podolsky and N.Rosen;Phys.Rev.47,777(1935)

2) A.Einstein in Albert Einstein:Philosopher Scientist,P.Schilpp ed., Library of Living Philosophers,Evanston,Ill.(1951)

3) V.Augelli,A.Garuccio and F.Selleri;Ann.Fond.L.de Broglie 1,154(1976)

4) N.Bohr;Phys.Rev.48,696 (1935)

5) J.S.Bell and B.d'Espagnat; Introductory talk to the THINKSHOP ON PHYSICS,April 18-23 (1976) held in Erice.

6) J.S.Bell;Physics 1, 195 (1965)

7) J.S.Bell;The Theory of Local Beables, Epistemological Letters, (March 1976)

8) J.F.Clauser and M.A.Horne;Phys.Rev.D10,526 (1974)

9) A.Garuccio and F.Selleri;Found.Phys.,to be published.

10) J.F.Clauser,M.A.Horne,A.Shimony and R.A.Holt;Phys.Rev.Lett.23,880 (1969)

11) S.J.Freedman and J.F.Clauser;Phys.Rev.Lett.28,938 (1972)

12) R.A.Holt and F.M.Pipkin;Harvard Univ.preprint (1973)

13) J.F.Clauser;Phys.Rev.Lett.36,1223 (1976)

14) E.S.Fry and R.C.Thompson;Phys.Rev.Lett.37,465 (1976)

15) L.R.Kasday,J.D.Ullman and C.S.WU;Nuovo Cimento 25B,633 (1975)

16) G.Faraci,S.Gutkowski,S.Notarrigo and A.R.Pennisi;Nuovo Cimento Lett.9,607 (1974)

17) A.R.Wilson,J.Lowe and D.K. Butt; Jour.Phys.G 2,613 (1976)

18) M.Bruno,M.d'Agostino and C.Maroni;Nuovo Cimento 40B,142 (1977)

19) R.Fox;Nuovo Cimento Lett. 2,565 (1971)

20) M.Lamehi-Rachti and W.Mittig; Phys.Rev.14,2543 (1976)

21) A.Aspect;Phys.Rev.D14,1944 (1976)

G. Bertolini,private communication

V.Rapisarda, private communication

22) J.F. Clauser and A.Shimony;Rep.Progr.Phys. 41,1881 (1978)

23) R.Livi;Nuovo Cimento 48B, 272 (1978)

24) A.Garuccio and F.Selleri;Paper presented at the Paris Conference on the Hundreth Anniversary of AlbertEistein,6-9 June 1979.

25) C.W. Rietdjik;Found.Phys. 8,615 (1978)
N.Herbert;preprint (1979)

26) D.Bohm and B.J.Hiley;Found.Phys. 5,93 (1978)

27) B.J.Hiley;Contemp.Phys.18,411 (1977)

28) J.P.Vigier;Nuovo Cimento Lett.24,258 (1979);N.Cufaro Petroni and J.P.Vigier;Breaking of restricted relativistic einsteinian causality by q.m.;H.Poincarè Inst. preprint (1978)

29) J.Rayski;Found.Phys.3,89 (1973) and 7,151 (1977)

30) F.Selleri and G.Tarozzi;Quantum Mechanics,reality and separability, sumbitted to Rivista del Nuovo Cimento.

31) H.P.Stapp, Phys.Rev.D13,947 (1976)

32) O.Costa de Beauregard;Found.Phys.6,539(1976) and Phys.Lett.67A,171 (1978)

33) A.S.Wightman;Proc.of Symposia in Pure Mathematics,28,147 (1976)

34) P.A.M. Dirac; Directions in Physics; Ed. by H.Hora and J.R. Shepansky;Wiley - Sidney (1976)

MIX
Papier aus verantwortungsvollen Quellen
Paper from responsible sources
FSC® C105338

If you have any concerns about our products,
you can contact us on
ProductSafety@springernature.com

In case Publisher is established outside the EU,
the EU authorized representative is:
Springer Nature Customer Service Center GmbH
Europaplatz 3, 69115 Heidelberg, Germany

Printed by Libri Plureos GmbH
in Hamburg, Germany